Electromagnetic Scintillation
II. Weak Scattering

Electromagnetic scintillation describes the phase and amplitude fluctuations imposed on signals that travel through the atmosphere. These volumes provide a modern reference and comprehensive tutorial for this subject, treating both optical and microwave propagation. Measurements and predictions are integrated at each step of the development. The first volume dealt with phase and angle-of-arrival measurement errors, which are accurately described by geometrical optics.

This second volume concentrates on amplitude and intensity fluctuations of the received signal. Diffraction plays a dominant role in this aspect of scintillation and one must use a full-wave description. The Rytov approximation provides the basis for describing weak-scattering conditions that characterize a wide range of important measurements. Astronomical observations in the optical, infrared and microwave bands fall in this category. So also do microwave signals received from earth-orbiting satellites and planetary spacecraft. Weak scattering describes microwave communication near the surface. Level fluctuations induced by atmospheric irregularities both in the troposphere and in the ionosphere are estimated for these applications and compared with experimental results. Laser signals on terrestrial paths are described by this approach if the transmission distance is less than approximately 300 m. Experiments and applications using longer paths involve strong scattering, which will be discussed in Volume III.

This book will be of particular interest to astronomers, applied physicists and engineers developing instruments and systems at the frontier of technology. It also provides a unique reference for atmospheric scientists and scintillation specialists. It can be used as a graduate textbook and is designed for self study. Extensive references to original work in English and Russian are provided.

DR ALBERT D. WHEELON has been a visiting scientist at the Environmental Technology Laboratory of NOAA in Boulder, Colorado, for the past decade. He holds a BSc degree in engineering science from Stanford University and a PhD in physics from MIT, where he was a teaching fellow and a research associate in the Research Laboratory for Electronics. He has published thirty papers on radio physics and space technology in learned journals.

He has spent his entire career at the frontier of technology. He made important early contributions to ballistic missile and satellite technology at TRW, where he was director of the Radio Physics Laboratory. While in government service, he was responsible for the development and operation of satellite and aircraft recon-naissance systems. He later led the development of communication and scientific satellites at Hughes Aircraft. This firm was a world leader in high technology and he became its CEO in 1986.

He has been a visiting professor at MIT and UCLA. He is a Fellow of the American Physical Society, the IEEE and the AIAA. He is also a member of the National Academy of Engineering and has received several awards for his contributions to technology and national security including the R. V. Jones medal. He has been a trustee of Cal Tech and the RAND Corporation. He was a member of the Defense Science Board and the Presidential Commission on the Space Shuttle Challenger Accident. He has been an advisor to five national scientific laboratories in the USA.

Electromagnetic Scintillation

II. Weak Scattering

Albert D. Wheelon

Environmental Technology Laboratory
National Oceanic and Atmospheric Administration
Boulder, Colorado, USA

CAMBRIDGE
UNIVERSITY PRESS

CAMBRIDGE UNIVERSITY PRESS
Cambridge, New York, Melbourne, Madrid, Cape Town, Singapore, São Paulo

Cambridge University Press
The Edinburgh Building, Cambridge CB2 2RU, UK

Published in the United States of America by Cambridge University Press, New York

www.cambridge.org
Information on this title: www.cambridge.org/9780521801997

First published 2003
This digitally printed first paperback version 2006

A catalogue record for this publication is available from the British Library

ISBN-13 978-0-521-80199-7 hardback
ISBN-10 0-521-80199-0 hardback

ISBN-13 978-0-521-02425-9 paperback
ISBN-10 0-521-02425-0 paperback

These volumes are
dedicated to Valerian Tatarskii who taught us all

Contents

Contents

Preface

History

Quivering of stellar images can be observed with the naked eye and was noted by ancient peoples. Aristotle tried but failed to explain it. A related phenomenon noted by early civilizations was the appearance of shadow bands on white walls just before solar eclipses. When telescopes were introduced, scintillation was observed for stars but not for large planets. Newton correctly identified these effects with atmospheric phenomena and recommended that observatories be located on the highest mountains practicable. Despite these occasional observations, the problem did not receive serious attention until modern times.

How It Began

Electromagnetic scintillation emerged as an important branch of physics following the Second World War. This interest developed primarily in response to the needs of astronomy, communication systems, military applications and atmospheric forecasting. The last fifty years have witnessed a growing, widespread interest in this field, with considerable resources being made available for measurement programs and theoretical research.

Radio signals coming from distant galaxies were detected as this era began, thereby creating the new field of radio astronomy. Microwave receivers developed by the military radar program were used with large apertures to detect these faint signals. Their amplitude varied randomly with time and it was initially suggested that the galactic sources themselves might be changing. Comparison of signals measured at widely separated receivers showed that the scintillation was uncorrelated, indicating that the random modulation was imposed by ionized layers high in the earth's atmosphere. Careful study of this scintillation now provides an important tool for examining ionospheric structures that influence reflected short-wave signals and transionospheric propagation.

Vast networks of microwave relay links were established to provide wideband communications over long distances soon after the Second World War. The effect of scintillation on the quality of such signals was investigated and found not to be important for the initial systems. The same question arose later in connection with the development of communication satellites and gave rise to careful research. These questions are now being revisited as terrestrial and satellite links move to higher frequencies and more complicated modulation schemes.

Large optical telescopes were being designed after the war in order to refine astronomical images. It became clear that the terrestrial atmosphere places an un-welcome limit on the accuracy of position and velocity measurements. The same medium limits the collecting area for coherent signals to areas that are considerably smaller than the apertures of large telescopes. A concerted effort to understand the source of this optical noise was begun in the early 1950s. Temperature fluctuations in the lower atmosphere were identified as the source. When high-resolution earth-orbiting reconnaissance satellites were introduced in 1960 it was feared that the same mechanism might limit their resolution.

Development of long-range ballistic missiles began in 1953 and early versions relied on radio guidance. Astronomical experience suggested that microwave quiv-ering would limit their accuracy. This concern encouraged numerous terrestrial experiments to measure phase and amplitude fluctuations induced by the lower atmosphere. The availability of controlled transmitters on earth-orbiting satellites after 1957 made possible a wide range of propagation experiments designed to investigate atmospheric structure.

The presence of refractive irregularities in the atmosphere suggested the possibil-ity of scattering microwave signals to distances well beyond the optical horizon. This was confirmed experimentally in 1955 and became the basis for scatter-propagation communications links using turbulent eddies both in the troposphere and in the iono-sphere. Because of its military importance, a considerable amount of research on the interaction of microwave signals with atmospheric turbulence was sponsored.

Understanding the Phenomenon

The measurement programs that explored these applications generated a large body of experimental data bearing directly on the scattering of electromagnetic waves in random media. There was an evident need to develop theoretical understandings that could explain these results. The first attempts used geometrical optics to describe the electromagnetic propagation combined with spatial correlation models for the turbulent atmosphere. Time variations of the field were included by assuming the existence of a frozen random medium carried past the propagation path on prevail-ing winds. These models were successful at describing phase and angle-of-arrival measurements, but failed to explain amplitude and intensity variations. The next step was to exploit the Rytov approximation to describe the influence of random media on electromagnetic waves. This technique includes diffraction effects and provides a reliable description for weak-scattering conditions.

Our understanding of refractive irregularities in the lower atmosphere benefited greatly from basic research on turbulent flow fields. Using dimensional arguments, Kolmogorov was able to explain the most important features of turbulent velocities. His approach was later used to describe the turbulent behavior of temperature and humidity, which directly influence electromagnetic waves. These models now provide a physical basis for describing many of the features observed at optical and microwave frequencies.

The Second Wave of Applications

The development of coherent light sources took scintillation research into a new and challenging regime. It is possible to form confined beams of optical radiation with laser sources. These beam waves find important applications in military target-location systems. The ability to deliver concentrated forms of optical energy onto targets at some distance soon led to laser weapons. Wave-front-tilt monitors and corrective mirror systems were combined to correct the angle-of-arrival errors experienced by such signals and later applied to large optical telescopes. Rapidly deformable mirrors were developed later in order to correct higher-order errors in the arriving wavefront. These applications stimulated research on many aspects of atmospheric structure and electromagnetic propagation.

Radio astronomy has moved from 100 MHz to over 100 GHz in the past forty years. Microwave interferometry has become a powerful technique for refining astronomical observations using phase comparison of signals received at separated antennas. The lower atmosphere defines the inherent limit of angular accuracy that can be achieved with earth-based arrays. Considerable effort has gone into programs to measure the phase correlation as a function of the separation between receivers and use it as a guide for the design of large interferometric arrays.

The development of precision navigation and location techniques using constellations of earth-orbiting satellites focused attention on phase fluctuations at 1550 MHz. Ionospheric errors are removed by scaling and subtracting the phase of two signals at nearby frequencies. The ultimate limit on position determination is thus set by phase fluctuations induced by the troposphere, which are the same errors that limit the resolution of interferometric arrays.

Coherent signals radiated by spacecraft sent to explore other planets have been used to examine the plasma distribution in our own solar system. Transmission of spacecraft signals through the atmospheres of planets and their moons provides a unique way to investigate the atmospheres of neighboring bodies. The discovery of microwave sources far out in the universe led to exploration of the interstellar plasma with scintillation techniques. Comparing different frequency components of pulsed signals that travel along the same path provides a unique tool with which to study this silent medium. One can use scintillation measurements to estimate the sizes of distant quasars with surprising accuracy.

An extension of scatter propagation occurred when radars became sensitive enough to measure backscattering by turbulent irregularities. The structure of atmospheric layers has been established from the troposphere to the ionosphere using high-power transmitters and large vertically pointed antennas. It was later found that scanning radars can detect turbulent conditions over considerable areas, thereby providing a valuable warning service to aircraft. The same phenomenon is now making an important contribution to meteorology. A network of phased-array radars has been installed in the USA to measure the vertical profiles of wind speeds and temperature by sensing the signal returned from irregularities and its Doppler shift.

Acoustic propagation is a complementary field to the one we will examine. Long-range acoustic detection programs sponsored by the military have supported important experimental and theoretical research. Controlled acoustic experiments that are not often possible with electromagnetic signals in the atmosphere can be done in the ocean. Theoretical descriptions of acoustic propagation have helped us to understand the strong scintillation observed at optical frequencies. Investigations of the acoustic and electromagnetic problems are mutually supporting endeavors.

These recent applications have stimulated further theoretical research. Laser systems often operate in the strong-scattering regime where the Rytov approximation is not valid. This encouraged a sustained effort to develop techniques that can describe saturation effects. It has taken three directions. The first is based on the Markov approximation, which results in differential equations for moments of the electric-field strength. The second approach is an adaptation of the path-integral method developed for quantum mechanics. The third approach relies on Monte Carlo simulation techniques in which the random medium is replaced by a succession of phase screens.

Resources for Learning About Scintillation

After fifty years of extensive experimental measurements and intense theoretical development this subject has become both deep and diverse. Many are asking "How can one learn about this expanding field?" Where does one go to find results that can be applied in the practical world? Despite its growing importance, one finds it difficult to establish a satisfactory understanding of the field without an enormous investment of time. That luxury is not available to most engineers, applied physicists and astronomers. They must find reliable results quickly and apply them.

There are few reference books in this field. A handful of early books were written in Russia where much of the basic work was done. These were influential in shaping research programs but subsequent developments now limit their utility. A later Russian series summarized the theory of strong scattering but made little contact with experimental data. Several books on special topics in random-medium propagation have appeared recently. Even with these references, it takes a great deal of time to establish a confident understanding of what is known and not known – even in small sectors of the field.

The Origin of this Series

This series on electromagnetic scintillation came about in a somewhat unusual way. It resulted from my return to a field in which I had worked as a young physicist. My life changed dramatically in 1962 and the demands of developing large radar, reconnaissance and communication systems at the frontier of technology took all of my energy for several decades. That experience convinced me how important research in this field has become. When I returned to scientific work in 1988, I resolved to explore the considerable progress that had been made during my absence.

I was immediately confronted with an enormous literature, scattered over many journals. Fortunately the Russian journals had been translated into English and were available at MIT where I was then teaching. As an aid to my exploration, I began to develop a set of notes with which I could navigate through the literature. My journal grew steadily as I added detail, made corrections and included new insights. It soon became several large notebooks. I was invited to work with the Environmental Technology Laboratories of NOAA in 1990. This coincided with the arrival from Moscow of several leaders in this field, which has made Boulder the premier center of research. I shared my notebooks with colleagues there who encouraged me to bring them into book form.

In reviewing the progress made over the past thirty years, I found a number of loose ends and apparent conflicts. To resolve these issues, I spent a good deal of time examining the field. Several areas needed clarification and this resulted in some original research, which is reported here for the first time.

Approach and Intended Users

The purpose of this series is to provide an understanding of the underlying principles of electromagnetic propagation through random media. We shall focus on transmission experiments in which small-angle forward scattering is the dominant mechanism.[1] The elements common to different applications are emphasized by focusing on fundamental descriptions that transcend their boundaries. I hope that this approach will serve the needs of a diversified community of technologists who need such information. Measurements and theoretical descriptions are presented together in an effort to build confidence in the final results. In each application, I have tried to identify critical measurements that confirm the basic expressions. These experiments are often summarized in the form of tables that readily lead one to the original sources. Actual data is occasionally reproduced so that the reader can judge the agreement for himself. In some cases, I give priority to the early experiments to recognize pioneering work and to provide a sense of historical development. In other cases, I have used the most recent and accurate data for comparison.

It is important to explain how the series is organized. The goal is always to present the simplest description for a measured quantity. We advance to more sophisticated explanations only when the simpler models prove inadequate. The first volume explores the subject with the most elementary description of electromagnetic radiation – geometrical optics. We find that it gives a valid description for phase and angle-of-arrival fluctuations for almost any situation. In the second volume we introduce the Rytov approximation which includes diffraction effects and provides a significant improvement on geometrical optics. With it one can describe weak fluctuations of signal amplitude and intensity over a wide range of applications. The third volume is devoted to strong scattering, which is encountered at optical and millimeter-wave frequencies. That regime presents a greater analytical challenge and one must lean more heavily on experimental results to understand it.

This presentation emphasizes scaling laws that show how the measured quantities vary with the independent variables; namely frequency, distance, aperture size, inter-receiver separation, time delay, zenith angle and frequency separation. It is often possible to rely on these scaling laws without knowing the absolute value for a measured quantity. That is important because the level of turbulent activity changes diurnally and seasonally. The scaling laws are expressed in closed form wherever possible. Numerical computations are presented when it is not. Brief descriptions of the special functions needed for these analyses are given in appendices and are referenced in the text. This should allow those who have studied mathematical physics to proceed rapidly, while providing a convenient reference for those less familiar with such techniques. Problems are included at the end of each chapter. They are designed to develop additional insights and to explore related topics.

---

[1] The original plan for the series included a volume on wide-angle forward scattering and backscattering of electromagnetic waves. This plan was deferred as the material relating to line-of-sight propagation expanded rapidly.

The turbulent medium itself is a vast subject about which much has been written. Each new work on propagation attempts to summarize the available information in order to lay a foundation for describing the electromagnetic response to it. I looked for ways to avoid that obligatory preamble. Alas, I could find no way out and a brief summary is included in the first volume, which identifies the basis on which we proceed. In doing so, I have tried to avoid promoting particular models of the turbulent medium. This is especially important for the very-large- and very-small-scale regions of the spectrum. Between these extremes, there is good reason to use the Kolmogorov model to characterize the inertial range. All too often, we find that large eddies or the dissipation range play an important – though subtle – role. We have no model based on physical understanding to describe these regions and they must be explored by experiment.

Any attempt to describe the real atmosphere must address the reality of anisotropy. We know that plasma irregularities in the ionosphere are elongated in the direction of the magnetic field. In the troposphere, irregularities near the surface are correlated over greater distances in the horizontal direction than they are in the vertical direction. That disparity increases rapidly with altitude. One cannot ignore the influence of anisotropy on signals that travel through the atmosphere and much of the new material included here is the result of recent attempts to include this effect.

Acknowledgments

This series is the result of many conversations with people who have contributed mightily to this field. Foremost among these is Valerian Tatarskii, who came to Boulder just as my exploration was taking on a life of its own. He has become my teacher and my friend. Valery Zavorotny also moved to Boulder and has been a generous advisor. Hal Yura reviewed the various drafts and suggested several ingenious derivations. Rich Lataitis has been a steadfast supporter, carefully reviewing my approach and suggesting important references. Reg Hill has subjected this work to searching examination for which I am truly grateful. Rod Frehlich helped by identifying important papers from the remarkable filing system that he maintains. Jim Churnside and Gerard Ochs have been generous reviewers and have paid special attention to the experiments I have cited. My friend Robert Lawrence did all the numerical computing and I owe him a special debt.

Steven Clifford extended the hospitality of the Environmental Technology Laboratory to me and has been a strong supporter from the outset. As a result of his initiative, I have enjoyed wonderful professional relationships with the people identified above. I cannot end without acknowledging the help of Mary Alice Wheelon who edited too many versions of the manuscript and double checked all the references. The drawings and figures were designed by Andrew Davies and Peter Wheelon.

Jane Watterson and her colleagues at the NOAA Library in Boulder have provided prompt and continuing reference support for my research.

At the end of the day, however, the work presented here is mine. I alone bear the responsibility for the choice of topics and their accuracy. My reward is to have taken this journey with wonderful friends.

Albert D. Wheelon
Santa Barbara, California

1

Introduction

The first volume on electromagnetic scintillation exploited geometrical optics to describe propagation in random media. That method represents an approximate solution for Maxwell's equations, which define the electromagnetic field. It was surprisingly successful in two important respects, even though it completely ignores diffraction effects.

Geometrical optics provides an accurate description for the signal-phase fluctuations imposed by a random medium. In this approximation, phase and range variations are caused by random speeding up and slowing down of the signal as it travels along the nominal ray trajectory. The phase variance estimated in this way is proportional to the distance traveled and to the first moment of the spectrum of refractive irregularities. It is therefore primarily sensitive to large eddies and diffraction effects can be safely ignored. This description is confirmed over an unusually wide range of wavelengths and propagation conditions.

The same technique was used to describe the phase difference measured between adjacent receivers. That result is needed in order to interpret observations made with microwave and optical interferometers. A similar expression characterizes the angular resolution of large telescopes in the limit of small separations. In this approach, angular errors are caused by random refractive bending of the rays as they travel through the random medium. The predicted resolution is proportional to the distance traveled and to the spectrum's third moment. Since aperture averaging suppresses the influence of irregularities smaller than the receiver, angular accuracy depends primarily on the inertial range. Diffraction effects are relatively unimportant for this reason. The resulting description is confirmed over a wide range of wavelengths and propagation conditions.

It was hoped that the same approximation would provide an accurate description for the amplitude and intensity fluctuations. These quantities are usually measured on terrestrial links and during astronomical observations. In the context of geometrical optics, amplitude fluctuations are caused by random bunching and

1

spreading of the ray trajectories. The scintillation level should be proportional
to the third power of the distance and to the spectrum's fifth moment in this description. The smallest eddies are therefore crucial for such measurements. In establishing the basic framework for geometrical optics, however, we had to insist
that all participating irregularities be large relative to the Fresnel length. That was
necessary in order that diffraction effects could be ignored. It means that the inner
scale length sets the following condition:

$$\sqrt{\lambda R} < \ell_0 \tag{1.1}$$

This limits the path length to distances less than 100 m for optical experiments and
far smaller values for microwave signals.

We are thus confronted with the reality that geometrical optics cannot describe amplitude fluctuations under the circumstances which are usually encountered. That conclusion is confirmed when one compares the predictions of
ray theory with a wide range of measurements. This makes clear the need for
a description of scintillation that includes diffraction as an essential feature.[1]
Such an account is necessarily three-dimensional because the amplitude of a
scattered electromagnetic wave is strongly influenced by interference effects in most
situations.

Maxwell's equations are the basis for all descriptions of electromagnetic propagation. We showed in Section 2.1 of Volume 1 that those relationships imply that
the electric field must satisfy the following vector wave equation:

$$\nabla^2 \mathbf{E} + k^2(1 + \delta\varepsilon)\mathbf{E} = -4\pi i k \mathbf{j}(r) - \nabla[\mathbf{E} \cdot \nabla(\delta\varepsilon)] \tag{1.2}$$

Changes in polarization induced by the random medium are characterized by the
last term. We will demonstrate in Chapter 11 that depolarization effects are far
below the measurement threshold for line-of-sight transmissions. This means that
one can describe the propagation by considering the individual components of the
electric-field vector, each of which satisfies a scalar differential equation:

$$\nabla^2 E + k^2[1 + \delta\varepsilon(\mathbf{r}, t)]E = -4\pi i k j(r) \tag{1.3}$$

Establishing solutions for this *random-wave equation* is our first challenge. We must
solve this equation for specified current sources and a variety of statistical models
for the dielectric variations. The same equation describes acoustic propagation in
the presence fluctuations in the speed of sound and research effort devoted to that
topic can be adapted to the electromagnetic problem.

---

[1] Astronomical telescopes and optical scintillometers often set a higher threshold for ignoring diffraction effects
because their aperture size replaces ℓ_0 in the condition (1.1).

1.1 The Born Approximation

The first descriptions of weak scattering that included a full account of the diffraction phenomenon were based on the Born approximation. This technique had been developed for solving scattering problems in quantum mechanics. In its applications to electromagnetic scattering by random media, one exploits the smallness of the dielectric fluctuations relative to unity:

$$\delta\varepsilon_{rms} \ll 1 \tag{1.4}$$

One then expands the electric-field strength in a series of terms that are proportional to successively higher powers of $\delta\varepsilon$:

$$E = E_0 + E_1 + E_2 + E_3 + \cdots \tag{1.5}$$

This approach has the advantage of providing a clear physical description of the process of scattering. The terms in this series are illustrated by the diagrams in Figure 1.1. The first term, E_0, represents the field strength that would be measured if the propagation medium contained no irregularities. The term E_1 represents single scattering of the incident plane wave by an irregularity at the point $\mathbf{r}$, with the scattered component reaching the receiver as a spherical wave. The term E_2 represents double scattering of the transmitted wave and is illustrated by the third

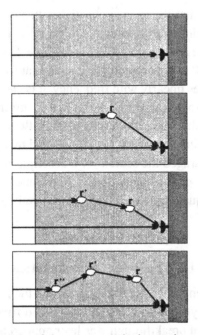

Figure 1.1: A diagrammatic description of plane-wave scattering by a random medium. The first panel represents the unperturbed plane wave. Subsequent panels portray typical single-, double- and triple-scattering sequences.

panel. The fourth term represents triple scattering and is suggested in the last panel of Figure 1.1. Differential equations that define these terms emerge when we substitute the series (1.5) into the random-wave equation (1.3) and group the results according to ascending powers of $\delta\varepsilon$.

The Born-approximation expression for the electric field is considerably more complicated than the corresponding geometrical optics description. Line integrals of the dielectric variation are replaced by volume integrations and the analytical challenge is considerably greater. With the spectrum method, however, we are able to separate the electromagnetic features of the problems from the description of turbulent irregularities. This approach also simplifies the propagation calculations and the measured quantities can be expressed as weighted averages of the wavenumber spectrum for most applications.

Despite its intuitive advantages, the Born approximation suffers from a serious limitation when it is applied to electromagnetic scattering by random media. One would like to be able to use the first few terms in the series to estimate the signal's phase and amplitude. For that to be possible, we must be sure that each term in the Born expansion is smaller than the preceding term. In Chapter 7 we show that the sum of the phase and amplitude variances must satisfy the following condition for the Born series (1.5) to converge rapidly:

$$\text{Born condition:} \qquad \left\langle \varphi^2 \right\rangle + \left\langle \chi^2 \right\rangle < 1 \qquad (1.6)$$

We can readily agree that the term $\left\langle \chi^2 \right\rangle$ satisfies this condition because we are considering weak scattering in this volume. On the other hand, the phase variance presents a major obstacle. In Section 4.1.5 of Volume 1 we found that measured phase variations are small only for microwave signals at frequencies below 5 GHz. Since the rms phase scales linearly with frequency, it is clear that the Born approximation cannot be used to describe millimeter-wave, infrared and optical signals. To make matters worse, the distribution of amplitude fluctuations predicted by the Born approach is contradicted both by microwave and by optical measurements.

These problems encouraged the search for a description of weak scattering that places a looser limit on phase fluctuations – as geometrical optics does. Fortunately, an approach adequate to this challenge was available for application to the electromagnetic problem.

1.2 The Rytov Approximation

A significant improvement on geometrical optics and the Born approximation was discovered in 1937 by Rytov, who was analyzing the diffraction of light by sound waves [1]. That analytical breakthrough was later applied by Obukhov to describe the propagation of electromagnetic waves in random media [2]. This technique is now known alternatively as the *method of smooth perturbations* or the *Rytov approximation*. It is widely used to describe line-of-sight propagation in turbulent media when the amplitude variations are small [3][4].

The Rytov approximation is fundamentally an enlargement of geometrical optics that includes diffraction effects. The essence of this method is to express the field strength as the product of the unperturbed field and the exponential of a surrogate function, which must be determined. That function is similar to the eikonal of geometrical optics but represents a far more complete physical picture. It is a complex function that describes the important influence of diffraction because it is derived from the random-wave equation. To solve specific transmission problems one expands the surrogate function in powers of $\delta\varepsilon$. The first term in this expansion is simply the single-scattering integral generated by the Born series. Most descriptions of propagation rely on this basic solution. Some estimates require the second-order solution and a few depend on higher-order terms. These solutions can be expressed as algebraic combinations of comparable terms from the Born series, which now appear as building blocks in the more powerful theory.

It was initially hoped that this approach would provide a description of multiple scattering and thus describe electromagnetic propagation for optical and infrared frequencies at large distances. Considerable debate regarding the applicability of the Rytov approximation ensued. One group maintained that it is no better than the Born method for describing transmission through random media. Others felt sure that it described a much wider class of problems. This debate was remarkable both for its intensity and for its duration. It took almost two decades to answer the following simple question: *What are the limits of applicability for the Rytov method?* Using qualitative arguments, Pisareva demonstrated that different restrictions are placed on the phase and amplitude of the propagating field [5]. She showed that the variance of logarithmic amplitude variations must be less than unity in all situations:

$$\text{Rytov condition:} \qquad \langle \chi^2 \rangle < 1 \qquad\qquad (1.7)$$

Notice that this is the same condition as that placed on the logarithmic amplitude by geometrical optics by Equation (3.45) in Volume 1. Tatarskii confirmed this with explicit calculations [3]. Pisareva also showed that the phase is unbounded for the usual case of Fresnel scattering. These conditions gives us the flexibility required to characterize weak scattering.

In the hierarchy of propagation theories we are developing, the Rytov approximation represents a natural stopping point between geometrical optics and modern theories of strong fluctuations. It describes some features of multiple scattering, just as geometrical optics does. On the other hand, it can describe weak fluctuations in amplitude and intensity – which geometrical optics cannot. We shall show that its results reduce to those of geometrical optics when the influential scatterers are concentrated near the ray path of the unperturbed field. Like Born theory, it captures the influence of diffraction phenomena. The log-normal distribution emerges as a natural consequence of the new method. An important difference with Born theory is that phase fluctuations appear naturally in the exponent of the Rytov solutions and can be very large for most applications. This amounts to a significant improvement over Born theory and means that one can treat both microwave and optical propagation in an even-handed way.

1.3 The Plan for this Volume

This volume emphasizes fluctuations in amplitude and intensity imposed on electromagnetic signals that propagate through random media. It does so because geometrical optics provides a valid description for phase and angle-of-arrival variations over a wide range of applications.

In the second chapter we develop the Rytov approximation as the method of choice for describing scintillation phenomena. We will rely primarily on the basic or first-order Rytov solution. That solution reduces to the geometrical-optics description when diffraction effects can be ignored. Moreover, it reduces to the Born approximation when the phase and amplitude variations are both small. One needs the second-order solution in some applications and two equivalent versions are established for it.

Expressions for the variance of fluctuations in amplitude and intensity are established in Chapter 3. The results for spherical and beam waves are developed for horizontal transmission paths. These predictions compare favorably with experimental data. Since amplitude fluctuations are determined primarily by small eddies, the dissipation region of the spectrum is important for interpreting optical measurements made on short paths. Aperture averaging is always a central consideration for terrestrial and astronomical measurements. This feature is readily included in our descriptions. Plane waves arriving from stellar sources are analyzed in order to understand the scintillation imposed on astronomical signals. The predicted scaling of scintillation level with zenith angle and telescope size agrees with astronomical observations. Source averaging can also influence scintillation levels and this coupling provides an important way to explore distant galaxies. Microwave signals from spacecraft and radio-astronomical sources are examined last, recognizing that the ionosphere often plays an influential role for frequencies below 10 GHz.

The correlation of amplitude fluctuations measured at adjacent receivers is addressed in Chapter 4. From the first observations of galactic radio sources, spatial correlations of radio-astronomical signals have provided important information about the ionosphere. The spatial correlation depends only on the ratio of the inter-receiver separation and the Fresnel length. That general conclusion is applied to spherical and beam waves traveling close to the surface. Aperture-averaging and inner-scale corrections are sometimes important for such links. Diffraction of plane waves passing through the atmosphere generates shadow patterns at the surface. These shadow patterns are readily observed at the focal plane of a telescope and techniques for inverting them in order to reconstruct the profile of atmospheric irregularities are reviewed.

The time correlation and power spectrum provide equivalent descriptions of the rapid fluctuations in amplitude imposed by atmospheric irregularities. In Chapter 5 we emphasize the power spectrum because it can be combined more easily with other system characteristics to estimate performance in the presence of the time-varying scintillations. We use Taylor's frozen-random-medium hypothesis to calculate the amplitude power spectrum for plane and spherical waves. Those predictions agree with atmospheric measurements in the weak-scattering regime.

The wavelength correlation for amplitude fluctuations is considered in Chapter 6. It should be the same for plane and spherical waves traveling along terrestrial paths when inner-scale and aperture-averaging effects can be ignored. This simple model agrees with microwave measurements. By contrast, the wavelength correlation of optical signals is confirmed only when the inner-scale region is properly modeled. Astronomical measurements of bichromatic correlations agree with a plane-wave model of atmospheric transmission. The possibility of inverting scintillation data recorded at different wavelengths is evaluated. This chapter also examines the frequency correlation of radio-astronomical and satellite sources, which are strongly influenced by the ionosphere at the VHF frequencies usually employed.

The discussion turns to phase fluctuations in Chapter 7, using the Rytov approximation to incorporate diffraction effects. The phase variance calculated in this way is little different than the geometrical-optics result. However, the Rytov approach is needed for estimating the cross correlation of phase and amplitude fluctuations. Diffraction expressions are also used to evaluate the phase structure function, angle-of-arrival errors and power spectrum of phase-difference measurements. The Rytov method suggests that the probability density function for phase variations is Gaussian, as has been suggested by geometrical optics and confirmed by experiments.

Double scattering of waves by refractive irregularities is important for developing a complete description of the field strength. The average value of the second order Rytov solution is required in order to estimate field-strength moments and is addressed in Chapter 8. This average can be evaluated exactly for spherical waves traveling along horizontal paths and for plane waves passing vertically through the atmosphere. These expressions include wide-angle scattering and therefore describe microwave propagation with the same fidelity as that normally associated with optical expressions. An important relationship is established between these double-scattering averages and the combined variances of phase and amplitude. Beam waves traveling along terrestrial paths can be analyzed only by assuming that the individual scattering events are described with the paraxial approximation. The results of this chapter are primarily analytical and intended to provide modules needed later. The reader who is more interested in applications may elect to pass over this material.

The description of propagation in random media is significantly enlarged in Chapter 9, where various moments of the electric-field strength are derived. One needs both the first- and the second-order Rytov solution in order to generate accurate descriptions of these moments. The average field strength or mean field decays exponentially with distance for all three types of wave. By contrast, the mean irradiance is everywhere equal to its free-space value for plane and spherical waves. This is not true for beam waves, which are broadened by scattering in the random medium. Conservation of energy is demonstrated using both diagrammatic and analytical methods. The mutual coherence function provides a fundamental description of the electromagnetic field and is calculated for the three types of wave. These expressions agree generally with astronomical and laser-link measurements. Irradiance fluctuations are often measured with logarithmic amplifiers to compress the large dynamic range of scintillations encountered at optical and infrared wavelengths. The mean logarithmic irradiance and its variance are simply related to familiar path and signal parameters when the scattering is weak. These predictions are confirmed by optical experiments over a limited range. The predictions and measured quantities rapidly depart from one another above a certain level. The experimental data saturate as the path length or structure constant increases. This behavior represents the onset of multiple scattering and is addressed in Volume 3. It cannot be explained with the Rytov approximation and such experiments establish an experimental boundary for our approach to weak scattering.

The probability density function for amplitude fluctuations is discussed in Chapter 10. The basic Rytov solution predicts a log-normal distribution for short-term variations, which is confirmed over a wide range of frequencies and propagation conditions. The second order Rytov solution suggests that this distribution should be slightly skewed and that prediction too is confirmed. The bivariate distribution of rapid fluctuations measured with separated receivers or displaced times is examined. Intermittent atmospheric structures exert a strong influence on the measured distribution when the path length is short or the sample size small. The amplitude distribution measured over much longer time scales is important for predicting the performance of a communication system. Diurnal and seasonal variations of signal-level distributions are closely related to atmospheric conditions and are best described by phenomenological models. Satellite-signal fluctuations are examined separately and seem to fit a somewhat different pattern.

The polarization of electromagnetic waves is altered very slightly as they travel through random media. This depolarization is explored in Chapter 11 and found to be so small that it should not be measurable on line-of-sight paths. That prediction is confirmed by experiments.

In Chapter 12 we return to the following important question: *Under what circumstances is the Rytov approximation valid?* The answer emerges when one compares the first- and second-order terms in the Rytov series. This conclusion is confirmed by the measurements summarized in Chapter 9.

An extensive review of refractive irregularities in the troposphere and ionosphere was provided in Chapter 2 of Volume 1. That material is not repeated in this volume. Instead, frequent references are made in the text and in footnotes to the relevant figures and descriptions in Volume 1.

The appendices are a combination of new material and subjects presented at the end of the first volume. The glossary is enlarged to include new symbols that have been introduced in this volume. The previous appendices on probability distributions and Kummer functions in Volume 1 have been expanded to provide additional results that are needed here. New appendices have been added to cover Green's function, cumulant analysis, diffraction integrals and Feynman formulas.

References

[1] S. M. Rytov, "Diffraction of Light by Ultrasonic Waves," *Izvestiya Akademii Nauk SSSR, Seriya Fizicheskaya (Bulletin of the Academy of Sciences of the USSR, Physical Series)*, No. 2, 223–259 (1937). (No English translation is available.)

[2] A. M. Obukhov, "On the Influence of Weak Atmospheric Inhomogeneities on the Propagation of Sound and Light," *Izvestiya Akademii Nauk SSSR, Seriya Geofizicheskaya (Bulletin of the Academy of Sciences of the USSR, Geophysical Series)*, No. 2, 155–165 (1953). (English translation by W. C. Hoffman, published by U. S. Air Force Project RAND as Report T-47, Santa Monica, CA, 28 July 1955.)

[3] V. I. Tatarskii, *The Effects of the Turbulent Atmosphere on Wave Propagation* (translated from the Russian and issued by the National Technical Information Office, U. S. Department of Commerce, Springfield, VA 22161, 1971), 218–258.

[4] Yu. N. Barabanenkov, Yu. A. Kravtsov, S. M. Rytov and V. I. Tatarskii, "Status of the Theory of Propagation of Waves in a Randomly Inhomogeneous Medium," *Uspekhi Fizicheskikh Nauk (Soviet Physics – Uspekhi)*, **13**, No. 5, 551–580 (March–April, 1971).

[5] V. V. Pisareva. "Limits of Applicability of the Method of 'Smooth' Perturbations in the Problem of Radiation Propagation through a Medium Containing Inhomogeneities," *Akusticheskii Zhurnal (Soviet Physics – Acoustics)*, **6**, No. 1, 81–86 (July–September 1960).

2

The Rytov Approximation

The purpose of this chapter is to lay the foundation for describing phase and amplitude fluctuations on an equal footing. That foundation rests on the *Rytov approximation* or *method of smooth perturbations* that was introduced in Chapter 1. This method provides a powerful technique with which one can solve the random-wave equation (1.3) subject to the condition (1.7). For weak scattering of waves in random media, the Rytov approximation provides a significant improvement both on geometrical optics and on the Born approximation.

There are two steps in this approach. The first and most important step is to completely change the structure of the problem. Rytov discovered a transformation of the electric-field strength that changes the random-wave equation to one that can be solved in many interesting situations [1]. The second step is to expand the transformed field strength into a series of terms proportional to successively higher powers of the dielectric variations [2].

The same approach is often applied to the parabolic-wave equation, which represents a special case of (1.3) that describes small-angle forward scattering [3]. We shall use the more general Helmholtz equation here so as to include microwave propagation which often includes wide-angle scattering. We specialize these results using the paraxial approximation for optical and infrared applications – where the scattering angle is usually small. This chapter is necessarily analytical and the reader who is primarily interested in applications may wish to proceed directly to those topics which begin with the next chapter.

2.1 Rytov's Transformation

The starting point for our development of the Rytov approximation is the random-wave equation (1.3) without the polarization term, which we repeat here for easy reference:

$$\nabla^2 E + k^2[1 + \delta\varepsilon(\mathbf{r}, t)]E = -4\pi i k j(\mathbf{r}) \tag{2.1}$$

The challenge is to solve this equation for a specified current source and various statistical models of the dielectric variations. That is a difficult analytical task even if $\delta\varepsilon$ is a simple algebraic function. The challenge is infinitely greater in our application because $\delta\varepsilon(\mathbf{r}, t)$ is a stochastic function. This function is specified only by its moments and correlations estimated over an ensemble of possible atmospheric configurations. Since preemptive surrender is not an option, we must find a way to simplify the task.

The fundamental problem in solving (2.1) is that the random function $\delta\varepsilon(\mathbf{r}, t)$ multiplies the function $E(\mathbf{r})$ we are trying to discover. In an ideal world, we would like to separate these two functions and have $\delta\varepsilon(\mathbf{r}, t)$ appear as an additional source term. It is quite astonishing that *Rytov's transformation* achieves precisely that goal. His idea is to express the solution as the product of two terms: (a) the coherent field strength E_0 that would be measured in the absence of irregularities, and (b) the exponent of a *surrogate function* that must be discovered:

$$E(\mathbf{r}) = E_0(\mathbf{r}) \exp[\Psi(\mathbf{r})] \tag{2.2}$$

When this trial solution is substituted into the random-wave equation, one finds that the surrogate function $\Psi(\mathbf{r})$ must satisfy a nonlinear partial differential equation in which the dielectric variation appears only as an additional source term.

Let us pause to note the similarity between this approach and geometrical optics. The surrogate function $\Psi(\mathbf{r})$ is similar to the eikonal of ray theory[1] and we will find a good deal of common ground in the two approaches. However, the eikonal is purely imaginary and is an effective descriptor only of phase changes. The surrogate function introduced here is assumed to be complex so that it can describe both the phase and the amplitude of a wave. A second difference is that the Laplacian term is discarded in geometrical optics but plays a major role in the Rytov approximation. The influence of diffraction is included by keeping the Laplacian and allowing Ψ to be complex. In this way one develops a description of propagation in random media that is superior both to geometrical optics and to the Born approximation.

We substitute the field-strength transformation (2.2) into the random-wave equation (2.1) and use the following identity from Appendix J:

$$\nabla^2(FG) = F\,\nabla^2 G + 2\,\nabla F \cdot \nabla G + G\,\nabla^2 F$$

to demonstrate that

$$E_0\,\nabla^2(e^\Psi) + 2\,\nabla E_0 \cdot \nabla(e^\Psi) + k^2\,\delta\varepsilon\,E_0 e^\Psi$$
$$= -4\pi i k j(\mathbf{r}) - e^\Psi\,\nabla^2 E_0 - k^2 E_0 e^\Psi$$

---

[1] See Chapter 3 of Volume 1.

The terms on the right-hand side are each proportional to the transmitter current density since the unperturbed field strength satisfies the simple wave equation

$$\nabla^2 E_0 + k^2 E_0 = -4\pi i k j(\mathbf{r}) \tag{2.3}$$

This means that

$$\nabla^2(e^\Psi) + 2\,\nabla(\ln E_0)\cdot\nabla(e^\Psi) + k^2 e^\Psi\,\delta\varepsilon = -4\pi i k j(\mathbf{r})\left(\frac{e^{-\Psi}-1}{E_0}\right)$$

We can drop the current-density term on the right-hand side because we are primarily interested in solving this equation far from the transmitter. Using the differential identity

$$\nabla^2(e^\Psi) = \nabla\cdot(e^\Psi\,\nabla\Psi) = e^\Psi(\nabla\Psi)^2 + e^\Psi\,\nabla^2\Psi$$

we find that the surrogate for the field strength must satisfy the following equation:

$$\nabla^2\Psi + (\nabla\Psi)^2 + 2\,\nabla(\ln E_0)\cdot\nabla(\Psi) = -k^2\,\delta\varepsilon(\mathbf{r}, t) \tag{2.4}$$

We see now that the goal of Rytov's transformation has been achieved. The stochastic function $\delta\varepsilon$ no longer multiplies the function we are seeking. Instead, it appears as an artificial source term.

The transmitter current density influences the solution of (2.4) through the unperturbed field strength which satisfies (2.3). It is convenient to represent this field as the exponential of a dimensionless function that is similar to the surrogate function:

$$E_0(\mathbf{r}) = \exp[\psi_0(\mathbf{r})] \tag{2.5}$$

We shall use these equivalent descriptions of the unperturbed field interchangeably. The total field which travels through the random medium can now be written in exponential form:

$$E(\mathbf{r}) = \exp[\psi_0(\mathbf{r}) + \Psi(\mathbf{r})] \tag{2.6}$$

When we substitute this expression into (2.4) we see that the surrogate function must satisfy a nonlinear partial differential equation:

$$\nabla^2\Psi + (\nabla\Psi)^2 + 2\,\nabla\psi_0\cdot\nabla\Psi + k^2\,\delta\varepsilon = 0 \tag{2.7}$$

2.1.1 The Relation to Riccati's Equation

The equation defining the surrogate function bears a close resemblance to a differential equation that has been studied by mathematicians for almost three hundred years. To see this connection we represent the surrogate's gradient by

$$\mathbf{U} = \nabla\Psi \tag{2.8}$$

which allows us to rewrite (2.7) in the following form:

$$\nabla \cdot \mathbf{U} + \mathbf{U} \cdot \mathbf{U} + 2\mathbf{U} \cdot \nabla \psi_0 + k^2 \, \delta\varepsilon = 0 \tag{2.9}$$

This is the three-dimensional version of *Riccati's differential equation*, whose general form is given by

$$\frac{dy}{dx} + b(x)y^2 + a(x)y + c(x) = 0 \tag{2.10}$$

This nonlinear differential equation plays an influential role in several fields of mathematics. It is important in differential geometry and there are twenty references to Riccati's equation in Darboux's *Leçons sur la théorie générale des surfaces* [4]. It also plays a central role in the theory of Bessel functions [5]. More recently, the same equation has emerged in the description of soliton waves [6]. The considerable mathematical literature associated with this equation is summarized by Watson [5].

It has long been known that one can transform Riccati's equation into a linear second-order differential equation with variable coefficients. One does so by introducing

$$y = \frac{1}{bw} \frac{dw}{dx} \tag{2.11}$$

into (2.10), which yields

$$\frac{d^2 w}{dx^2} + \left(a - \frac{1}{b} \frac{db}{dx} \right) \frac{dw}{dx} + cbw = 0 \tag{2.12}$$

This change is traditionally used to find solutions of Riccati's equation by solving the resulting linear second-order equation.

Rytov's approach reverses this process. He begins with a second-order differential equation with variable coefficients (2.1) and transforms it into the nonlinear first-order equation (2.7). In this revised description, the variable coefficients appear as a driving term and are separated from the function that is being sought. The enormous contribution of Rytov and Obukhov was to recognize this possibility and to note that the transformation works equally well when one of the coefficients is a stochastic function of position.

2.1.2 The Series-expansion Solution

What remains is to solve the transformed wave equation (2.7). We do so by expanding the surrogate function in powers of the dielectric variation:

$$\Psi(\mathbf{r}) = \psi_1(\mathbf{r}) + \psi_2(\mathbf{r}) + \psi_3(\mathbf{r}) + \psi_4(\mathbf{r}) + \cdots \tag{2.13}$$

The terms ψ_n are proportional to the nth power of $\delta\varepsilon$. When this expansion is substituted into the nonlinear equation (2.7) and the terms separated according to ascending powers of $\delta\varepsilon$, one finds that the following equations determine the successive approximations to the surrogate function [7]. We also include the equation satisfied by ψ_0 which represents the unperturbed field strength. In doing so, we consider only points in the random medium away from the transmitter:

$$\nabla^2\psi_0 + (\nabla\psi_0)^2 + k^2 = 0$$

$$\nabla^2\psi_1 + 2\nabla\psi_0\cdot\nabla\psi_1 + k^2\,\delta\varepsilon = 0$$

$$\nabla^2\psi_2 + 2\nabla\psi_0\cdot\nabla\psi_2 + (\nabla\psi_1)^2 = 0$$

$$\nabla^2\psi_3 + 2\nabla\psi_0\cdot\nabla\psi_3 + 2\nabla\psi_1\cdot\nabla\psi_2 = 0 \qquad (2.14)$$

$$\nabla^2\psi_4 + 2\nabla\psi_0\cdot\nabla\psi_4 + 2\nabla\psi_1\cdot\nabla\psi_3 + (\nabla\psi_2)^2 = 0$$

$$\nabla^2\psi_n + 2\nabla\psi_0\cdot\nabla\psi_n + \sum_{p=1}^{n-1}\nabla\psi_p\cdot\nabla\psi_{n-p} = 0$$

The vacuum field defined by (2.3) influences each succeeding approximation through the coupling term $2\,\nabla\psi_0\cdot\nabla\psi_n$.

2.2 The Basic Rytov Solution

The *basic Rytov solution* is used in the vast majority of applications to describe propagation in random media. It is defined by the term ψ_1 in the expansion (2.13) and is the solution of the second field equation in (2.14). That equation depends only on the unperturbed field surrogate ψ_0 and the dielectric variation $\delta\varepsilon$. We rewrite it as follows:

$$\nabla^2\psi_1 + 2\nabla\psi_0\cdot\nabla\psi_1 = -k^2\,\delta\varepsilon \qquad (2.15)$$

To solve this equation we make the following substitution:

$$\psi_1(\mathbf{r}) = Q(\mathbf{r})\exp[-\psi_0(\mathbf{r})] \qquad (2.16)$$

When this form is introduced into (2.15) that equation becomes

$$e^{-\psi_0}\nabla^2 Q - 2e^{-\psi_0}\nabla\psi_0\cdot\nabla Q + Q[-e^{-\psi_0}\nabla^2\psi_0 + e^{-\psi_0}(\nabla\psi_0)^2]$$

$$+ 2\nabla\psi_0\cdot(e^{-\psi_0}\nabla Q - e^{-\psi_0}Q\nabla\psi_0) = -k^2\,\delta\varepsilon(\mathbf{r}, t)$$

which simplifies to

$$\nabla^2 Q + Q[-\nabla^2\psi_0 - (\nabla\psi_0)^2] = -k^2 e^{\psi_0}\,\delta\varepsilon(\mathbf{r}, t)$$

The coefficient of Q in square brackets is just k^2 in view of the first equation in (2.14). The auxiliary function therefore satisfies a normal *Helmholtz equation* with a random source term:

$$\nabla^2 Q + k^2 Q = -k^2 \, \delta\varepsilon \, \exp[\psi_0(\mathbf{r})]$$

This is solved using Green's function, which is explained in Appendix L:

$$Q(\mathbf{R}) = -k^2 \int d^3r \, G(\mathbf{R}, \mathbf{r}) \, \delta\varepsilon(\mathbf{r},t) \exp[\psi_0(\mathbf{r})] \qquad (2.17)$$

The basic Rytov solution emerges when we combine this with (2.15) and replace the exponential of $\psi_0(\mathbf{r})$ by $E_0(\mathbf{r})$:

$$\psi_1(\mathbf{R}) = B_1 = -k^2 \int d^3r \, G(\mathbf{R}, \mathbf{r}) \, \delta\varepsilon(\mathbf{r},t) \frac{E_0(\mathbf{r})}{E_0(\mathbf{R})} \qquad (2.18)$$

This is just the single-scattering term generated by the Born approximation, which we designate by B_1. It is normalized by the unperturbed field strength. The basic Rytov approximation becomes

$$E_1(\mathbf{R}) = E_0(\mathbf{R}) \exp\left(-k^2 \int d^3r \, G(\mathbf{R}, \mathbf{r}) \, \delta\varepsilon(\mathbf{r},t) \frac{E_0(\mathbf{r})}{E_0(\mathbf{R})}\right) \qquad (2.19)$$

or

$$E_1(\mathbf{R}) = E_0(\mathbf{R}) \exp(B_1) \qquad (2.20)$$

When the scattering is weak the exponential can then be expanded and this expression replicates the first term in the Born approximation:

$$B_1 \ll 1 \qquad E_1(\mathbf{R}) = E_0(\mathbf{R})(1 + B_1 + \cdots)$$

The first Born term is large in our applications for most cases of interest and the scattering integral must remain in the exponent. This has a profound effect on the theory and its agreement with experiments.

Let us pause to make our result explicit for two common cases. If the unperturbed field is a plane wave coming from a distant source, the basic solution can be written

$$E_1(\mathbf{R}) = \mathcal{E}_0 \exp\left(ikR - k^2 \int d^3r \, G(\mathbf{R}, \mathbf{r}) \, \delta\varepsilon(\mathbf{r}) \exp[ik\,(z - R)]\right) \qquad (2.21)$$

If the transmitted field is a spherical wave, the basic solution becomes

$$E_1(\mathbf{R}) = \frac{\mathcal{E}_0}{4\pi R} \exp\left(ikR - k^2 \int d^3r \, G(\mathbf{R}, \mathbf{r}) \frac{R}{r} \delta\varepsilon(\mathbf{r}) \exp[ik(r - R)]\right)$$

$$(2.22)$$

2.2.1 Phase and Amplitude Expressions

One can identify the phase and amplitude of the received signal in the expression (2.19) for the basic Rytov solution. If we write the field strength as

$$E_1(\mathbf{R}) = A(\mathbf{R}) \exp\{i[\phi_0 + \varphi(\mathbf{R})]\} \tag{2.23}$$

the fluctuating part of the signal phase is simply the imaginary part of the scattering integral:

$$\varphi(\mathbf{R}) = -k^2 \int d^3r\, \delta\varepsilon(\mathbf{r})\, \Im\left(G(\mathbf{R}, \mathbf{r}) \frac{E_0(\mathbf{r})}{E_0(\mathbf{R})}\right) \tag{2.24}$$

This expression provides a surprisingly complete description of phase fluctuations but is only slightly better than geometrical optics. It provides a vast improvement on Born theory, whose series expansion does not converge when the phase variations are greater than unity. The imaginary part of the scattering integral now occurs in the exponent of (2.19) where one expects to find a phase term. In this location it can assume large values and the Rytov phase expression can describe the phase of optical and millimeter-wave signals.

The signal amplitude is related logarithmically to the real part of the scattering integral:

$$A(\mathbf{R}) = |E_0(\mathbf{R})| \exp\left[-k^2 \int d^3r\, \delta\varepsilon(\mathbf{r})\, \Re\left(G(\mathbf{R}, \mathbf{r}) \frac{E_0(\mathbf{r})}{E_0(\mathbf{R})}\right)\right] \tag{2.25}$$

This formulation is fundamentally different than the Born approximation. It leads naturally to the log-normal distribution for fluctuations in amplitude and intensity, thereby remedying a major defect of Born theory. In doing so it fills the void left by geometrical optics. We usually deal with the logarithmic amplitude, which is written as

$$\chi = \log\left(\frac{A}{E_0}\right) = -k^2 \int d^3r\, \delta\varepsilon(\mathbf{r})\, \Re\left(G(\mathbf{R}, \mathbf{r}) \frac{E_0(\mathbf{r})}{E_0(\mathbf{R})}\right) \tag{2.26}$$

It is helpful to introduce a simplified notation when one is calculating field quantities with the basic Rytov solution. We designate the real and imaginary parts of the normalized scattering integral by two dimensionless random variables:

$$\mathbf{B}_1 = a + ib = -k^2 \int d^3r\, \delta\varepsilon(\mathbf{r})\, G(\mathbf{R}, \mathbf{r}) \frac{E_0(\mathbf{r})}{E_0(\mathbf{R})} \tag{2.27}$$

To first order the fluctuations in logarithmic amplitude and phase are identical to these variables:

$$\chi = a \qquad \text{and} \qquad \varphi = b \tag{2.28}$$

The moments of these stochastic variables are the primary quantities that must be compared with microwave and optical measurements. The average values of a and b vanish because $\langle \delta \varepsilon \rangle = 0$ in view of the partitioning of dielectric changes outlined in Section 2.2.1 of Volume 1. This means that the following mean values are realized if the sample length is sufficiently long relative to the important harmonic components of the dielectric variations:

$$\langle \chi \rangle = \langle a \rangle = 0 \quad \text{and} \quad \langle \varphi \rangle = \langle b \rangle = 0 \tag{2.29}$$

The variances of phase and logarithmic amplitude are the first measures of the signal's variability. To calculate these quantities we decompose the spatial weighting function in the first Born term into its real and imaginary parts:

$$G(\mathbf{R}, \mathbf{r}) \frac{E_0(\mathbf{r})}{E_0(\mathbf{R})} = A(\mathbf{R}, \mathbf{r}) + i B(\mathbf{R}, \mathbf{r}) \tag{2.30}$$

We can now express the phase and amplitude fluctuations as

$$\chi = a = -k^2 \int d^3r \, A(\mathbf{R}, \mathbf{r}) \, \delta \varepsilon(\mathbf{r}, t) \tag{2.31}$$

$$\varphi = b = -k^2 \int d^3r \, B(\mathbf{R}, \mathbf{r}) \, \delta \varepsilon(\mathbf{r}, t) \tag{2.32}$$

and the variances are given by

$$\langle \chi^2 \rangle = k^4 \int d^3r \, A(\mathbf{R}, \mathbf{r}) \int d^3r' \, A(\mathbf{R}, \mathbf{r}') \langle \delta \varepsilon(\mathbf{r}, t) \, \delta \varepsilon(\mathbf{r}', t) \rangle \tag{2.33}$$

and

$$\langle \varphi^2 \rangle = k^4 \int d^3r \, B(\mathbf{R}, \mathbf{r}) \int d^3r' \, B(\mathbf{R}, \mathbf{r}') \langle \delta \varepsilon(\mathbf{r}, t) \, \delta \varepsilon(\mathbf{r}', t) \rangle \tag{2.34}$$

The spatial correlation of $\delta \varepsilon$ couples the two volume integrations in a complicated way. Authors of early studies of propagation in random media used analytical models for the spatial correlation and struggled to calculate the double volume integrals in these expressions.

2.2.1.1 Homogeneous Random Media

A far more versatile approach is based on introducing the Fourier wavenumber decomposition of the spatial correlation. For a homogeneous random medium this simplifies the analytical problem dramatically. We use the equivalent description for the spatial correlation of $\delta \varepsilon$ that was introduced in Section 2.2.5 of Volume 1:

$$\langle \delta \varepsilon(\mathbf{r}, t) \, \delta \varepsilon(\mathbf{r}', t) \rangle = \int d^3\kappa \, \Phi_\varepsilon(\kappa) \exp[i\kappa \cdot (\mathbf{r} - \mathbf{r}')] \tag{2.35}$$

This substitution separates the wave-scattering features of the problem from the description of the turbulent medium. When it is combined with (2.33) we see that the volume integrations are uncoupled and can be interchanged with the wavenumber integration:

$$\langle \chi^2 \rangle = k^4 \int d^3\kappa \, \Phi_\varepsilon(\kappa) \int d^3r \, A(\mathbf{R}, \mathbf{r}) \exp(i\boldsymbol{\kappa} \cdot \mathbf{r})$$
$$\times \int d^3r' \, A(\mathbf{R}, \mathbf{r}') \exp(-i\boldsymbol{\kappa} \cdot \mathbf{r}')$$

If we define the *amplitude weighting function* by

$$D(\kappa) = k^2 \int d^3r \, A(\mathbf{R}, \mathbf{r}) \exp(i\boldsymbol{\kappa} \cdot \mathbf{r}) \tag{2.36}$$

we can express the amplitude variance as a weighted wavenumber integral of the spectrum of irregularities:

$$\langle \chi^2 \rangle = \int d^3\kappa \, \Phi_\varepsilon(\kappa) D(\kappa) D(-\kappa) \tag{2.37}$$

In a similar way the phase variance can be expressed as

$$\langle \varphi^2 \rangle = \int d^3\kappa \, \Phi_\varepsilon(\kappa) E(\kappa) E(-\kappa) \tag{2.38}$$

where the *phase weighting function* is defined by

$$E(\kappa) = k^2 \int d^3r \, B(\mathbf{R}, \mathbf{r}) \exp(i\boldsymbol{\kappa} \cdot \mathbf{r}) \tag{2.39}$$

The cross correlation between amplitude and phase can also be expressed in terms of these weighting functions:

$$\langle \chi\varphi \rangle = \int d^3\kappa \, \Phi_\varepsilon(\kappa) D(\kappa) E(-\kappa) \tag{2.40}$$

The electromagnetic and geometrical features of the propagation are isolated to the weighting functions $D(\kappa)$ and $E(\kappa)$. By contrast, conditions in the random medium enter the calculation only through the turbulence spectrum $\Phi_\varepsilon(\kappa)$. The two influences are merged through the wavenumber integration as the last step in the description. This is clearly a desirable outcome and makes it possible to combine modules from different attacks on the problem.

Much of the development in this volume is devoted to estimating the weighting functions $D(\kappa)$ and $E(\kappa)$ for various types of signals and propagation geometries. That task is simplified by combining these two functions as the real and imaginary parts of a *complex weighting function* [8]:

$$\Lambda(\kappa) = D(\kappa) + iE(\kappa) \tag{2.41}$$

If we introduce the definitions (2.36) and (2.39), we see that this function can be written as follows:

$$D(\kappa) + iE(\kappa) = k^2 \int d^3r \, (A(\mathbf{R}, \mathbf{r}) + iB(\mathbf{R}, \mathbf{r})] \exp(i\kappa \cdot \mathbf{r})$$

In view of the earlier separation (2.30) of the spatial weighting terms, we can write the complex weighting function as follows:

$$\Lambda(\kappa) = k^2 \int d^3r \left(G(\mathbf{R}, \mathbf{r}) \frac{E_0(\mathbf{r})}{E_0(\mathbf{R})} \right) \exp(i\kappa \cdot \mathbf{r}) \tag{2.42}$$

This means that we need not separate the combination in large parentheses into its real and imaginary parts before evaluating the volume integration. Once the complex weighting function has been established, $D(\kappa)$ and $E(\kappa)$ can be calculated easily from the following relationships:

$$D(\kappa) = \tfrac{1}{2}\big[\Lambda(\kappa) + \Lambda^*(-\kappa)\big] \tag{2.43}$$

and

$$E(\kappa) = \tfrac{1}{2i}\big[\Lambda(\kappa) - \Lambda^*(-\kappa)\big] \tag{2.44}$$

A single calculation thus leads directly to estimates of the amplitude and phase variances, and to their cross correlation. We shall find that $\Lambda(\kappa)$ also determines the correlations of phase and amplitude with distance, time delay and frequency offset. The complex weighting function therefore provides a powerful unifying device for our work. It is determined by the propagation geometry and the properties of the radiated wave; namely frequency, power, signal waveform, antenna pattern, etc. Unfortunately, this elegant method has a limitation.

2.2.1.2 Inhomogeneous Random Media

The separation of turbulent and electromagnetic effects presented above depends squarely on the representation (2.35) for the spatial covariance of $\delta\varepsilon$. That expression is valid only for homogeneous random media. In this volume we will often consider transmission of electromagnetic waves through the entire atmosphere. That situation occurs for all optical and radio-astronomy observations. It also represents the propagation of microwave signals transmitted by communication and navigation satellites, which must pass through the atmosphere if they are to be useful.

We learned in Section 2.3.3 of Volume 1 that the parameters which characterize atmospheric turbulence change dramatically as one ascends from ground level to the outer boundary of the atmosphere. One evidently cannot assume that a random medium is homogeneous in these applications. A procedure for describing inhomogeneous media was presented in Section 2.2.8 of Volume 1. In that system the spatial covariance is represented by a wavenumber integral but the turbulence spectrum now depends on the average position of the two points:

$$\langle \delta\varepsilon(\mathbf{r}_1)\,\delta\varepsilon(\mathbf{r}_2)\rangle = \int d^3\kappa \; \Phi_\varepsilon\left(\kappa, \frac{\mathbf{r}_1+\mathbf{r}_2}{2}\right) \exp[i\kappa\cdot(\mathbf{r}_1-\mathbf{r}_2)] \qquad (2.45)$$

When this expression is substituted into the amplitude-variance description (2.33) one finds that

$$\langle \chi^2\rangle = k^4 \int d^3\kappa \int d^3r \, A(\mathbf{R}, \mathbf{r}) \exp(i\kappa\cdot r)$$

$$\times \int d^3r' \, A(\mathbf{R}, \mathbf{r}') \exp(-i\kappa\cdot r') \, \Phi_\varepsilon\left(\kappa, \frac{\mathbf{r}_1+\mathbf{r}_2}{2}\right) \qquad (2.46)$$

The location dependence of the spectrum Φ_ε couples the volume integrals and prevents one from representing the electromagnetic features of the problem by an amplitude weighting function. Instead, one must deal directly with the nine-fold integral presented here. We will suggest approximate ways to do so in Chapter 3 when we first consider astronomical observations.

There is a second way to describe random media that are not uniform. In this approach, one represents the spatial covariance of two dielectric fluctuations by the product of three terms. The first is a true spatial correlation function that depends only on the separation between the two points. This is multiplied by the rms values of the dielectric variation measured at both points:

$$\langle \delta\varepsilon(\mathbf{r})\,\delta\varepsilon(\mathbf{r}')\rangle = \delta\varepsilon_{\mathrm{rms}}(\mathbf{r})\delta\varepsilon_{\mathrm{rms}}(\mathbf{r}')C(\mathbf{r}-\mathbf{r}') \qquad (2.47)$$

In the context of tropospheric irregularities the rms values are proportional to the local value of C_n^2. We now introduce a *turbulence-profile function* defined by the local values of this important parameter:[2]

$$\wp(\mathbf{r}) = \sqrt{\frac{C_n^2(\mathbf{r})}{C_n^2(0)}} \qquad (2.48)$$

With this dimensionless ratio we can describe the covariance as a product of the profile functions for the two points and a spatial correlation function that depends on the difference in location:

$$\langle \delta\varepsilon(\mathbf{r})\delta\varepsilon(\mathbf{r}')\rangle = \wp(\mathbf{r})\wp(\mathbf{r}')\langle \delta\varepsilon^2\rangle C(\mathbf{r}-\mathbf{r}')$$

---

[2] This definition does not recognize the variation of inner and outer scale lengths with height, but their variabilities are usually small compared with the large changes in C_n^2 that are observed.

The last two terms can be represented by the wavenumber integral of the spectrum introduced in (2.35):

$$\langle \delta\varepsilon(\mathbf{r}, t)\delta\varepsilon(\mathbf{r}', t)\rangle = \wp(\mathbf{r})\wp(\mathbf{r}') \int d^3\kappa \, \Phi_\varepsilon(\kappa) \exp[i\kappa \cdot (\mathbf{r} - \mathbf{r}')] \qquad (2.49)$$

When this representation is introduced into the variance expression (2.33), the electromagnetic and turbulent features of the problem are separated. The amplitude weighting function is now defined by a volume integral that is weighted by the profile function:

$$D(\kappa) = k^2 \int d^3r \, A(\mathbf{R}, \mathbf{r}) \exp(i\kappa \cdot \mathbf{r})\wp(\mathbf{r}) \qquad (2.50)$$

The same approach leads to a more general version of the complex weighting function:

$$\Lambda(\kappa) = \int d^3r \, G(\mathbf{R}, \mathbf{r}) \frac{E_0(\mathbf{r})}{E_0(\mathbf{R})} \exp(i\kappa \cdot \mathbf{r}) \wp(\mathbf{r}) \qquad (2.51)$$

The profile function $\wp(\mathbf{r})$ varies slowly with position and it is usually a function only of height above the earth's surface. We will exploit this feature to simplify the calculations that follow. It will turn out that both approaches lead to the same conclusion in the approximations that we will use.

2.2.2 Reduction to Geometrical Optics

Let us pause to show how the basic Rytov solution reduces to geometrical optics. Consider a plane wave falling on a turbulent medium. We use the rectangular coordinates shown in Figure 2.1 to write the first-order solution:

$$E_1(R) = \mathcal{E}_0 \exp\left(ikR - k^2 \int_0^R dz \int_{-\infty}^\infty dx \int_{-\infty}^\infty dy \, \delta\varepsilon(x, y, z) \right.$$
$$\times \frac{\exp\left(ik\sqrt{x^2 + y^2 + (R-z)^2}\right)}{4\pi\sqrt{x^2 + y^2 + (R-z)^2}} \exp[ik(z - R)] \Bigg)$$

$$(2.52)$$

Since geometrical optics depends primarily on values of the dielectric variation close to the nominal ray path, one can expand $\delta\varepsilon$ in powers of x and y:

$$\delta\varepsilon(x, y, z) = \delta\varepsilon(0, 0, z) + x\left(\frac{\partial\delta\varepsilon}{\partial x}\right)_0 + y\left(\frac{\partial\delta\varepsilon}{\partial y}\right)_0$$
$$+ \frac{1}{2}\left[x^2\left(\frac{\partial^2}{\partial x^2}\delta\varepsilon\right)_0 + xy\left(\frac{\partial^2}{\partial x \partial y}\delta\varepsilon\right)_0 + y^2\left(\frac{\partial^2}{\partial x^2}\delta\varepsilon\right)_0 \right]$$

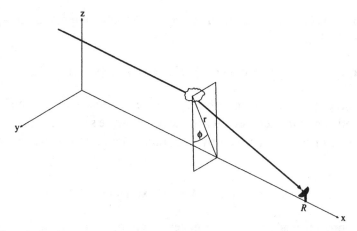

Figure 2.1: The geometry for analyzing the reduction of the Rytov approximation to geometrical optics for a plane wave incident on a random medium.

When we substitute this expansion into (2.52) and convert to cylindrical coordinates, the azimuth integration eliminates linear and cross terms in $\sin\phi$ and $\cos\phi$:

$$E_1(R) = \mathcal{E}_0 \exp\left(ikR - k^2 \int_0^R dz\, e^{ik(z-R)} \int_0^\infty dr\, r\right.$$

$$\times \frac{\exp\left[ik\sqrt{r^2 + (R-z)^2}\right]}{\sqrt{r^2 + (R-z)^2}}$$

$$\left.\times \left[\tfrac{1}{2}\delta\varepsilon(0,0,z) + \tfrac{1}{8}r^2(\nabla_\perp^2\delta\varepsilon)_{x=y=0}\right]\right) \qquad (2.53)$$

The radial integrations are barely convergent but they can be evaluated by introducing a small attenuation term and using Lamb's integral, which is discussed in Appendix D:

$$\lim_{\epsilon\to 0}\int_0^\infty dr\, r\, J_0(r\epsilon)\frac{\exp\left[ik\sqrt{r^2+(R-z)^2}\right]}{\sqrt{r^2+(R-z)^2}}$$

$$= \lim_{\epsilon\to 0}\left(\frac{ie^{i|R-z|\sqrt{k^2-\epsilon^2}}}{\sqrt{k^2-\epsilon^2}}\right) = \frac{i}{k}e^{ik|R-z|}$$

$$\lim_{\epsilon\to 0}\int_0^\infty dr\, r^3\, J_0(r\epsilon)\frac{\exp\left[ik\sqrt{r^2+(R-z)^2}\right]}{\sqrt{r^2+(R-z)^2}}$$

$$= \lim_{\epsilon\to 0}\left(-\frac{2}{\epsilon}\frac{\partial}{\partial\epsilon}\frac{ie^{i|R-z|\sqrt{k^2-\epsilon^2}}}{\sqrt{k^2-\epsilon^2}}\right) = -2\left(\frac{|R-z|^2}{k^2} + \frac{i}{k^3}\right)e^{ik|R-z|}$$

The imaginary term in the second result is small relative to the leading term because the path length is usually many wavelengths long. A familiar result emerges when we combine these results with (2.53):

$$E_1(R) = \mathcal{E}_0 \exp\left[ikR - \left(\frac{1}{2}ik \int_0^R dz\, \delta\varepsilon(0,0,z)\right.\right.$$
$$\left.\left. -\frac{1}{4}\int_0^R dz\,(R-z)(\nabla_\perp^2 \delta\varepsilon)_0\right)\right] \qquad (2.54)$$

The second term in the exponent is just the geometrical-optics description for the phase fluctuation. The third term is the amplitude fluctuation calculated with ray theory. Tatarskii demonstrated the same reduction starting from the paraxial approximation version of the field strength [9].

2.3 The Second-order Solution

There are several reasons for developing an explicit expression for the second term in the series expansion (2.13). We will establish the range of validity for the Rytov approximation by comparing the first- and second-order terms. Conservation of energy depends both on the first- and on the second-order solution. We also need the second term when we address moments of the field strength in Chapter 9. Only by including this solution can one establish the correct descriptions for the average field strength, the mean irradiance and the mutual coherence function. The second-order solution predicts that the probability density function of amplitude fluctuations should be skewed and this feature is confirmed by numerical simulations. Two alternative expressions for ψ_2 have been developed and we will show later that they are equivalent.

2.3.1 Tatarskii's Version

The second term in the Rytov series is the solution of the third field equation in (2.14), which we rewrite as follows:

$$\nabla^2 \psi_2 + 2\nabla\psi_0 \cdot \nabla\psi_2 = -(\nabla\psi_1)^2 \qquad (2.55)$$

The right-hand side of this equation is known since ψ_1 is the normalized single scattering term in (2.18). The exponential substitution (2.16) allows one to solve equations of this type:

$$\psi_2(\mathbf{R}) = -\int d^3r\, G(\mathbf{R},\mathbf{r})[\nabla\psi_1(\mathbf{r})]^2 \exp[\psi_0(\mathbf{r}) - \psi_0(\mathbf{R})]$$

This can be written in terms of the unperturbed field strength by using the connection (2.5):

$$\psi_2(\mathbf{R}) = -\int d^3r \, G(\mathbf{R}, \mathbf{r})[\nabla \psi_1(\mathbf{r})]^2 \frac{E_0(\mathbf{r})}{E_0(\mathbf{R})} \qquad (2.56)$$

Tatarskii estimated the average value of ψ_2 and its variance using this formulation [10]. With those results he found conditions on the amplitude variance for propagation conditions under which the Rytov approximation is valid [11]. The electric-field strength can be written to second order as follows:

$$E_2(\mathbf{R}) = E_0(\mathbf{R}) \exp\left(-\int d^3r \, G(\mathbf{R}, \mathbf{r}) \frac{E_0(\mathbf{r})}{E_0(\mathbf{R})}[k^2 \, \delta\varepsilon(\mathbf{r}) + (\nabla \psi_1(\mathbf{r}))^2]\right)$$

$$(2.57)$$

2.3.2 Yura's Version

A different but equivalent version of the second-order Rytov solution was discovered later by Yura [12][13]. The essential step is to express the surrogate function in terms of a new unknown function B_2 and the basic Rytov solution:

$$\psi_2 = B_2 - \tfrac{1}{2}B_1^2 \qquad (2.58)$$

When this assumed form is substituted into the third field equation in (2.14) the successive versions of that equation become

$$\nabla^2(B_2 - \tfrac{1}{2}B_1^2) + 2\,\nabla\psi_0 \cdot (\nabla B_2 - \tfrac{1}{2}\nabla B_1^2) = -(\nabla B_1)^2$$

$$\nabla^2 B_2 - \tfrac{1}{2}[2B_1\,\nabla_1^2 B + 2(\nabla B_1)^2] + 2\,\nabla\psi_0 \cdot (\nabla B_2 - B_1\,\nabla B_1) = -(\nabla B_1)^2$$

$$\nabla^2 B_2 + 2\,\nabla\psi_0 \cdot \nabla B_2 - B_1(\nabla_1^2 B_1 + 2\,\nabla\psi_0 \cdot \nabla B_1) = 0$$

In the last version the combination in parentheses is equal to $-k^2 \, \delta\varepsilon$ in view of the second field equation in (2.14). The unknown function B_2 must therefore satisfy the following equation:

$$\nabla^2 B_2 + 2\,\nabla\psi_0 \cdot \nabla B_2 = -k^2 \, \delta\varepsilon \, \psi_1$$

The solution of this familiar equation was given in (2.17):

$$B_2(\mathbf{R}) = -k^2 \int d^3r_1 \, G(\mathbf{R}, \mathbf{r}_1) \frac{E_0(\mathbf{r})}{E_0(\mathbf{R})} \, \delta\varepsilon(\mathbf{r}) \, \psi_1(\mathbf{r}) \qquad (2.59)$$

On substituting the expression (2.18) for $\psi_1(\mathbf{r})$ one finds an explicit solution for the unknown function:

$$B_2(\mathbf{R}) = k^4 \int d^3r_1 \, G(\mathbf{R}, \mathbf{r}_1) \, \delta\varepsilon(\mathbf{r}_1) \int d^3r_2 \, G(\mathbf{r}_1, \mathbf{r}_2) \, \delta\varepsilon(\mathbf{r}_2) \frac{E_0(\mathbf{r}_2)}{E_0(\mathbf{R})}$$

$$(2.60)$$

This is recognized as the second term in the Born series. It describes double scattering of the incident wave normalized by the unperturbed field and is illustrated in Figure 1.1.

The field strength which is valid to second order can now be written in terms of the solution (2.60) and the single-scattering term as follows:

$$E_2(\mathbf{R}) = E_0(\mathbf{R}) \exp\left[B_1(\mathbf{R}) + B_2(\mathbf{R}) - \tfrac{1}{2}B_1^2(\mathbf{R})\right] \qquad (2.61)$$

When the exponential terms are small this can be expanded and generates the first three terms in the Born series:

$$\begin{aligned} E_2 &= E_0\left[1 + \left(B_1 + B_2 - \tfrac{1}{2}B_1^2\right) + \tfrac{1}{2}\left(B_1 + B_2 - \tfrac{1}{2}B_1^2\right)^2\right] \\ &= E_0(1 + B_1 + B_2) \end{aligned} \qquad (2.62)$$

When the phase variations are large – as they are in optical and millimeter-wave propagation – one cannot expand the exponential in this way. The Rytov method provides a description that is vastly superior to the Born approximation in those cases.

To make estimates with Yura's version it is helpful to break the double-scattering solution into real and imaginary parts, as we did in writing the first Born term as $\psi_1 = a + ib$.

$$B_2 = c + id = k^4 \int d^3r\, G(\mathbf{R}, \mathbf{r}) \int d^3r'\, G(\mathbf{r}, \mathbf{r}') \frac{E_0(\mathbf{r}')}{E_0(\mathbf{R})} \delta\varepsilon(\mathbf{r})\, \delta\varepsilon(\mathbf{r}') \qquad (2.63)$$

Two new random variables c and d are introduced in this way. The second-order Rytov expression for the field strength can be written in terms of the dimensionless quantities a, b, c and d in the following form:

$$E_2(\mathbf{R}) = E_0(\mathbf{R}) \exp\left[a + c - \tfrac{1}{2}a^2 + \tfrac{1}{2}b^2 + i(b + d - ab)\right] \qquad (2.64)$$

Yura's version of the second-order solution has several advantages. The real and imaginary parts of the first two Rytov terms in the exponent are separated, in contrast to (2.57). This means that we can immediately identify the phase fluctuation

$$\varphi_2 = b + d - ab \qquad (2.65)$$

and the corresponding logarithmic amplitude

$$\chi_2 = a + c - \tfrac{1}{2}a^2 + \tfrac{1}{2}b^2 \qquad (2.66)$$

The moments and correlations of a, b, c and d are sometimes known from calculations made with the Born approximation. When this is the case, they can be introduced as modules in estimating amplitude and intensity moments to second order. We will base our moment predictions in Chapter 9 on this approach.

On the other hand, the Yura version shares a fundamental difficulty with the Tatarskii version. The problem is that ψ_2 is not a Gaussian random variable (GRV) since it depends on the product of two GRVs: $\nabla \psi_1^2$ in the Tatarskii version and $\delta\varepsilon_1 \delta\varepsilon_2$ in the Yura version. This product need not be a Gaussian random variable – even if the components are.[3] This means that one *cannot assume* that B_2 is a GRV. One is often compelled to use the method of cumulants summarized in Appendix M to evaluate ensemble averages of signal components to second order.

2.3.3 The Equivalence of the Two Versions

It is important to demonstrate that the Tatarskii and Yura versions of the second-order solution are equivalent. We compare the two expressions for ψ_2 after multiplying both by the unperturbed field:

$$-\tfrac{1}{2}E_0(\mathbf{R})\psi_1^2(\mathbf{R}) - k^2 \int d^3r\, G(\mathbf{R},\mathbf{r})E_0(\mathbf{r})\,\delta\varepsilon(\mathbf{r})\,\psi_1(\mathbf{r}) \qquad (2.67)$$

$$\overset{?}{=} -\int d^3r\, G(\mathbf{R},\mathbf{r})E_0(\mathbf{r})[\nabla\psi_1(\mathbf{r})]^2$$

The wave operator $\nabla_R^2 + k^2$ is applied to both sides and the terms rearranged:

$$\int d^3r\left(\nabla_R^2 + k^2\right)G(\mathbf{R},\mathbf{r})\{[\nabla\psi_1(\mathbf{r})]^2 - k^2 E_0(\mathbf{r})\,\delta\varepsilon(\mathbf{r})\,\psi_1(\mathbf{r})\}$$

$$\overset{?}{=} \tfrac{1}{2}\left(\nabla_R^2 + k^2\right)\left[E_0(\mathbf{R})\psi_1^2(\mathbf{R})\right]$$

Acting on Green's function the wave operator produces a delta function:[4]

$$\left(\nabla_R^2 + k^2\right)G(\mathbf{R},\mathbf{r}) = \delta(\mathbf{R} - \mathbf{r})$$

This collapses the volume integral on the left-hand side and the relationship

$$\nabla^2(AB) = A\,\nabla^2 B + 2\,\nabla A \cdot \nabla B + B\,\nabla^2 A$$

from Appendix J simplifies the right-hand side:

$$(\nabla\psi_1)^2 - k^2 E_0\,\delta\varepsilon\,\psi_1 \overset{?}{=} \tfrac{1}{2}\{\psi_1^2(\nabla^2 + k^2)E_0 + 4\psi_1\,\nabla E_0 \cdot \nabla\psi_1$$

$$+ 2E_0[\psi_1\nabla^2\psi_1 + (\nabla\psi_1)^2]\}$$

The wave equation (2.3) for the unperturbed field simplifies the combination in curly brackets still further:

$$E_0\psi_1[\nabla^2\psi_1 + 2\,\nabla \ln E_0 \cdot \nabla\psi_1 - k^2\,\delta\varepsilon] \overset{?}{=} 0 \qquad (2.68)$$

---

[3] As noted in Problem 3 of Chapter 2 in Volume 1.
[4] In view of the defining equation for Green's function given in Appendix L.

The term in the square brackets vanishes because ψ_1 satisfies the second equation in (2.14).

This calculation shows that the wave operator acting on both solutions gives the same result. The two expressions for ψ_2 are therefore the same except for a function that satisfies the vacuum-wave equation:

$$(\nabla^2 + k^2)F = 0$$

This *function of integration* must vanish since both solutions must approach the unperturbed wave as the dielectric variation goes to zero.[5] The two versions are therefore the same.

2.4 Higher-order Solutions

A new problem occurs if one attempts to use the first two Rytov solutions to estimate intensity fluctuations. It is similar to the problem one encounters if one uses only the first-order solution to calculate the mean irradiance or mutual coherence function. Those measurements depend on the product of two field-strength expressions and we shall find in Chapter 9 that these quantities are properly described only if one uses the first two terms in the Rytov series. The variance of intensity fluctuations depends on four field-strength expressions averaged over the ensemble of atmospheric conditions. To calculate intensity moments properly we need four terms in the Rytov series:

$$E_4(\mathbf{R}) = E_0(\mathbf{R})\exp(\psi_1 + \psi_2 + \psi_3 + \psi_4) \tag{2.69}$$

We have already solved for ψ_1 and ψ_2. It is surprisingly easy to find the third and fourth terms using Yura's approach.

2.4.1 The Third-order Solution

The third term in the Rytov series is completely determined by the fourth equation in (2.14):

$$\nabla^2\psi_3 + 2\,\nabla\psi_0\cdot\nabla\psi_3 = -2\,\nabla\psi_1\cdot\nabla\psi_2 \tag{2.70}$$

Since we now have explicit expressions for ψ_1 and ψ_2 we can use the solution technique (2.17) to find ψ_3. It is more revealing to follow Yura's path and represent

---

[5] The demonstration is actually more complicated because the field strength is a *functional* of the stochastic function $\delta\varepsilon(\mathbf{r})$. We have used a general measure of the dielectric fluctuations here because we have not yet laid a proper foundation for treating functionals. That will be done in Volume 3 and this issue should then become clearer.

ψ_3 in terms of the first two Born terms and a new unknown function:

$$\psi_3 = B_3 + \tfrac{1}{3}B_1^3 - B_1B_2 \qquad (2.71)$$

With this expression and the previous results for ψ_1 and ψ_2 we can write the third-order field equation as follows:

$$\nabla^2\left(B_3 + \tfrac{1}{3}B_1^3 - B_1B_2\right) + 2\,\nabla\psi_0 \cdot \left(\nabla B_3 + B_1^2\,\nabla B_1 - B_1\,\nabla B_2 - B_2\,\nabla B_1\right)$$
$$+ 2\,\nabla B_1 \cdot \left(\nabla B_2 - B_1\,\nabla B_1\right) = 0$$

Several terms cancel out and the remaining elements can be grouped as follows:

$$\nabla^2 B_3 + 2\,\nabla\psi_0 \cdot \nabla B_3 - B_1(\nabla^2 B_2 + 2\,\nabla\psi_0 \cdot \nabla B_2)$$
$$+ \left(B_1^2 - B_2\right)\left(\nabla^2 B_1 + 2\,\nabla\psi_0 \cdot \nabla B_1\right) = 0$$

The combination which multiplies B_1 in the first set of parentheses is just $-k^2\,\delta\varepsilon\,B_1$ in view of the equation for B_2 given above (2.59). The group of terms in the second bracket is $-k^2\,\delta\varepsilon$ because of (2.15). After combining these elements there emerges an equation for B_3 that is now familiar:

$$\nabla^2 B_3 + 2\,\nabla\psi_0 \cdot \nabla B_3 = -k^2\,\delta\varepsilon\,B_2$$

This can be solved using the technique described by (2.17). The expression (2.59) for B_2 is then introduced to yield the unknown function:

$$B_3(\mathbf{R}) = -k^6 \int d^3r_1\, G(\mathbf{R}, \mathbf{r}_1)\,\delta\varepsilon(\mathbf{r}_1) \int d^3r_2\, G(\mathbf{r}_1, \mathbf{r}_2)\,\delta\varepsilon(\mathbf{r}_2)$$
$$\times \int d^3r_3\, G(\mathbf{r}_2, \mathbf{r}_3)\,\delta\varepsilon(\mathbf{r}_3)\,\frac{E_0(\mathbf{r}_3)}{E_0(\mathbf{R})} \qquad (2.72)$$

This is just the normalized Born term which describes triple scattering of the incident field and is illustrated in Figure 1.1. This result and the previous relations for ψ_1 and ψ_2 allow us to write the surrogate function to third order in $\delta\varepsilon$ as follows:

$$\Psi = B_1 - \tfrac{1}{2}B_1^2 + \tfrac{1}{3}B_1^3 + B_2(1 - B_1) + B_3 \qquad (2.73)$$

It is more convenient to use the following dimensionless random variables to describe the real and imaginary parts of the solution (2.72):

$$B_3 = e + if \qquad (2.74)$$

Using these new variables and those introduced previously, we can write the third-order surrogate in terms of its orthogonal components:

$$\Psi_3 = e + \tfrac{1}{3}a^3 - ab^2 - ac + bd + i\left(a^2b - \tfrac{1}{3}b^3 - bc - ad\right) \qquad (2.75)$$

2.4.2 The Fourth-order Solution

The procedure described above also yields the fourth Rytov approximation. The equation for ψ_4 is the penultimate one in (2.14):

$$\nabla^2\psi_4 + 2\,\nabla\psi_0\cdot\nabla\psi_4 + 2\,\nabla\psi_1\cdot\nabla\psi_3 + (\nabla\psi_2)^2 = 0 \qquad (2.76)$$

The important step in solving this is to make a good assumption for ψ_4. On the basis of previous experience it is not hard to guess the form of the trial solution,

$$\psi_4 = B_4 - B_3 B_1 + B_2\big(B_1^2 - \tfrac{1}{2}B_2\big) - \tfrac{1}{4}B_1^4 \qquad (2.77)$$

where B_4 is an undetermined function and the other terms have their previous meanings. When this expression is substituted into the field equation, one finds that it can be progressively simplified using the equations satisfied by the lower-order solutions. The undetermined function then satisfies the familiar equation

$$\nabla^2 B_4 + 2\,\nabla\psi_0\cdot\nabla B_4 = -k^2\,\delta\varepsilon\,B_3$$

The solution of this equation is the normalized Born term which describes quadruple scattering of the incident wave:

$$B_4(\mathbf{R}) = k^8 \int d^3 r_1\, G(\mathbf{R}, \mathbf{r}_1)\,\delta\varepsilon(\mathbf{r}_1) \int d^3 r_2\, G(\mathbf{r}_1, \mathbf{r}_2)\,\delta\varepsilon(\mathbf{r}_2)$$

$$\times \int d^3\mathbf{r}_3\, G(\mathbf{r}_2, \mathbf{r}_3)\,\delta\varepsilon(\mathbf{r}_3) \int d^3\mathbf{r}_4\, G(\mathbf{r}_3, \mathbf{r}_4)\,\delta\varepsilon(\mathbf{r}_4)\,\frac{E_0(\mathbf{r}_4)}{E_0(\mathbf{R})} \qquad (2.78)$$

Characterizing the fourth Born term by two dimensionless random variables,

$$B_4 = g + ih \qquad (2.79)$$

one can express the fourth surrogate as

$$\psi_4 = \big[g - ea + fb + c(a^2 - b^2) - 2dab - \tfrac{1}{2}c^2 + \tfrac{1}{2}d^2$$
$$- \tfrac{1}{4}(a^4 - 6a^2b^2 + b^4)\big] + i[h - af - be + d(a^2 - b^2)$$
$$+ 2ca - cd - a^3 b + ab^3] \qquad (2.80)$$

and the field strength to fourth order becomes

$$E_4(\mathbf{R}) = E_0(\mathbf{R})\exp\big\{a + c - \tfrac{1}{2}a^2 + \tfrac{1}{2}b^2 + e + \tfrac{1}{3}a^3 - ab^2 - ac + bd + g$$
$$- ea + fb + c(a^2 - b^2) - 2dab - \tfrac{1}{2}c^2 + \tfrac{1}{2}d^2$$
$$- \tfrac{1}{4}(a^4 - 6a^2b^2 + b^4) + i\big[b + d - ab + a^2 b - \tfrac{1}{3}b^3 - bc$$
$$- ad + h - af - be + d(a^2 - b^2) + 2ca$$
$$- cd - a^3 b + ab^3\big]\big\} \qquad (2.81)$$

The real and imaginary terms in the exponential define the logarithmic amplitude and phase fluctuations to fourth order.

2.4.3 Perspective

With the development of these higher-order solutions, a new perspective on the Rytov method emerges that is not apparent from the basic solution. One can write the Rytov solution and Born series to fourth order as follows:

$$\text{Born:} \quad E_4 = E_0(1 + B_1 + B_2 + B_3 + B_4)$$
$$\text{Rytov:} \quad E_4 = E_0 \exp(\psi_1 + \psi_2 + \psi_3 + \psi_4)$$

$$(2.82)$$

One can expand the exponential if the terms ψ_n in the Rytov expression are small. On equating the two expressions for E_4 and grouping terms of the same order in $\delta\varepsilon$ one finds that the surrogate functions and Born terms should be connected by algebraic equations:

$$\psi_1 = B_1$$
$$\psi_2 = B_2 - \tfrac{1}{2}B_1^2$$
$$\psi_3 = B_3 + \tfrac{1}{3}B_1^3 - B_1 B_2$$
$$\psi_4 = B_4 - B_1 B_3 + B_2\left(B_1^2 - \tfrac{1}{2}B_2\right) - \tfrac{1}{4}B_1^4$$

$$(2.83)$$

The same relationships were obtained when we solved the successive field equations.

This correspondence suggests to some that the Rytov approximation is little more than an exponential version of the Born series. However, there are important differences. The imaginary components of the surrogate function represent signal phase and now occur naturally in the exponent where they belong. When the phase fluctuations are large – as they are in optical and infrared propagation – one cannot expand the exponential and the Rytov method provides a superior description. In addition, signal amplitude and the real part of the surrogate are connected by logarithmic relations and that leads to the observed distribution for fluctuations in signal amplitude and intensity. It is clear that the Rytov approximation should be preferred over the Born series for describing propagation in random media.

2.5 Problems

Problem 1

Use the basic Rytov approximation and a Taylor-series expansion of the dielectric variation about the nominal ray path to show that the angle-of-arrival components deduced from the field-gradient expressions,

$$\delta\theta_x = \frac{\partial E}{\partial x} \bigg/ \frac{\partial E}{\partial z} \quad \text{and} \quad \delta\theta_y = \frac{\partial E}{\partial y} \bigg/ \frac{\partial E}{\partial z}$$

lead naturally to the geometrical-optics results for these quantities derived in Chapter 3 of Volume 1.

Problem 2

Using the approach described in Section 2.4, show that the Rytov approximation for the field strength to fifth order is given by

$$E_5(\mathbf{r}) = E_0(\mathbf{r}) \exp\big[B_1 - \tfrac{1}{2}B_1^2 + \tfrac{1}{3}B_1^3 - \tfrac{1}{4}B_1^4 + \tfrac{1}{5}B_1^5 + B_2\big(1 - B_1 + B_1^2$$
$$- B_1^3 - \tfrac{1}{2}B_2^2 - B_1 B_2\big) + B_3\big(1 - B_1 + B_1^2 - B_2\big)$$
$$+ B_4\big(1 - B_1\big) + B_5 \big]$$

where B_5 is the Born term that describes five-fold scattering of the incident wave. Can you sum the series of terms that seem to be emerging? How do these summations compare with the Born expansion? What do they tell one about the relationship between the Born and Rytov approaches?

References

[1] S. M. Rytov, "Diffraction of Light by Ultrasonic Waves," *Izvestiya Akademii Nauk SSSR, Seriya Fizicheskaya (Bulletin of the Academy of Sciences of the USSR, Physical Series)*, No. 2, 223–259 (1937). (No English translation is available.)

[2] A. M. Obukhov, "On the Influence of Weak Atmospheric Inhomogeneities on the Propagation of Sound and Light," *Izvestiya Akademii Nauk SSSR, Seriya Geofizicheskaya (Bulletin of the Academy of Sciences of the USSR, Geophysical Series)*, No. 2, 155–165 (1953). (English translation by W. C. Hoffman, published by U. S. Air Force Project RAND as Report T-47, Santa Monica, CA, 28 July 1955.)

[3] V. I. Tatarskii, *The Effects of the Turbulent Atmosphere on Wave Propagation* (translated from the Russian and issued by the National Technical Information Office, U. S. Department of Commerce, Springfield, VA 22161, 1971), 218–253.

[4] H. T. H. Piaggio, *An Elementary Treatise on Differential Equations and Their Applications* (G. Bell and Sons, London, 1946), 201–205.

[5] G. N. Watson, *A Treatise on the Theory of Bessel Functions* (Cambridge University Press, Cambridge, 1952), 85–94.

[6] G. L. Lamb, *Elements of Soliton Theory* (John Wiley and Sons, New York, 1980), 80–83.

[7] R. A. Schmeltzer, "Means, Variances and Covariances for Laser Beam Propagation Through a Random Medium," *Quarterly Journal of Applied Mathematics*, **24**, No. 4, 339–354 (January 1967).

[8] S. M. Rytov, Yu. A. Kravtsov and V. I. Tatarskii, *Principles of Statistical Radiophysics 4, Wave Propagation Through Random Media* (Springer-Verlag, Berlin, 1989), 49–50.

[9] See [3], pages 224–225.

[10] V. I. Tatarskii, "Second Approximation to the Problem of Propagation of Waves in a Medium with Random Inhomogeneities," *Izvestiya Vysshikh Uchebnykh Zavedenii, Radiofizika (Soviet Radiophysics)*, **5**, No. 3, 164–202 (1962).

[11] See [3], pages 253–258.

[12] H. T. Yura, "Optical Propagation Through a Turbulent Medium," *Journal of the Optical Society of America*, **59**, No. 1, 111–112 (January 1969).

[13] H. T. Yura, C. C. Sung, S. F. Clifford and R. J. Hill, "Second-Order Rytov Approximation," *Journal of the Optical Society of America*, **73**, No. 4, 500–502 (April 1983).

3

Amplitude Variance

We turn now to the task of describing the variance of amplitude and intensity fluctuations. We avoided this problem in Volume 1 and it is important to recall why we did so. That development was based entirely on solutions for Maxwell's equations generated by the geometrical-optics approximation. The electromagnetic wavelength is completely absent from the eikonal and transport equations that define those solutions.[1] On the other hand, astronomical observations and terrestrial experiments show that the level of scintillation *does* depend on wavelength. The fundamental problem is that geometrical optics completely ignores diffraction. Yet we know that amplitude fluctuations are caused primarily by small atmospheric irregularities. Diffraction by these small eddies is the principal cause of the amplitude fluctuations that characterize atmospheric scintillation.

Our task in this chapter is to estimate the level of scintillation for optical and microwave propagation. To do so we will depend on the Rytov approximation developed in Chapter 2 that includes diffraction in a rigorous way. It is limited to weak-scattering situations but that covers a broad range of applications. The Rytov approximation gives a complete description for terrestrial microwave links and can often describe millimeter-wave propagation. The same approach characterizes the scintillation experienced by optical and infrared signals that travel near the surface on relatively short paths. By contrast, astronomical observations are characterized by weak scattering unless the source is close to the horizon. Electro-optical observations of the earth's surface by scientific and reconnaissance spacecraft are described by this approach. It also describes microwave signals that are received from earth-orbiting satellites and from galactic radio sources if the frequency is above 1000 MHz. These applications provide ample motivation to develop weak-scattering expressions for the amplitude variance, and to test them against the many measurements that have been made.

---

[1] See Chapter 3 in Volume 1.

The variance of logarithmic amplitude fluctuations must be less than unity in those situations for which weak-scattering expressions can be employed, as we shall demonstrate in Chapter 12. Since the average value of the logarithmic amplitude vanishes to first order we can write

$$\langle \chi^2 \rangle < 1 \tag{3.1}$$

This condition allows one to relate the variations of amplitude and intensity directly. Recall that the voltage induced in a receiver is proportional to the magnitude of the field strength of the arriving signal:

$$V = \mathcal{G}|E| = \mathcal{G}|E_0|(1 + \chi)$$

The corresponding intensity or irradiance is therefore described by

$$I = \mathcal{G}^2|E_0|^2(1 + \chi)^2 \simeq I_0(1 + 2\chi)$$

This establishes a basic connection between intensity fluctuations and the logarithmic amplitude:

$$\frac{\delta I}{I_0} = \frac{I - I_0}{I_0} = 2\chi$$

Their variances are related by

$$\left\langle \left(\frac{\delta I}{I_0} \right)^2 \right\rangle = 4\langle \chi^2 \rangle \tag{3.2}$$

We shall concentrate here on $\langle \chi^2 \rangle$ with the confidence that it accurately leads to the intensity variance when the scattering is weak.

The descriptions that we will develop express $\langle \chi^2 \rangle$ as weighted integrals of the wavenumber spectrum for turbulent irregularities. Those weighting functions depend on the propagation geometry, the type of signal and its wavelength. They have several features that are common to all applications. The logarithmic amplitude is determined primarily by small eddies. This means that the inertial range of the spectrum is primarily responsible for amplitude and intensity scintillation. That influence is usually bounded by aperture averaging – and sometimes by the transition to the dissipation range. Large eddies play a relatively minor role, in contrast to the dominant influence they exert on phase fluctuations.

This chapter is quite long and it may help to specify in advance the ground we will cover. The first step is to explain the paraxial approximation that is used throughout this volume. It describes small-angle forward scattering and we will explain it by examining a familiar problem in optical astronomy. We turn then to propagation along links near the surface. Our discussion is separated according to the type of transmitted signal (plane, spherical and beam waves). Aperture averaging plays a central role in understanding nearly all amplitude measurements. The wavelength scaling of scintillation measured at optical and microwave frequencies provides a crucial test for these descriptions. We will find that the noninertial regions of the spectrum are important in some applications. The influences of a finite data-sample length and anisotropy are also explored.

Astronomical measurements made in the visible and infrared bands are considered next. The inhomogeneous nature of tropospheric irregularities is an important consideration for this type of scintillation, which complicates the description. The arriving signals are always plane waves and that lightens our load. The scaling laws needed to interpret such observations are the variation of $\langle \chi^2 \rangle$ with source elevation, telescope diameter and wavelength.

The scintillation imposed on radio-astronomy signals is generated in the ionosphere by field-aligned variations in electron density. Our description of these irregularities is quite general. The measured frequency scaling of $\langle \chi^2 \rangle$ is our best guide for defining the power-law exponent in those models. Source averaging of microwave scintillation provides a new way to explore the detailed structure of distant galaxies.

Microwave signals transmitted by communication satellites are influenced by the troposphere or by the ionosphere – and sometimes by both. The important questions are the same as those asked in the astronomical context: how does the scintillation level depend on the angle of elevation, wavelength and size of receiver? When we have completed this program, we will have laid a strong foundation for the problems that will be encountered in succeeding chapters.

3.1 The Paraxial Approximation

Most of the results derived in this volume will be based on the paraxial or small-scattering-angle approximation. In the context of propagation in random media, this approach was first used by Obukhov [1] and later developed by Tatarskii [2][3] and Chernov [4]. It is based on a simple physical principle that is best illustrated by a specific example.

Consider a plane-wave optical signal from an overhead star falling on the terrestrial atmosphere, as illustrated in Figure 3.1. This type of propagation is accurately described by weak scattering. The single-scattering event shown in Figure 3.1 is completely defined by the complex weighting function (2.51). If the origin coincides with the receiver, the downcoming wave in that expression is given by

$$E(\mathbf{r}) = \mathcal{E}_0 \exp(-ikz) \tag{3.3}$$

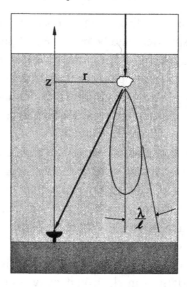

Figure 3.1: Forward scattering by an eddy that is large relative to the electromagnetic wavelength. The pattern of reradiated energy is quite narrow in this case and the eddy must be close to the vertical axis for its scattered field to influence the signal measured at the receiver.

Green's function is naturally expressed in terms of the cylindrical coordinates defined by Figure 3.1:

$$G(\mathbf{R}, \mathbf{r}) = \frac{\exp\left(ik\sqrt{z^2 + r^2}\right)}{4\pi\sqrt{z^2 + r^2}} \tag{3.4}$$

The profile function depends only on the height z above ground level and the complex weighting function can be written as

$$\Lambda(\kappa) = k^2 \int_0^\infty dz\, \wp(z) \exp(-ikz) \int_0^\infty dr\, r \int_0^{2\pi} d\phi\, \frac{\exp\left(i\sqrt{r^2 + z^2}\right)}{4\pi\sqrt{r^2 + z^2}}$$
$$\times \exp\{i[\kappa_z z + \kappa_r r \cos(\phi - \omega)]\} \tag{3.5}$$

Here the wavenumber vector is also described by cylindrical components:

$$\kappa = \mathbf{i}_z \kappa_z + \mathbf{i}_x \kappa_r \cos \omega + \mathbf{i}_y \kappa_r \sin \omega \tag{3.6}$$

Studies of electromagnetic scattering by a spherical object with an index of refraction different than unity show that the scattered energy is concentrated in a cone of angular width [5]

$$\theta = \lambda/\ell \tag{3.7}$$

A turbulent eddy behaves like a semi-transparent sphere and scatters most of the incident energy in the forward direction. This spreading is the feature of diffraction that influences propagation in random media. It is helpful to make a preliminary estimate of the effect. Traditional astronomical observations use visible light with wavelengths less than 1 μm. By contrast, the smallest eddies in atmospheric turbulence are of magnitude a few millimeters so that

$$\lambda \ll \ell$$

for all eddies and the scattering angle θ should be less than one milliradian.

The paraxial approximation exploits the smallness of λ relative to ℓ in the following way. The influential scattering eddies are evidently quite close to the vertical axis. From Figure 3.1 we note that the radial distance to such eddies must satisfy the inequality

$$r < z\lambda/\ell$$

for its scattered wave to influence the output at the receiver. This condition allows one to expand the scalar distance in Green's function (3.5) as follows:

$$k\sqrt{z^2 + r^2} \simeq kz + \pi\left(\frac{r^2}{\lambda z}\right) - \frac{\pi^2}{2kz}\left(\frac{r^2}{\lambda z}\right)^2 \qquad (3.8)$$

The first term sets the phase reference of the vacuum field and is enormous everywhere in the medium. By the same token the third term is trivial. The second term is the important one for our purposes. It depends only on the *Fresnel length*,

$$\text{Fresnel length} = \sqrt{\lambda z} \qquad (3.9)$$

which will play a central role in all our descriptions. We must keep the first two terms in the exponent of the Green function in order to capture the phase-dependent interference effects that characterize diffraction. On the other hand, we need only the first term in the scalar denominator of (3.5). With these approximations the complex weighting function becomes

$$\Lambda(\kappa) = \frac{k^2}{4\pi} \int_0^\infty dz \, \frac{\wp(z)}{z} \int_0^\infty dr \, r \int_0^{2\pi} d\phi \, \exp\left[i\left(z\kappa_z + r\kappa_r \cos(\phi - \omega) + \frac{r^2 k}{2z}\right)\right]$$

The azimuth integration gives the zeroth-order Bessel function:

$$\Lambda(\kappa) = \frac{k^2}{2} \int_0^\infty dz \, \frac{\wp(z)}{z} \exp(iz\kappa_z) \int_0^\infty dr \, r \, J_0(r\kappa_r) \exp\left(\frac{ir^2 k}{2z}\right)$$

We shall find that distant eddies are more influential than those nearby, so we need not worry about the apparent singularity at the receiver. The vertical integration is extended to infinity in the expectation that the profile function will limit its reach. Although the radial integration also runs to infinity, we expect that the narrow scattering diagrams of the eddies will suppress the contributions of volume elements that are far from the line of sight. With these conventions the radial integration can be done using an integral found in Appendix D:

$$\text{Plane wave:} \qquad \Lambda(\kappa) = \frac{ik}{2} \int_0^\infty dz \, \wp(z) \exp\left[iz\left(\kappa_z - \frac{\kappa_r^2}{2k}\right)\right] \qquad (3.10)$$

This result was developed for a plane wave penetrating the atmosphere at vertical incidence. On the other hand, the integration variable can be interpreted more generally. It could just as well represent the distance along any line-of-sight path. This means that the same expression can be used to describe the horizontal or inclined propagation of a plane wave. To remind ourselves of this generality, we change the integration variable from the vertical height z to the distance s measured along the nominal propagation direction. The phase and amplitude weighting functions are constructed from (3.10) by using the connections (2.43) and (2.44):

$$\text{Plane wave:} \qquad D(\kappa) = \frac{k}{2} \int_0^\infty ds \, \wp(s) \sin\left(\frac{s\kappa_r^2}{2k}\right) \exp(is\kappa_s) \qquad (3.11)$$

$$\text{Plane wave:} \qquad E(\kappa) = \frac{k}{2} \int_0^\infty ds \, \wp(s) \cos\left(\frac{s\kappa_r^2}{2k}\right) \exp(is\kappa_s) \qquad (3.12)$$

These expressions relate the measured quantities to Fourier transforms of the turbulent profile. It is a simple matter to compute $D(\kappa)$ and $E(\kappa)$ when the profile is known in analytical or numerical form. It is important to note that these expressions for $D(\kappa)$ and $E(\kappa)$ are valid only for plane-wave signals. In the next section, we will use the paraxial approximation to develop expressions for spherical and beam waves that illustrate the important differences.

When the turbulent eddies are large relative to the wavelength, most of the energy will be scattered in the forward direction. That energy is concentrated in a cone of angular width λ/ℓ as illustrated in Figure 3.1. We have observed that this ratio is small for optical wavelengths and the paraxial approximation provides a consistently accurate description of optical propagation. The situation for microwaves is more complicated because the wavelengths are larger than the inner scale length but smaller than the outer scale length. The paraxial approximation should describe the influence of large eddies but probably not the small ones. However, the falling spectrum reminds us that the large eddies are much stronger than the small ones. This suggests that the paraxial approximation may also describe microwave propagation. The exact solutions developed later will confirm this intuitive argument.

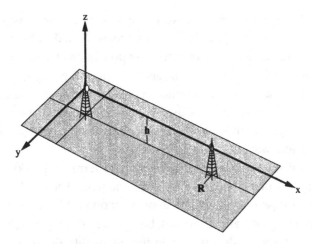

Figure 3.2: A microwave communication relay link operating on a typical path near the surface. The line-of-sight path coincides with the x axis at a height h above the surface.

3.2 Terrestrial Links

Many applications that are influenced by scintillation involve horizontal paths near the surface. The microwave relay link shown in Figure 3.2 is widely used in communication networks and illustrates this type of propagation. Scientific experiments are often conducted near the surface. In these tests, a controlled transmitter generates a signal that is monitored by one or more receivers some distance away. Laser ranging systems are widely used for surveying and precision geodesy. Similar techniques are exploited to locate and illuminate military targets. Optical and millimeter-wave radars are used for battlefield surveillance.

These applications are relatively easy to describe for several reasons. One can assume that the turbulence parameters are nearly constant along the path if the transmitter and receiver are both close to the surface. Turbulent eddies are relatively isotropic near the surface, in contrast to their significant elongation in the free atmosphere. One is always dealing with a controlled source whose signal characteristics are precisely known in these applications. The transmitted signal is often coherent and that too is helpful. Moreover, the path length is accurately known or can be measured by the system itself.

The type of wave that is radiated depends on the transmitter design. Three basic wave types are commonly used in terrestrial experiments: (a) plane waves, (b) spherical waves and (c) beam waves. There are significant differences among the levels of scintillation imposed on these types of wave and hence we must address them separately. Diverging waves are radiated by small transmitters and are widely used because they are the simplest to implement experimentally. We shall therefore spend considerable effort on studying the amplitude fluctuations imposed on spherical waves. Plane waves are relatively easy to analyze but must be simulated by collimated beams, at considerably greater expense. The development of coherent optical sources has created a wide range of applications based on beam waves. Collimated and focused beams are important examples of this general family. The description of beam-wave signals presents a considerable analytical challenge. To lighten that load we shall consistently use the paraxial approximation. We return to this question in Chapter 8 and develop descriptions for the wide-angle scattering of plane and spherical waves.

The fundamental question we face in describing terrestrial links concerns the type of scattering that occurs near the tropospheric path. The results developed in this volume are all based on the Rytov approximation which assumes that weak scattering occurs along the path. Their suitability is determined by the fundamental condition (3.1). The amplitude variance in that inequality depends on conditions along the path and on the frequency of the transmitted signal. It increases roughly as the carrier frequency and the square of the distance traveled. Microwave transmissions near the surface are thus described well by the Rytov approximation. That is often the case for millimeter-wave signals too. By contrast, optical signals traveling over a few hundred meters usually encounter strong-scattering conditions, for which the Rytov description of amplitude fluctuations is not valid.[2]

3.2.1 Plane Waves

The scintillation induced on terrestrial links was first analyzed for plane-wave signals [1][2][4]. That choice simplifies the analysis but is difficult to realize experimentally, primarily because of transmitter-size requirements and problems associated with reflection from the ground. It was soon recognized that a collimated beam wave is similar to a plane wave near the beam axis. Several optical experiments exploiting this similarity were performed using large collimators.

The complex weighting function defined by (2.42) is the starting point for understanding such experiments. One could use the expression established in the last section for astronomical observations (3.10) and modify it to fit the situation in Figure 3.3. It is somewhat clearer if we start afresh. We use the cylindrical coordinates (x, r, ϕ) illustrated in Figure 2.1 to define the scattering point and receiver locations, letting the coordinate origin coincide with the exit plane of the transmitter. The vacuum electric-field strength depends only on the downrange distance:

$$E(r) = \mathcal{E}_0 \exp(ikx) \qquad (3.13)$$

---

[2] Notice that phase fluctuations *are* described accurately by the Rytov approximation well into the strong-scattering regime. We will develop this topic in Chapter 7.

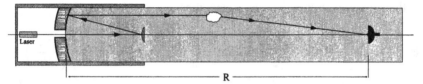

Figure 3.3: Small-angle forward scattering by a typical turbulent irregularity within the volume illuminated by a collimated beam. The signal can be approximated by a plane wave if the scattering eddy is close to the axis.

Green's function depends on the radial distance from the beam axis and the distance from the scattering eddy to the receiver:

$$G(\mathbf{R}, \mathbf{r}) = \frac{\exp\left(ik\sqrt{(R-x)^2 + r^2}\right)}{4\pi\sqrt{(R-x)^2 + r^2}} \tag{3.14}$$

With these conventions the complex weighting function is defined by the following volume integral:

$$\Lambda(\kappa) = k^2 \int_0^R dx \int_0^\infty dr\, r \int_0^{2\pi} d\phi \, \frac{\exp\left(ik\sqrt{(R-x)^2 + r^2}\right)}{4\pi\sqrt{(R-x)^2 + r^2}} \tag{3.15}$$
$$\times \exp[ik(R-x)]\exp\{i[x\kappa_x + r\kappa_r \cos(\phi - \omega)]\}$$

In writing this expression we have run the x integration from the transmitter to the receiver. We have therefore ignored irregularities that lie beyond the receiver because we consider only small-angle forward scattering. The cylindrical wavenumber components are defined by (3.6).

The paraxial approximation explained previously allows one to expand the square root in Green's function:

$$\Lambda(\kappa) = \frac{k^2}{4\pi} \int_0^R dx \, \frac{\exp(ix\kappa_x)}{R-x} \int_0^\infty dr\, r \int_0^{2\pi} d\phi \exp\left(\frac{ikr^2}{2(R-x)}\right)$$
$$\times \exp[ir\kappa_r \cos(\phi - \omega)]$$

The azimuth and radial integrations can be done using the techniques that were noted above (3.10):

Plane wave: $$\Lambda(\kappa) = \frac{ik}{2} \int_0^R dx \exp(ix\kappa_x)\exp\left(-i\frac{R-x}{2k}\kappa_r^2\right) \tag{3.16}$$

With this result we calculate the amplitude weighting function from (2.43):

Plane wave: $$D(\kappa) = \frac{k}{2} \int_0^R dx \exp(ix\kappa_x)\sin\left(\frac{R-x}{2k}\kappa_r^2\right) \tag{3.17}$$

or, with the substitution $x = R - x'$, it becomes

Plane wave: $$D(\kappa) = \frac{k}{2} \exp(i R \kappa_x) \int_0^R dx' \exp(-i x' \kappa_x) \sin\left(\frac{\kappa_r^2 x'}{2k}\right)$$

$$(3.18)$$

This is the equivalent to the general expression (3.11) but with the axis reversed to agree with the experimental situation illustrated in Figure 3.3.

There are two ways to proceed now that we have the weighting function. One can complete the integrations in (3.18) and write

$$D(\kappa) = \frac{k}{4}\left(\frac{\exp[i R \kappa_r^2/(2k)] - \exp(i R \kappa)}{\kappa_x - \kappa_r^2/(2k)} - \frac{\exp[-i R \kappa_r^2/(2k)] - \exp(i R \kappa)}{\kappa_x + \kappa_r^2/(2k)}\right)$$

$$(3.19)$$

The logarithmic amplitude variance is defined by (2.37) and one can proceed to do the three-fold wavenumber integration required there. In doing so, we convert to the refractivity spectrum using $\Phi_\varepsilon = 4\Phi_n$ to find

$$\langle \chi^2 \rangle = 4 \int d^3\kappa \, \Phi_n(\kappa) D(\kappa) D(-\kappa)$$

$$(3.20)$$

It is reasonable to assume that the tropospheric irregularities are isotropic near the surface. It is natural to use cylindrical wavenumber coordinates defined by (3.6) to express the amplitude variance:

$$\langle \chi^2 \rangle = 4 \int_{-\infty}^{\infty} d\kappa_x \int_0^\infty d\kappa_r \, \kappa_r \int_0^{2\pi} d\omega \, \Phi_n\left(\sqrt{\kappa_x^2 + \kappa_r^2}\right)$$
$$\times \, D(\kappa_x, \kappa_r) D(-\kappa_x, -\kappa_r)$$

$$(3.21)$$

This approach is developed in Problem 1, using the relationship of the Fresnel length to the wavenumber to simplify the task.

A second method exploits the relationships among physical quantities much earlier in the process and has become the standard approach. The integral expression (3.18) for $D(\kappa)$ is used to form the product

$$D(\kappa) D(-\kappa) = \frac{k^2}{4} \int_0^R dx_1 \int_0^R dx_2 \exp[i\kappa_x(x_2 - x_1)] \sin\left(\frac{x_1 \kappa_r^2}{2k}\right) \sin\left(\frac{x_2 \kappa_r^2}{2k}\right)$$

$$(3.22)$$

which is central to estimating the amplitude variance. We introduce the sum and difference coordinates

$$u = x_2 - x_1 \quad \text{and} \quad x = \tfrac{1}{2}(x_1 + x_2)$$

and use a standard formula for the product of two sine functions to give

$$D(\kappa)D(-\kappa) = \frac{k^2}{4} \int_0^R dx \int_{-x}^x du \, \exp(iu\kappa_x)\left[\sin^2\left(\frac{x\kappa_r^2}{2k}\right) - \sin^2\left(\frac{u\kappa_r^2}{4k}\right) \right] \quad (3.23)$$

The second term is negligible for most experimental situations. This assertion is based on the observation that the important contribution to the double integral in (3.23) comes from the region of the integration square where x_1 and x_2 are nearly the same. Along this diagonal line the inequality

$$\kappa(x_2 - x_1) < 1 \quad (3.24)$$

is valid because the fluctuations in refractive index are uncorrelated along the path if the points are separated by more than the eddy size [6][7]. This means that the argument in the second term can be estimated as follows:

$$\frac{u\kappa_r^2}{4k} < |\kappa(x_2 - x_1)|\frac{\kappa}{4k} < \frac{\kappa}{4k}$$

This ratio is very small at optical frequencies and the second term in (3.23) can be neglected entirely. The situation is more complicated at microwave frequencies. When we develop an exact theory of plane-wave propagation, we shall find that the contributions of eddies that are comparable in magnitude to the wavelength are not influential and the second term is properly discarded for that case also. The inequality (3.24) plays a minor role here in evaluating the amplitude–weighting-function product. It will play a much more important role in establishing the Markov approximation which describes strong fluctuations and multiple scattering.

The first term in (3.23) is therefore much larger than the second and describes most of the relevant physics. The difference coordinate integration can be evaluated to give the following expression for the amplitude–weighting-function product:

$$D(\kappa)D(-\kappa) = \frac{k^2}{2} \int_0^R dx \, \sin^2\left(\frac{x\kappa_r^2}{2k}\right)\frac{\sin(x\kappa_x)}{\kappa_x} \quad (3.25)$$

At this point, we exploit the relationship between the downrange distance and the turbulence wavenumber. The distance to almost every point along the path is greater than the size of the eddies:

$$x\kappa_x \gg 1$$

This means that the term in square brackets can be replaced by the Dirac delta function described in Appendix F:

$$D(\boldsymbol{\kappa})D(-\boldsymbol{\kappa}) = \frac{\pi k^2}{2}\delta(\kappa_x) \int_0^R dx \, \sin^2\left(\frac{x\kappa_r^2}{2k}\right) \tag{3.26}$$

When we combine this result with the basic definition of the amplitude variance (3.20) we find that

$$\langle \chi^2 \rangle = 2\pi k^2 \int d^3\kappa \, \Phi_n(\boldsymbol{\kappa})\delta(\kappa_x) \int_0^R dx \, \sin^2\left(\frac{x\kappa_r^2}{2k}\right) \tag{3.27}$$

The usual assumption takes the irregularities to be isotropic and one then casts this expression in spherical wavenumber coordinates:

$$\langle \chi^2 \rangle = 2\pi k^2 \int_0^\infty d\kappa \, \kappa^2 \int_0^\infty d\psi \, \sin\psi \int_0^{2\pi} d\omega \, \Phi_n(\kappa)\delta(\kappa\cos\psi)$$
$$\times \int_0^R dx \, \sin^2\left(\frac{x\kappa^2\sin^2\psi}{2k}\right)$$

or, finally,

$$\langle \chi^2 \rangle = 4\pi^2 Rk^2 \int_0^\infty d\kappa \, \kappa\Phi_n(\kappa) \int_0^R dx \, \sin^2\left(\frac{x\kappa^2}{2k}\right) \tag{3.28}$$

The horizontal integration is performed, yielding

$$\langle \chi^2 \rangle = 4\pi^2 Rk^2 \int_0^\infty d\kappa \, \kappa\Phi_n(\kappa)F_\chi(\kappa) \tag{3.29}$$

where

$$\text{Plane wave:} \quad F_\chi(\kappa) = \frac{1}{2}\left(1 - \frac{\sin\left(R\kappa^2/k\right)}{R\kappa^2/k}\right) \tag{3.30}$$

is the *spectral weighting function for amplitude variance* which describes the diffraction process. It depends only on the *scattering parameter*

$$\zeta = \frac{R\kappa^2}{k} = \frac{2\pi R\lambda}{\ell^2} \tag{3.31}$$

which relates the Fresnel length to the variable eddy size ℓ in the turbulence hierarchy. The weighting function is plotted in Figure 3.4. It starts at zero for small values of the scattering parameter, begins to rise for $\zeta = 1$ and eventually settles down to an asymptotic value of 0.5.

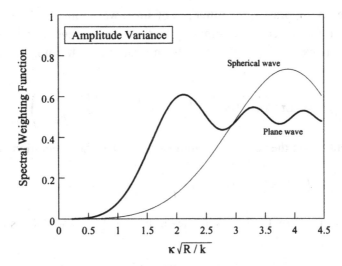

Figure 3.4: Wavenumber-weighting functions for logarithmic amplitude variance predicted for weak-scattering situations. The expressions for plane and spherical waves are plotted as functions of the scattering parameter defined by the wavenumber and Fresnel length.

3.2.1.1 Reduction to Geometrical Optics

Let us pause to compare the diffraction result for $\langle \chi^2 \rangle$ with the description provided by geometrical optics. We know that ray theory is valid only when the influential scattering eddies are large relative to the Fresnel length. If this were true for all eddies, one could expand the spectral weighting function in powers of the scattering parameter

$$\frac{1}{2}\left(1 - \frac{\sin(R\kappa^2/k)}{R\kappa^2/k}\right) \approx \frac{1}{12}\left(\frac{R\kappa^2}{k}\right)^2$$

to find

$$\langle \chi^2 \rangle = \frac{1}{3}\pi^2 R^3 \int_0^\infty d\kappa\, \kappa^5 \Phi_n(\kappa)$$

This is the geometrical-optics description for a plane wave and is independent of wavelength [8][9]. The two methods agree in this case. The problem is that all eddies are *not* large relative to the Fresnel length. The two weighting functions indicated above differ markedly when the scattering parameter is large and this difference has a profound influence in many applications.

3.2.1.2 Amplitude Variance for the Kolmogorov Model

Now let us return to (3.29) and with it calculate the variance for logarithmic amplitude fluctuations in terms of quantities that are measured for terrestrial paths. To do so we must specify the turbulence spectrum. The weighting function plotted in

Figure 3.4 clearly suppresses the influence of small wavenumbers. The amplitude variance is therefore determined primarily by the inertial range. If one retains the inner- and outer-scale wavenumber cutoffs one can use the Kolmogorov model to estimate it:

$$\langle \chi^2 \rangle = 2\pi^2 R k^2 0.033 C_n^2 \int_{\kappa_0}^{\kappa_s} \frac{d\kappa \, \kappa}{\kappa^{\frac{11}{3}}} \left(1 - \frac{\sin(R\kappa^2/k)}{R\kappa^2/k} \right) \tag{3.32}$$

It is convenient to use the scattering parameter (3.30) as the integration variable:

$$\langle \chi^2 \rangle = 0.326 C_n^2 R^{\frac{11}{6}} k^{\frac{7}{6}} \int_{\zeta_0}^{\zeta_m} \frac{d\zeta}{\zeta^{\frac{11}{6}}} \left(1 - \frac{\sin \zeta}{\zeta} \right)$$

The lower limit is set by the largest eddies in the spectrum and defines the *outer-scale scattering parameter*:

$$\zeta_0 = 2\pi R\lambda / L_0^2 \tag{3.33}$$

The upper limit is defined by the inner scale length as follows:

$$\zeta_m = \frac{R\kappa_m^2}{k} = 5.56 \frac{R\lambda}{\ell_0^2} \tag{3.34}$$

Now let us examine the upper and lower limits for typical propagation parameters and meteorological conditions. The Fresnel length is 2.24 cm for $R = 1$ km and $\lambda = 0.6$ μm. The lower limit η is essentially zero in this case because the outer scale length L_0 is much larger. This type of optical propagation is consistently described by Fresnel scattering. The inner scale length ℓ_0 is only a few millimeters and the upper limit is very large. We assume here that ζ_m is infinite but will reconsider its influence later. With these observations and a definite integral listed in Appendix B the important result first established by Tatarskii [10] emerges:

$$\text{Fresnel:} \qquad \langle \chi^2 \rangle = 0.307 C_n^2 R^{\frac{11}{6}} k^{\frac{7}{6}} \tag{3.35}$$

This frequency and distance dependence is dramatically different than that predicted by geometrical optics. We need to test this expression against experimental data.

The first confirmation came from experiments performed in the USSR by Gurvich and his colleagues [11]. They used He–Ne laser signals transmitted over distances ranging from 250 to 1750 m. The path length and wavelength were thus known precisely. The refractive-index structure constant C_n^2 was estimated from temperature-fluctuation data using the connection established in Section 2.3.1 of Volume 1. With these values the amplitude variance could be predicted using (3.35). Values of logarithmic intensity were measured and related to $\langle \chi^2 \rangle$ by the formula

$$\sigma_{\log I}^2 = \langle |\log I - \langle \log I \rangle|^2 \rangle = 4 \langle \chi^2 \rangle$$

The actual data from these experiments is reproduced later in this volume as Figure 9.11. These measurements agreed quite well with the prediction when $\langle \chi^2 \rangle < 1$. This confirms the general correctness of Tatarskii's result (3.35) and indirectly showed that the expected limitation (3.1) on the Rytov approximation is appropriate. Subsequent experiments have continued to confirm these conclusions.

It is also important to test the curious frequency and distance dependence of (3.35). It is difficult to verify the distance dependence of (3.35) because different path lengths necessarily encounter different regions of the atmosphere. If one compares amplitude fluctuations measured at receivers placed along the same path, one cannot be sure that C_n^2 is the same – as we have assumed. This problem evidently worsens as the path length increases since intermittent structures can intrude on one part of the path but not another. One *could* do an optical experiment using multiple transits back and forth over a short path, but this seems not to have been attempted.

On the other hand, it should be a straightforward matter to verify the frequency scaling of (3.35) by comparing the intensity fluctuations measured at different wavelengths. To do so, it is necessary to include the effects of inner- and outer-scale corrections plus aperture averaging – all of which depend on wavelength.

3.2.1.3 The Inner-scale Influence

The upper limit in the amplitude-variance expression (3.32) is no longer large when the Fresnel length is comparable to the inner scale length ℓ_0. That situation occurs for short optical paths, which are often employed in order to ensure that only weak scattering is encountered. The inner scale length is several millimeters near the surface and the Fresnel length is only 1 cm for a 200-m path. In this case the important eddies fall near the energy-loss region of the turbulence spectrum. One cannot capture this influence with a simple cutoff at the inner-scale wavenumber because the result is sensitive to the spectrum's actual behavior beyond that point. We found in Section 2.2.6 of Volume 1 that the turbulence spectrum can be described by the Kolmogorov model multiplied by a universal function – which we call the *energy-dissipation function*:

$$\Phi_n(\kappa) = \frac{0.033 C_n^2}{\kappa^{\frac{11}{3}}} \mathcal{F}(\kappa \ell_0) \qquad \text{for} \qquad \kappa_0 < \kappa < \infty \qquad (3.36)$$

If the Fresnel length is comparable to the inner scale length one can be confident that it is much smaller than the outer scale length. This means that one can take the lower limit to zero in the amplitude-variance expression (3.32) and write

$$\langle \chi^2 \rangle = 2\pi^2 R k^2 0.033 C_n^2 \int_0^\infty \frac{d\kappa\,\kappa}{\kappa^{\frac{11}{3}}} \mathcal{F}(\kappa \ell_0) \left(1 - \frac{\sin(R\kappa^2/k)}{R\kappa^2/k} \right) \qquad (3.37)$$

To make further progress one must specify $\mathcal{F}(\kappa \ell_0)$.

The Gaussian model. Early estimates of the influence of the inner scale length used a Gaussian model to describe the dissipation range [12][13]:

$$\mathcal{F}(\kappa \ell_0) = \exp\left(-\kappa^2/\kappa_m^2\right) \quad \text{with} \quad \kappa_m = 5.91/\ell_0 \qquad (3.38)$$

This model was chosen primarily for its analytical tractability and the corresponding amplitude variance is represented by

$$\langle \chi^2 \rangle = 2\pi^2 R k^2 0.033 C_n^2 \int_0^\infty \frac{d\kappa}{\kappa^{\frac{8}{3}}} \left(1 - \frac{\sin(R\kappa^2/k)}{R\kappa^2/k}\right) \exp\left(\frac{-\kappa^2}{\kappa_m^2}\right)$$

Using the identity

$$1 - \frac{\sin z}{z} = z \int_0^1 du \, (1 - u) \sin(uz)$$

one can write

$$\langle \chi^2 \rangle = 2\pi^2 R k^2 0.033 C_n^2 \int_0^1 du \, (1 - u) \mathcal{J}(u)$$

where

$$\mathcal{J}(u) = \int_0^\infty \frac{d\kappa}{\kappa^{\frac{2}{3}}} \sin\left(\frac{u R\kappa^2}{k}\right) \exp\left(\frac{-\kappa^2}{\kappa_m^2}\right)$$

$$= \Im\left\{\int_0^\infty \frac{d\kappa}{\kappa^{\frac{2}{3}}} \exp\left[-\kappa^2\left(\frac{1}{\kappa_m^2} - i\frac{u R\kappa^2}{k}\right)\right]\right\}$$

$$= \frac{1}{2}\Gamma\left(\frac{1}{6}\right)(\kappa_m)^{\frac{1}{3}} \Im\left[\left(1 - \frac{i u R\kappa_m^2}{k}\right)^{-\frac{1}{6}}\right]$$

The result is given by

$$\langle \chi^2 \rangle = \Gamma\left(\frac{1}{6}\right)\pi^2 R k^2 0.033 C_n^2 (\kappa_m)^{\frac{1}{3}} \Im\left[\int_0^1 du \, (1 - u)\left(1 - \frac{i u R\kappa_m^2}{k}\right)^{-\frac{1}{6}}\right]$$

The remaining integration can be expressed as a hypergeometric function, using its basic definition given in Appendix H:

$$\langle \chi^2 \rangle = \left(0.307 R^{\frac{11}{6}} k^{\frac{7}{6}} C_n^2\right) \mathcal{I}_{pl}\left(\frac{R\kappa_m^2}{k}\right) \qquad (3.39)$$

The *inner scale factor* for the Gaussian model is defined by

$$\mathcal{I}_{pl}(\zeta_m) = 2.593 \zeta_m^{-\frac{1}{6}} \Im\left[{}_2F_1\left(\frac{1}{6}, 1, 3; i\zeta_m\right)\right] \qquad (3.40)$$

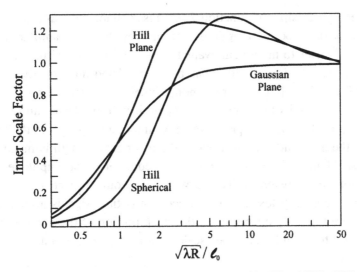

Figure 3.5: The inner scale factor for plane and spherical waves computed by Hill and Clifford [15]. These curves are based on the Hill bump model for the spectrum in the energy-loss region. The curve for a plane wave with a Gaussian dissipation function is shown for reference.

and depends on the dimensionless variable ζ_m defined by (3.34). One can verify that the inner scale factor goes to unity when ℓ_0 vanishes.[3] The function defined by (3.40) is plotted in Figure 3.5 as a function of the Fresnel length divided by the inner scale length. It rises steadily from zero to unity as this ratio increases. We will not use this model because meteorological measurements have since shown that the Gaussian model is not a good description for the dissipation region.[4]

The Hill bump model. A more accurate description of the spectrum in the energy-loss region was developed later by Hill [14]. His model for $\mathcal{F}(\kappa \ell_0)$ is the solution of a second-order differential equation and is plotted in Figure 2.7 of Volume 1. It includes the removal of eddies caused both by diffusion and by viscosity in the turbulent velocity field. It provides a substantially different physical description than the Gaussian model. More to the point, it agrees with numerous meteorological and electromagnetic propagation measurements.

---

[3] One uses the linear transformation property of hypergeometric functions

$$_2F_1(a, b, c; z) = (1 - z)^{-a} {}_2F_1\left(a, c - b, c; \frac{z}{z - 1}\right)$$

to handle the large value of z and the special value

$$_2F_1(a, b, c; 1) = \frac{\Gamma(c)\Gamma(c - a - b)}{\Gamma(c - a)\Gamma(c - b)}$$

[4] This problem is discussed in Section 2.2.6 of Volume 1.

The Hill energy-dissipation function $\mathcal{F}(\kappa \ell_0)$ is available only as a computer file and one cannot describe the amplitude variance in terms of familiar functions. This poses no barrier to numerical evaluation – indeed it is a virtue. The amplitude variance that corresponds to it has been calculated numerically by several authors [15][16][17][18]. The corresponding inner scale factor for a plane wave is plotted in Figure 3.5 where one can compare it with the Gaussian result. This curve also approaches an asymptotic value of unity and reproduces the basic expression for the amplitude variance when the inner scale length is much smaller than the Fresnel length. It predicts a substantial reduction of the amplitude variance when that situation is reversed. This is what one would expect if the important eddies fall in the energy-loss region. Notice that optical signals used with short path lengths are typically operating on the steep portion of these curves and one is well advised to include the inner scale factor when one is interpreting such measurements.

3.2.1.4 The Outer-scale Influence

The outer-scale region of the spectrum can also influence the amplitude variance in some situations of interest. For instance, millimeter-wave links sometimes exhibit this effect [19]. Their Fresnel lengths are usually only a few meters and the outer scale length can be that small for paths close to the surface. The outer-scale scattering parameter can thus be greater than unity and one should expect Fraunhofer scattering:

$$\lambda R/L_0^2 \geq 1$$

This means that one must not take the lower limit to zero in the amplitude-variance expression (3.32). On the other hand, one can then be confident that the upper limit there is enormous because $\ell_0 \ll L_0$ for the large Reynolds numbers that characterize atmospheric flow fields. This allows one to write the amplitude variance as the product of two terms:

$$\langle \chi^2 \rangle = 0.307 C_n^2 R^{\frac{11}{6}} k^{\frac{7}{6}} \mathbb{O}\left(\frac{2\pi R\lambda}{L_0^2} \right) \tag{3.41}$$

The first combination is just the basic expression for the amplitude variance.

The second term defines the *outer scale factor*. It depends both on L_0 and on the actual behavior of the turbulence spectrum for small wavenumbers. To show the second feature we examine two models that are often used to describe this region: the Kolmogorov model with a cutoff and the von Karman model.[5] The outer scale

---

[5] These models are pictured in Figure 2.8 of Volume 1.

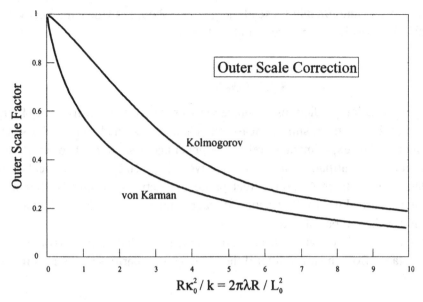

Figure 3.6: Outer scale factors for amplitude variance that are predicted by the von Karman model and the Kolomogorov inertial model with a sharp cutoff.

factors for these models are defined by the following expressions:

Kolmogorov model: $\mathbb{O}(\eta) = 1.060 \int_{\eta}^{\infty} \dfrac{d\zeta}{\zeta^{\frac{11}{6}}} \left(1 - \dfrac{\sin \zeta}{\zeta} \right)$

von Karman model: $\mathbb{O}(\eta) = 1.060 \int_{0}^{\infty} \dfrac{d\zeta}{(\zeta + \eta)^{\frac{11}{6}}} \left(1 - \dfrac{\sin \zeta}{\zeta} \right)$

(3.42)

These functions are plotted together in Figure 3.6. They are normalized to unity at zero at the origin but then separate noticeably. This shows that the correction is sensitive to the shape of the spectrum near the outer-scale wavenumber.

For large values both functions fall gradually and their asymptotic behavior

$$\lim_{\eta \to \infty} \mathbb{O}(\eta) = 1.272/\eta^{\frac{5}{6}}$$

indicates that the variance should have a different form in the Fraunhofer regime [20].

Fraunhofer: $\langle \chi^2 \rangle = 0.391 \, R k^2 C_n^2 (\kappa_0)^{-\frac{5}{3}}$ (3.43)

The frequency and distance scaling here is different than that predicted by the Fresnel expression (3.35) and the consequences for measurements will be discussed in Section 3.2.4. Microwave-propagation transmissions sometimes fall in the range between Fresnel and Fraunhofer scattering, so one should use the numerical values in Figure 3.6 to estimate their scintillation levels.

3.2.2 Spherical Waves

Testing scintillation predictions for plane waves required the construction of large collimators. By contrast, simple measurements could be made with point-source transmitters. This experimental preference generated a surge of theoretical work to describe the scintillation of spherical waves [21][22][23]. Geometrical-optics calculations in Volume 1 showed that phase fluctuations are quite different for plane and spherical waves. It was judged that amplitude fluctuations might also be different for these types of wave.

A point source should generate an electric-field strength that depends only on the scalar distance from the transmitter if there are no irregularities in the transmission volume:

$$E_0(\mathbf{r}) = \mathcal{E}_0 \frac{\exp(ik|\mathbf{r} - \mathbf{T}|)}{|\mathbf{r} - \mathbf{T}|} \tag{3.44}$$

We have omitted the static and induction field components here because the vast majority of the scattering irregularities are many wavelengths from the transmitter. In the units we are using the constant $\mathcal{E}_0$ is related to the transmitted power by

$$P_{\mathrm{T}} = \tfrac{1}{2}c|\mathcal{E}_0|^2 \tag{3.45}$$

and this connection will be examined in Problem 3.

To evaluate the amplitude variance for fluctuations imposed on a spherical wave we return to the complex weighting function defined by (2.42). It is convenient to use the cylindrical coordinates identified in Figure 3.7 and place the transmitter at the coordinate origin. The unperturbed fields at the eddy and at the receiver are given in these coordinates by

$$E_0(\mathbf{r}) = \mathcal{E}_0 \frac{\exp\left(ik\sqrt{x^2 + r^2}\right)}{\sqrt{x^2 + r^2}} \qquad \text{and} \qquad E_0(\mathbf{R}) = \mathcal{E}_0 \frac{\exp(ikR)}{R}$$

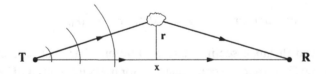

Figure 3.7: The geometry employed to describe the scattering of a spherical wave by atmospheric irregularities.

In the same coordinates Green's function is

$$G(\mathbf{R}, \mathbf{r}) = \frac{\exp\left(ik\sqrt{(R-x)^2 + r^2}\right)}{4\pi\sqrt{(R-x)^2 + r^2}}$$

and the complex weighting function then becomes

$$\Lambda(\kappa) = k^2 \int dx \int_0^\infty dr\, r \int_0^{2\pi} d\phi\, \frac{\exp\left(ik\sqrt{(R-x)^2 + r^2}\right)}{4\pi\sqrt{(R-x)^2 + r^2}} \frac{R}{\sqrt{x^2 + r^2}}$$
$$\times \exp\left[ik\left(\sqrt{x^2 + r^2} - R\right)\right] \exp\{i\,[x\kappa_x + r\kappa_r \cos(\phi - \omega)]\} \quad (3.46)$$

We will rely on the paraxial approximation to estimate this quantity.[6] We can ignore eddies that lie to the left of the transmitter and to the right of the receiver since they would require back scattering to influence the received signal. We therefore need to integrate on the line-of-sight variable only in the limited range $0 < x < R$. However, the small-angle-scattering assumption has a more important effect. As explained in Section 3.1, it means that one can expand the square roots which occur in our expression for the complex weighting function:

$$\sqrt{(R-x)^2 + r^2} \approx R - x + \frac{r^2}{2(R-x)} + \cdots$$

and

$$\sqrt{x^2 + r^2} \approx x + \frac{r^2}{2x} + \cdots$$

We require the first two terms in these expansions to describe the phase of the vacuum field strength and Green's function. In the denominators of these functions we need only the first-order terms:

$$\Lambda(\kappa) = \frac{k^2}{4\pi} \int_0^R dx\, \frac{R}{x(R-x)} \int_0^\infty dr\, r \int_0^{2\pi} d\phi\, \exp\left(r^2 \frac{ikR}{2x(R-x)}\right)$$
$$\times \exp\{i[x\kappa_x + r\kappa_r \cos(\phi - \omega)]\}$$

The azimuth integration gives the zeroth-order Bessel function:

$$\Lambda(\kappa) = \frac{k^2}{2} \int_0^R dx\, \frac{R\exp(ix\kappa_x)}{x(R-x)} \int_0^\infty dr\, r\, J_0(r\kappa_r) \exp\left[-r^2\left(\frac{-ikR}{2x(R-x)}\right)\right]$$

---

[6] We will show later that $\Lambda(\kappa)$ can be evaluated exactly so as to describe the wide-angle scattering of microwave signals.

The radial integration was encountered in the plane-wave case and can be found in Appendix D:

$$\Lambda(\kappa) = \frac{ik}{2} \int_0^R dx \, \exp(ix\kappa_x) \exp\left(\frac{i\kappa_r^2 x(R-x)}{2kR}\right) \tag{3.47}$$

The corresponding amplitude weighting function is computed from (2.43):

$$D(\kappa) = \frac{k}{2} \int_0^R dx \, \exp(ix\kappa_x) \sin\left(\frac{\kappa_r^2 x(R-x)}{2kR}\right) \tag{3.48}$$

Let us pause to compare this result with expression (3.17) for a plane wave. They are the same but for the replacement

$$x \rightarrow x\left(1 - \frac{x}{R}\right)$$

This change causes measurable differences in the scintillation levels imposed on the two types of wave.

With the amplitude weighting function in hand, we are now ready to calculate the amplitude variance with the relationship (2.37). We convert to the spectrum of refractive-index irregularities because we are examining propagation close to the surface:

$$\langle\chi^2\rangle = k^2 \int d^3\kappa \, \Phi_n(\kappa) \int_0^R dx \int_0^R dx' \exp[ix\kappa_x(x-x')]$$
$$\times \sin\left(\frac{\kappa_r^2 x(R-x)}{2kR}\right) \sin\left(\frac{\kappa_r^2 x'(R-x')}{2kR}\right)$$

The double path integration generates a delta function, as we learned in the plane-wave case:

$$\langle\chi^2\rangle = 2\pi k^2 \int d^3\kappa \, \Phi_n(\kappa)\delta(\kappa_x) \int_0^R dx \sin^2\left(\frac{\kappa_r^2 x(R-x)}{2kR}\right)$$

It is reasonable to assume that the irregularities are isotropic for such paths and use the spherical wavenumber coordinates (3.28). With the rescaling $x = uR$ we find that

$$\langle\chi^2\rangle = 2\pi Rk^2 \int_0^\infty d\kappa \, \kappa^2 \Phi_n(\kappa) \int_0^\pi d\psi \sin\psi \int_0^{2\pi} d\omega \, \delta(\kappa \cos\psi)$$
$$\times \int_0^1 du \sin^2\left(\frac{\kappa^2 \sin^2\psi \, Ru(1-u)}{2k}\right)$$

This can be reduced to a familiar form:

$$\langle \chi^2 \rangle = 4\pi^2 Rk^2 \int_0^\infty d\kappa\, \kappa\, \Phi_n(\kappa) F_\chi(\kappa) \tag{3.49}$$

where the *spectral weighting function of amplitude variance for spherical waves* is defined by

Spherical wave:
$$F_\chi(\kappa) = \frac{1}{2} \int_0^1 du \left[1 - \cos\left(\frac{\kappa^2 Ru(1-u)}{k} \right) \right] \tag{3.50}$$

By setting $2u = 1 - t$ one finds the result first established by Tatarskii [24]:

Spherical wave:
$$F_\chi(\kappa) = \frac{1}{2} \left\{ 1 - \sqrt{\frac{2\pi k}{R\kappa^2}} \left[\cos\left(\frac{R\kappa^2}{4k} \right) C\left(\sqrt{\frac{R\kappa^2}{2\pi k}} \right) \right. \right.$$
$$\left. \left. + \sin\left(\frac{R\kappa^2}{4k} \right) S\left(\sqrt{\frac{R\kappa^2}{2\pi k}} \right) \right] \right\} \tag{3.51}$$

where the Fresnel integrals defined by

$$C(x) = \int_0^x dt \cos\left(\frac{\pi t^2}{2} \right) \quad \text{and} \quad S(x) = \int_0^x dt \sin\left(\frac{\pi t^2}{2} \right) \tag{3.52}$$

are tabulated in standard references [25]. The weighting function in (3.51) is also plotted in Figure 3.4 so that it can be compared with the plane-wave result. The two functions are considerably different in the way they approach the common asymptotic value. The spherical weighting rises more slowly than does the plane-wave version and reaches a higher peak. On the other hand, the spectrum is falling rapidly and the small-wavenumber region is decisive for optical applications. The plane wave is therefore picking up larger contributions from the spectrum than is the spherical wave. The plane-wave variance should be larger than the spherical result for identical values of the parameters C_n^2, R and k. Let us find out by how much.

We now know that amplitude fluctuations are caused by small eddies in Fresnel scattering and the result should depend on the inertial range of the spectrum. We can rely on the Kolmogorov model, ignoring the inner- and outer-scale limitations in that case. The easiest way to calculate the variance is to use (3.50) and reverse the order of integration in (3.49):

$$\langle \chi^2 \rangle = 0.033\pi^2 Rk^2 C_n^2 \int_0^1 du \int_0^\infty \frac{d\kappa}{\kappa^{\frac{8}{3}}} \left[1 - \cos\left(\frac{\kappa^2 Ru(1-u)}{k} \right) \right]$$

The integrations are separated if we define a new variable w by the argument of the cosine term:

$$\langle \chi^2 \rangle = 0.033\pi^2 R^{\frac{11}{6}} k^{\frac{7}{6}} C_n^2 \int_0^1 du\, [u(1-u)]^{\frac{5}{6}} \int_0^\infty \frac{dw}{w^{\frac{11}{6}}} (1 - \cos w)$$

Both integrals are found in Appendix B and the amplitude variance is [24]

$$\text{Spherical wave:} \qquad \langle \chi^2 \rangle = 0.124 R^{\frac{11}{6}} k^{\frac{7}{6}} C_n^2 \qquad (3.53)$$

This is only 40% of the plane-wave result, demonstrating the strong influence of the type of wave transmitted on the amplitude variance.

The dissipation region of the turbulence spectrum can also influence the spherical-wave variance on short paths. The Hill bump model for the energy-dissipation function yields

$$\text{Spherical wave:} \qquad \langle \chi^2 \rangle = 0.124 R^{\frac{11}{6}} k^{\frac{7}{6}} C_n^2 \mathcal{I}_{\text{sph}}\left(\frac{\sqrt{\lambda R}}{\ell_0}\right) \qquad (3.54)$$

This inner scale factor is also plotted in Figure 3.5 and is similar to the plane-wave factor for the Hill bump model.

One could also evaluate the influence of the outer-scale region on the amplitude variance, as we did for plane waves in Figure 3.6. The required calculations are clear cut and we will not pause to develop them. We are anxious to press on to see whether our amplitude-variance expressions agree with optical and microwave measurements. Before we can make those comparisons, we must include an essential factor that is still missing.

3.2.3 Aperture Averaging of Scintillation

Our description thus far has considered only the amplitude variance measured at a single point. This corresponds to a small optical or microwave detector. By contrast, most communication systems and scientific experiments employ a good deal of antenna gain. They do so to enhance the signal-to-noise ratio and to eliminate multipath signals reflected from terrain and off-axis objects.

The first task is to relate the incoming field to the receiver output. How one does so depends on the type of measurement that is being made. If the receiver is an array of photo diodes, the measured intensity is simply the irradiance of the arriving wave integrated over the array:

$$\bar{I} = \frac{1}{A} \iint_A d^2\sigma\, I(\sigma) \qquad (3.55)$$

One should next ask what happens when the lens of an astronomical telescope concentrates the arriving waves at its focal plane. Because the image wanders as a result of random deflections in the turbulent atmosphere,[7] a matrix of photo diodes or film plate is usually placed at the focal plane to capture all the available light. In this case, one can use (3.55) to describe the recorded signal provided that A is identified with the area of the telescope's primary mirror or lens. The situation is no different when a parabolic reflector is used to concentrate incoming microwave signals at a focal feed. The variance of irradiance fluctuations measured by such receivers depends on the spatial covariance of intensity fluctuations at adjacent points on the aperture surface:

$$\overline{\langle \delta I^2 \rangle} = \frac{1}{A^2} \iint_A d^2\sigma_1 \iint_A d^2\sigma_2 \langle [I(\sigma_1) - \langle I \rangle][I(\sigma_2) - \langle I \rangle] \rangle \tag{3.56}$$

For weak scattering we can use the relationship between intensity variations and the logarithmic amplitude developed above (3.2) to express the result in terms of the spatial covariance of χ:

$$\overline{\langle \delta I^2 \rangle} = \frac{4I_0^2}{A^2} \iint_A d^2\sigma_1 \iint_A d^2\sigma_2 \langle \chi(\sigma_1)\chi(\sigma_2) \rangle \tag{3.57}$$

This expression is widely used in the literature to address aperture averaging [26][27][28][29] and we will adopt it as our point of departure.

We should pause to note that a different type of measurement is made with telescopes or optical transmission systems when it is important to recover phase information. If one places a point detector at the focus of a parabolic concentrator, the instantaneous response in the receiver is proportional to the electric-field strength averaged over the area of the reflector. Provided that the aperture size is large relative to the electromagnetic wavelength,

$$\overline{E} = \frac{1}{A} \iint_A d^2\sigma \, E(\sigma) \tag{3.58}$$

The absolute mean-square value of the aperture-averaged field now depends on the *mutual coherence function*, which plays a key role in describing image transmission:

$$\langle |\overline{E}|^2 \rangle = \frac{1}{A^2} \iint_A d^2\sigma_1 \iint_A d^2\sigma_2 \langle E(\sigma_1)E^*(\sigma_2) \rangle \tag{3.59}$$

The integrand depends both on the phase and on the amplitude of the field strength:[8]

$$\langle E_1 E_2^* \rangle = \langle A_1 A_2 \exp[i(\varphi_1 - \varphi_2)] \rangle \tag{3.60}$$

---

[7] An experimental record of image wander in the focal plane of an astronomical telescope is reproduced in Figure 7.1 of Volume 1.
[8] This combination is discussed further in Section 9.4 of this volume.

The following approximation can be justified:

$$\langle |\overline{E}|^2 \rangle = \frac{1}{A^2} \iint_A d^2\sigma_1 \iint_A d^2\sigma_2 \, \langle A_1 A_2 \exp[i(\varphi_1 - \varphi_2)] \rangle$$

$$\simeq \frac{1}{A^2} \iint_A d^2\sigma_1 \iint_A d^2\sigma_2 \, \langle A_1 A_2 \rangle \langle \exp[i(\varphi_1 - \varphi_2)] \rangle$$

so the second ensemble average is proportional to the exponential of the phase structure function:

$$\langle |\overline{E}|^2 \rangle = \frac{1}{A^2} \iint_A d^2\sigma_1 \iint_A d^2\sigma_2 \, \langle A_1 A_2 \rangle \exp\left[-\tfrac{1}{2} D_\varphi(|\sigma_1 - \sigma_2|)\right] \qquad (3.61)$$

Phase fluctuations are correlated only over distances that are small relative to the coherence length of amplitude fluctuations. The measured signal is therefore determined primarily by phase fluctuations across the aperture.

We will concentrate here on the more common situation described by (3.57). We use the aperture-averaged variance of the logarithmic amplitude as our basic reference:

$$\overline{\langle \chi^2 \rangle} = \frac{1}{A^2} \iint_A d^2\sigma_1 \iint_A d^2\sigma_2 \, \langle \chi(\sigma_1)\chi(\sigma_2) \rangle \qquad (3.62)$$

To complete the calculation we must anticipate the results of Chapter 4 to describe the spatial covariance of χ measured at adjacent points in a plane normal to the line of sight.

3.2.3.1 Plane Waves

If the irregularities are isotropic and the spectrum does not change along the path, the spatial covariance for plane waves can be written as

$$\langle \chi(\sigma)\chi(\sigma') \rangle_{pl} = 2\pi^2 R k^2 \int_0^\infty d\kappa \, \kappa \, \Phi_n(\kappa) J_0(\kappa|\sigma - \sigma'|) \left(1 - \frac{\sin(R\kappa^2/k)}{R\kappa^2/k}\right)$$

$$(3.63)$$

In this application, we identify $|\sigma - \sigma'|$ with the scalar distance between two surface elements on the receiving aperture.

The antenna patterns of receivers are defined by the reflector–lens combination and by the arrangement of the detectors. The reflector must be large relative to the wavelength in order to produce appreciable gain. Optical and microwave receivers invariably use circular apertures to enhance symmetry and that simplifies the analysis enormously. We assume that the receiver is a uniformly illuminated disk and

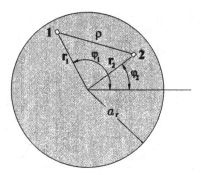

Figure 3.8: Cylindrical coordinates used to locate the surface elements on a circular receiver of radius a_r.

use the circular geometry of Figure 3.8 to make the aperture averages explicit:

$$\overline{\langle \chi^2 \rangle}_{pl} = 2Rk^2 \frac{1}{a_r^4} \int_0^\infty d\kappa \, \kappa \, \Phi_n(\kappa) \left(1 - \frac{\sin(R\kappa^2/k)}{R\kappa^2/k} \right) \int_0^{a_r} dr_1 \, r_1 \int_0^{2\pi} d\phi_1$$

$$\times \int_0^{a_r} dr_2 \, r_2 \int_0^{2\pi} d\phi_2 \, J_0 \left(\kappa \sqrt{r_1^2 + r_2^2 - 2r_1 r_2 \cos(\phi_1 - \phi_2)} \right)$$

The double surface integral was evaluated in estimating aperture-averaged phase fluctuations:[9]

$$\overline{\langle \chi^2 \rangle}_{pl} = 2\pi^2 Rk^2 \int_0^\infty d\kappa \, \kappa \, \Phi_n(\kappa) \left(1 - \frac{\sin(R\kappa^2/k)}{R\kappa^2/k} \right) \left(\frac{2J_1(\kappa a_r)}{\kappa a_r} \right)^2 \tag{3.64}$$

The *aperture-averaging wavenumber weighting function* appears in the second set of large parentheses. It eliminates the contributions of wave-numbers greater than the reciprocal of the radius of the receiver.

The amplitude variance depends primarily on the similarity range of the spectrum if the receiver is large relative to the inner scale length. One can then take the upper wavenumber limit for the Kolmogorov model to infinity and ignore the outer-scale lower limit:

$$\overline{\langle \chi^2 \rangle}_{pl} = 0.651 Rk^2 C_n^2 \int_0^\infty \frac{d\kappa}{\kappa^{\frac{8}{3}}} \left(1 - \frac{\sin(R\kappa^2/k)}{R\kappa^2/k} \right) \left(\frac{2J_1(\kappa a_r)}{\kappa a_r} \right)^2$$

Using the scattering parameter as the integration variable, we find that

$$\overline{\langle \chi^2 \rangle}_{pl} = 0.307 R^{\frac{11}{6}} k^{\frac{7}{6}} C_n^2 G \left(a_r \sqrt{\frac{2\pi}{R\lambda}} \right) \tag{3.65}$$

[9] See Section 4.1.8 of Volume 1.

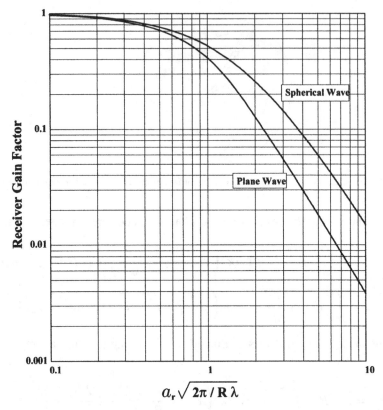

Figure 3.9: Receiver gain factors for plane and spherical waves.

where the *receiver gain factor* for plane waves is defined by the following integral:

$$\text{Plane wave:} \quad G(\eta) = 1.060 \int_0^\infty \frac{dx}{x^{\frac{11}{6}}} \left(1 - \frac{\sin x}{x}\right) \left(\frac{2J_1(\eta\sqrt{x})}{\eta\sqrt{x}}\right)^2 \quad (3.66)$$

Numerical values for this function are plotted in Figure 3.9. The following expression gives a close approximation to that curve and is adequate for engineering applications [30]:

$$G(\eta) = \left[1 + 1.07\left(\frac{2\pi a_r^2}{\lambda R}\right)^{\frac{7}{6}}\right]^{-1} \quad (3.67)$$

The gain factor is essentially unity when the radius of the receiver is less than the Fresnel length. In that case one can use the point-detector result (3.35) to interpret plane-wave measurements. The asymptotic expression

$$\lim_{\eta > 2} G(\eta) = \frac{0.934}{\eta^{\frac{7}{3}}}$$

indicates that the amplitude variance is described by

$$a_{\mathrm{r}} > \sqrt{\lambda R} \qquad \overline{\langle \chi^2 \rangle}_{\mathrm{pl}} = 0.287 C_n^2 R^3 (a_{\mathrm{r}})^{-\frac{7}{3}} \qquad (3.68)$$

when the aperture radius is much greater than the Fresnel length. This expression is independent of wavelength and identical to the prediction of geometrical optics. That coincidence occurs because amplitude fluctuations are caused by small eddies and their diffraction effects are eliminated by aperture smoothing. Expression (3.68) can be used to interpret observations made with large astronomical telescopes and optical scintillometers.

The plane-wave expression (3.65) gives the complete variation of $\overline{\langle \chi^2 \rangle}$ with distance, frequency and aperture size. The dependences are interconnected through the gain factor. This means that specific values of two variables will influence scaling with the third. We should like to test these predictions against actual measurements. Most of the data for horizontal propagation has been taken with diverging waves and we turn to that problem next.

3.2.3.2 Spherical Waves

The description of aperture averaging for spherical waves proceeds along the same route. We must again anticipate the results of Chapter 4 to express the spatial covariance for a spherical wave:

$$\langle \chi(\sigma_1)\chi(\sigma_2) \rangle_{\mathrm{sph}} = 4\pi^2 R k^2 \int_0^\infty d\kappa \, \kappa \, \Phi_n(\kappa) \int_0^1 du \, J_0(\kappa u |\sigma_1 - \sigma_2|)$$

$$\times \sin^2\left(\frac{R\kappa^2 u(1-u)}{2k} \right) \qquad (3.69)$$

The parametric u integration describes the variable separation of spherical wave components. We combine this result with (3.62) and use the cylindrical coordinates of Figure 3.8 to write the scalar separation between the individual surface elements:

$$\overline{\langle \chi^2 \rangle}_{\mathrm{sph}} = \frac{4 R k^2}{a_{\mathrm{r}}^4} \int_0^\infty d\kappa \, \kappa \, \Phi_n(\kappa) \int_0^1 du \, \sin^2\left(\frac{R\kappa^2 u(1-u)}{2k} \right) \int_0^{a_{\mathrm{r}}} dr_1 \, r_1 \int_0^{2\pi} d\phi_1$$

$$\times \int_0^{a_{\mathrm{r}}} dr_2 \, r_2 \int_0^{2\pi} d\phi_2 \, J_0\left(\kappa u \sqrt{r_1^2 + r_2^2 - 2r_1 r_2 \cos(\phi_1 - \phi_2)}\right)$$

The addition theorem for Bessel functions presented in Appendix D separates the aperture variables, and the surface integrals can then be performed [31]:

$$\overline{\langle \chi^2 \rangle}_{\mathrm{sph}} = 4\pi^2 R k^2 \int_0^\infty d\kappa \, \kappa \, \Phi_n(\kappa) \left[\int_0^1 du \, \sin^2\left(\frac{R\kappa^2 u(1-u)}{2k} \right) \left(\frac{2J_1(\kappa u a_{\mathrm{r}})}{\kappa u a_{\mathrm{r}}} \right)^2 \right]$$

$$(3.70)$$

The *aperture-averaging wavenumber weighting function* for spherical waves is contained in the square brackets.

We use the Kolmogorov model to describe the spectrum since the inertial range is usually the most influential for terrestrial experiments. The result can be expressed as the variance for a point receiver multiplied by the *receiver gain factor* for spherical waves:

$$\text{Spherical wave:} \quad \overline{\langle \chi^2 \rangle} = 0.124 R^{\frac{11}{6}} k^{\frac{7}{6}} C_n^2 G\left(a_r \sqrt{\frac{2\pi}{R\lambda}}\right) \tag{3.71}$$

where

$$\text{Spherical wave:} \quad G(\eta) = 5.246 \int_0^\infty \frac{dx}{x^{\frac{11}{6}}} \int_0^1 du \, \sin^2\left[\tfrac{1}{2}xu(1-u)\right]$$

$$\times \left(\frac{2J_1(\eta u \sqrt{x})}{\eta u \sqrt{x}}\right)^2 \tag{3.72}$$

This function is also plotted in Figure 3.9 so that one can compare it with the corresponding plane-wave result.

3.2.3.3 Experimental Tests

These predictions can be tested using laser signals on terrestrial paths. The receiver gain factor is equal to the ratio of the scintillation measured with a defined aperture to that which is sensed by a point detector:

$$G\left(a_r \sqrt{\frac{2\pi}{R\lambda}}\right) = \frac{\langle \chi^2(a_r) \rangle}{\langle \chi^2(0) \rangle} \tag{3.73}$$

If one compares the same signal measured with a variety of receiver openings simultaneously and that measured with a very small receiver, the resulting ratios should agree with the curves in Figure 3.9.

In two early experiments He–Ne lasers were used to perform this experiment. It was first done in Germany with path lengths of 4.5 and 14.5 km using aperture diameters between 5 and 80 mm [32]. A second experiment was performed on an 8-km path with diameters ranging from 1 to 100 mm [33]. The predictions of Rytov theory were not confirmed by these measurements. It was later realized that strong scattering was probably responsible for the scintillation experienced on these long links.

Some short-path optical experiments were then designed in order to test the Rytov predictions under weak-scattering conditions [34]. A diverging wave was generated by a He–Ne laser. Receivers with diameters ranging from 1 to 203 mm were deployed normal to the line of sight. The smallest receiver represented the point-source reference. The receiver gain factor was measured with four path lengths between

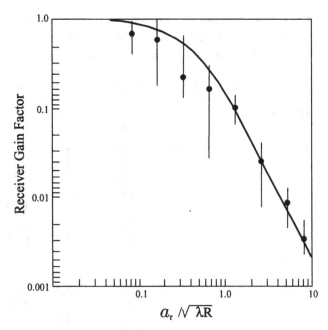

Figure 3.10: The receiver gain factor measured with a He–Ne laser signal on a 215.4-m path by Homstad, Strohbehn, Berger and Heneghan [34]. The error bars represent 90% confidence levels. The solid curve is the prediction for a spherical wave.

200 and 1600 m. The actual data taken for $R = 215.4$ m is reproduced in Figure 3.10 and indicates that good agreement with the spherical-wave prediction was found. In similar experiments performed at 430.8 and 861.6 m larger-than-expected values were found where $a_r/\sqrt{R\lambda} > 1$. This suggests the onset of strong scintillation in which the reduction in gain factor is softened by multiple scattering as discussed in Volume 3. That experiment was repeated with improved data-processing programs using an array of six apertures with diameters of 1, 2.75, 5, 10, 25 and 50 mm [30][35]. In this experiment good agreement was found for path lengths of 250 and 500 m if the prediction included the influences both of the aperture size and of the inner scale length.

3.2.4 Wavelength Scaling of the Scintillation Level

The problems encountered in correlating electromagnetic data with single-point meteorological measurements can be avoided if one compares signals at different frequencies transmitted simultaneously over the same path. The ratio of the scintillation measured at one wavelength to that measured at a second should be independent of the structure constant C_n^2. The relative importance of the various terms in the amplitude-variance expression depends on whether one is using microwave or optical waves.

Table 3.1: *A summary of results from experiments that examined the wavelength scaling of scintillation experienced by microwave and millimeter-wave signals*

Location	Year	R (km)	h (m)	f (GHz)	$\sqrt{\lambda R}$ (m)	a_r (m)	Ref.
Georgia	1971	43	40–50	4	56.8		36
				6	46.4		
Hawaii	1975	64	0–3000	9.55	44.8	1.35	37
				19.1	31.7	1.35	
				22.2	29.4	1.35	
				25.4	27.5	1.35	
				33.3	24.0	1.00	
London	1977	4.1	15–50	36	5.8	0.29	19, 38
				110	3.3	0.23	
White Sands	1978	2	1.5	94	2.5		39
				140	2.1		
Vermont	1983	1.2	4	35	3.2	0.30	39
				94	2.0	0.30	
				140	1.6	0.30	
				217	1.3		
		0.3	4	35	1.6		
				94	1.0		
				140	0.8		
				217	0.7		

3.2.4.1 Microwave Measurements

Several wavelength-scaling experiments have been performed using two or more microwave signals. The parameters that define these experiments are summarized in Table 3.1. The Fresnel lengths that result from the choices of path and wavelength are large compared with the inner scale length in each case. They are also large compared with the receiving antennas employed. On the other hand, the Fresnel lengths are sometimes comparable to the outer scale length L_0 that is estimated or measured for these paths. The scintillation ratio should therefore be uniquely determined by the wavelength ratio and the outer scale factor:

$$\text{Microwave:} \qquad \frac{\langle \chi_1^2 \rangle}{\langle \chi_2^2 \rangle} = \left(\frac{\lambda_2}{\lambda_1} \right)^{\frac{7}{6}} \frac{\mathbb{O}(\sqrt{\lambda_1 R}/L_0)}{\mathbb{O}(\sqrt{\lambda_2 R}/L_0)} \qquad (3.74)$$

The numerical values for $\mathbb{O}(\eta)$ plotted in Figure 3.6 indicate that one can omit this function when the Fresnel length is less than the outer scale length. In that case one would expect the following scaling law to apply:

$$\sqrt{\lambda R} < L_0 \qquad \frac{\langle \chi_1^2 \rangle}{\langle \chi_2^2 \rangle} = \left(\frac{\lambda_2}{\lambda_1} \right)^{\frac{7}{6}} \qquad (3.75)$$

By contrast, there are situations in which the Fresnel length is greater than the outer scale length. This usually occurs for millimeter-wave signals traveling along short paths and the variance expression (3.43) suggests the scaling law

$$L_0 < \sqrt{\lambda R} \qquad \frac{\langle \chi_1^2 \rangle}{\langle \chi_2^2 \rangle} = \left(\frac{\lambda_2}{\lambda_1} \right)^2 \qquad (3.76)$$

Intermediate situations evidently occur when L_0 is comparable to the Fresnel length and one would expect frequency-scaling exponents from 1.17 to 2.0. This is a surprisingly wide range and we should be sensitive to its effect on experimental results.

In an early test scintillation levels on a commercial communication relay link were measured [36]. The associated Fresnel lengths were greater than the outer scale length, which was estimated from the relationship

$$L_0 \approx 0.4h \qquad (3.77)$$

that is valid near the surface.[10] The corresponding estimate for L_0 suggests that the first scaling law should apply. The 7/6 prediction of (3.75) gave a rough fit to the ratios of scintillation levels measured at 3.910, 4.198 and 5.975 GHz.

A large-scale experiment was then mounted in order to measure amplitude and phase fluctuations on a 64-km path between Maui and Hawaii [37]. Five microwave frequencies were selected to represent bands being considered for new communication services. The transmission path rose from sea level to 3000 m and most of the scattering took place in the free atmosphere where the horizontal outer scale length is several kilometers. That is much greater than the Fresnel lengths for this path and one would expect the scintillation level to follow (3.75). Measurements of the amplitude variance taken at different frequencies were fitted to the expression

$$\langle \chi^2 \rangle = \text{constant} \times f^\mu$$

Inferred values of the exponent fell in the range $0.66 < \mu < 0.98$, which values are smaller than $\mu = 1.17$. Atmospheric attenuation may have distorted the 22-GHz signal which was intentionally chosen to be near the water-vapor resonance. Nonetheless, the result of this experiment did not agree with the prediction.

A long-term experiment was then conducted along a 4.1-km path over London using 36- and 110-GHz signals [19][38]. The outer scale length was estimated

---

[10] This expression is due to Tatarskii and is discussed in Section 2.3.5.1 of Volume 1.

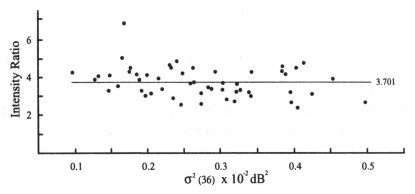

Figure 3.11: Ratios of amplitude variances for 110 and 36 GHz measured on a 4-km path near London by Ho, Mavrokoukoulakis and Cole [38]. The outer scale length was approximately 23 m during a 90-min period and the predicted ratio is indicated.

from temperature spectra measured near the path. The ratio of scintillation levels at these two frequencies is reproduced in Figure 3.11. The data is plotted as a function of the amplitude variance taken at the lower frequency. The outer scale length was measured during this test as $L_0 \approx 23$ m so one would expect (3.75) to apply. The predicted ratio 3.701 is a reasonable fit to the data scatter. A later test was conducted with the same equipment when the outer scale length was approximately $L_0 \approx 1.5$ m. In this case the measured intensity ratios clustered about the value 9.81 which is predicted by (3.76). In combination, these experiments support the basic description (3.74) and emphasize the important role played by the outer scale length.

The important role played by L_0 was also observed later in an ambitious experimental program using four different millimeter-wave signals [39]. Simultaneous measurements were made on overlapping folded paths with lengths of 120 and 300 m. The common height was $h = 4$ m and (3.77) suggests that $L_0 \approx 1.6$ m. The Fresnel lengths for the longer path listed in Table 3.1 are generally larger than this value and one would expect (3.75) to apply. The scintillation measurements made at 35, 94, 140 and 217 GHz on the long path were accurately fitted by the 7/6 scaling law. The situation on the shorter path was substantially different. The Fresnel length there was smaller than the estimated L_0 and the frequency scaling was best described by the second scaling law (3.76).

The role played by deviative absorption in setting scintillation levels has also been explored for infrared and millimeter-wave signals. Theoretical considerations suggest that intensity fluctuations should decrease by as much as 5 dB near absorption lines [40]. Measurements that confirmed this forecast were made in the USSR on terrestrial paths ranging from 360 m to 2.5 km using signals with wavelengths of 0.896, 0.920, 0.938 and 0.986 mm [41].

3.2.4.2 Optical Measurements

Signals generated by optical and infrared lasers have also been used to test the scaling of scintillation with wavelength. These experiments have usually been done on paths approximately 1 km long. The associated Fresnel lengths vary from 2 to 12 cm. These values are a good deal larger than the inner scale length and smaller than the outer scale length,

$$\ell_0 \ll \sqrt{\lambda R} \ll L_0$$

This means that the principal modification of the 7/6 scaling law arises from aperture averaging:

$$\text{Optical:} \qquad \frac{\langle \chi_1^2 \rangle}{\langle \chi_2^2 \rangle} = \left(\frac{\lambda_2}{\lambda_1}\right)^{\frac{7}{6}} \frac{G(a_{r1}/\sqrt{\lambda_1 R})}{G(a_{r2}/\sqrt{\lambda_2 R})} \qquad (3.78)$$

where the receiver-gain loss factors $G(\eta)$ are plotted in Figure 3.9.

A pioneering experiment was performed to test the expression (3.78). Diverging waves were generated by He–Ne and CO_2 lasers [42]. These signals were transmitted over the same reflected path and the total distance traveled was 1.2 km. The receiver's diameter was 0.62 cm for the $\lambda = 0.632\,8$ μm signal and 2.5 cm for $\lambda = 10.6$ μm. In fact, the two openings were adjusted to scale exactly as $\sqrt{\lambda}$ and the gain factors in (3.78) therefore canceled out. The 7/6 scaling law given by (3.75) should describe these measurements. The ratio of scintillation measured with the CO_2-laser signal to that measured with the He–Ne-laser signal is reproduced in Figure 3.12. The data points cover a 2-h period in December and they are plotted as a function of the amplitude variance experienced by the He–Ne-laser signal. The mean value was

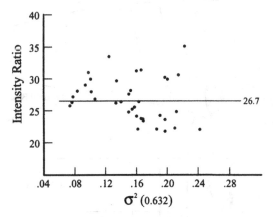

Figure 3.12: The ratio of irradiance measured at 0.632 μm to that measured at 10.6 μm by Fitzmaurice, Bufton and Minott [42]. The signals were generated by He–Ne and CO_2 lasers and transmitted over a 1.2-km path to scaled recievers. The horizontal line is the predicted ratio.

26.8 with a standard deviation of 3.3. This agrees with the value 26.7 predicted by the 7/6 scaling law. On the other hand, one should remember that our wavelength-scaling expressions are based on the weak-scattering assumption. Indeed, it is somewhat surprising that our prediction is verified over an optical path as long as 1 km.

To dispel that concern, the two-laser experiment was repeated using a 1-km path in Colorado [43]. A continuous-wave He–Ne laser provided single-mode outputs at 0.6328 and 1.084 μm. The receiving apertures were scaled as the square root of the wavelength so that the gain factors would cancel out. These measurements encountered a wide variety of turbulence conditions, ranging from weak to strong scattering. The 7/6 scaling law was confirmed when $\langle \chi^2 \rangle < 0.2$, which marked the beginning of saturation effects. Similar limitations were noted in subsequent experiments employing even longer paths [44][45].

3.2.5 Beam Waves

The development of lasers provided the practical means for creating spatially confined light beams, which have found a wide range of civilian and military applications. Three common types of beam waves are illustrated in Figure 3.13. Each is formed by an optical system of diameter $2w_0$ at the exit plane of the transmitter.

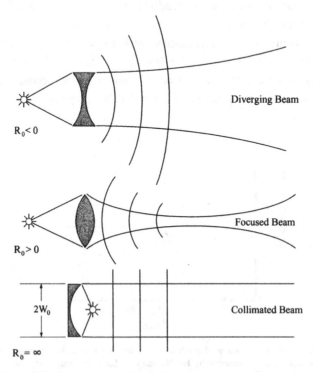

Figure 3.13: Diagrams illustrating the three common types of beam waves.

They are differentiated by the radius of curvature of the initial wave-front. A diverging beam is illustrated in the first panel and corresponds to a negative radius of curvature at the transmitter. A focused beam is created by converging optics and is characterized by a positive initial radius of curvature. A collimated beam corresponds to an infinitely large positive radius of curvature. Scattering by turbulent irregularities gradually destroys the spatial coherence of beam waves as they propagate through the atmosphere and thereby limits their utility in many applications.

3.2.5.1 Vacuum Field Strength for Beam Waves

The first task is to describe the electric field that would be measured if there were no refractive irregularities along the path. When we considered angle-of-arrival fluctuations in Volume 1, we used geometrical optics to characterize beam waves. In that approximation we described the energy flow and beam boundaries by straight lines converging at the focal point.

Diffraction is an important consideration for beam waves because it limits the extent to which their energy can be spatially confined. In examining the scattering of beam waves, we must use a wave-theory description for their vacuum field strength. Fortunately, a common description is available for the types of beam waves shown in Figure 3.13. The fundamental (TEM$_{00}$) mode is characterized by a Gaussian amplitude profile and a parabolic phase distribution at the exit plane of the transmitter:

$$E_0(\rho, x)|_{x=0} = \mathcal{E}_0 \exp\left[-r^2\left(\frac{1}{w_0^2} + i\frac{k}{2\mathcal{R}_0}\right)\right] \tag{3.79}$$

Here w_0 is the effective radius of the transmitter optics and $\mathcal{R}_0$ is the initial radius of curvature of the phase front. These parameters are easily adjusted to realize different beam shapes. This expression provides a good description for many lasers.

The next step is to describe the field strength as a function of distance from the transmitter. The Huygens–Fresnel method is the standard way to calculate the diffractive unfolding of the transmitted field strength [46]. In this procedure the expression (3.79) acts as the initial condition and yields

$$\text{Beam wave:} \quad E_0(r, x) = \frac{\mathcal{E}_0}{1 + i\alpha x} \exp\left(ikx - \frac{1}{2}k\alpha\frac{r^2}{1 + i\alpha x}\right) \tag{3.80}$$

The parameter α is a complex number defined by the initial beam radius and the radius of curvature of the phase front at the transmitter:

$$\alpha = \alpha_1 + i\alpha_2 = \frac{\lambda}{\pi w_0^2} + i\frac{1}{\mathcal{R}_0} \tag{3.81}$$

The description (3.80) is valid for distances

$$x < \pi^3 w_0^4 / \lambda^3$$

which is satisfied in all cases of practical interest. The basic field-strength expression (3.80) is consistently used to describe beam waves in the literature. Notice that it reduces to a plane wave if both the real part and the imaginary part of α vanish, but that occurs only when the aperture and radius of curvature of the transmitter are both very large.

The irradiance of the vacuum beam wave provides a useful reference. From the field-strength expression one calculates

$$I_0(r, x) = |E_0(r, x)|^2 = I_0 \left(\frac{w_0}{w(x)} \right)^2 \exp \left(-\frac{2r^2}{w^2(x)} \right) \tag{3.82}$$

where $w(x)$ is related to the real and imaginary parts of the parameter α by

$$w^2(x) = w_0^2 \left[\left(\frac{x\lambda}{\pi w_0^2} \right)^2 + \left(1 - \frac{x}{R_0} \right)^2 \right] \tag{3.83}$$

This function describes the *effective beam radius* at each point along the propagation path. Notice that the total transmitted power is independent of distance, as it should be:

$$P = \int_0^\infty dr\, r \int_0^{2\pi} d\phi\, I_0(r, x) = \frac{\pi}{2} w_0^2 I_0 \tag{3.84}$$

3.2.5.2 The Amplitude-variance Expression

We are interested to know how scattering by refractive irregularities modifies the irradiance of a beam wave. The first descriptions of this type of scintillation assumed that the receivers coincide with the centerline of the beam [47][48][49][50]. It was learned later that this assumption is unnecessary and that the scintillation level can be calculated for any combination of radial distance and path length [51]. To this end we must calculate the complex weighting function (2.42) using the cylindrical coordinates (r, ϕ, x) identified in Figure 3.14. The vacuum field strength (3.79) is already expressed in these coordinates. We must consider off-axis locations for the receiver since we need to explore modifications of the irradiance expression (3.82) throughout the beam structure:

$$E_0(\rho, R) = \frac{\mathcal{E}_0}{1 + i\alpha R} \exp \left(ikR - \frac{1}{2} k\alpha \frac{\rho^2}{1 + i\alpha R} \right) \tag{3.85}$$

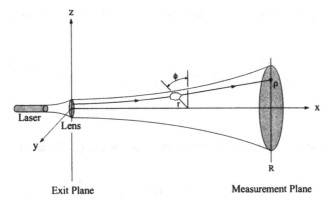

Figure 3.14: The geometry employed to describe the scattering of a beam wave by a refractive irregularity at (r, ϕ, x). The receiver is located in a normal plane at a radial distance ρ from the centerline of the beam.

Green's function connects the scattering eddy to the receiver,

$$G(\mathbf{R}, \mathbf{r}) = \frac{\exp\left(ik\sqrt{(R-x)^2 + r^2 + \rho^2 - 2r\rho\cos\phi}\right)}{4\pi\sqrt{(R-x)^2 + r^2 + \rho^2 - 2r\rho\cos\phi}} \quad (3.86)$$

and we use the paraxial approximation to simplify it:

$$G(\mathbf{R}, \mathbf{r}) = \frac{1}{4\pi(R-x)}\exp\left[ik\left((R-x) + \frac{r^2 + \rho^2 - 2r\rho\cos\phi}{2(R-x)}\right)\right]$$

With these preparations and the cylindrical wavenumber coordinates (3.6) we can express the complex weighting function as follows:

$$\Lambda(\kappa_r, \omega, \kappa_x) = \frac{k^2}{4\pi}\int_0^R \frac{dx}{R-x}\left(\frac{1+i\alpha R}{1+i\alpha x}\right)\exp(ix\kappa_x)$$

$$\times \exp\left[\frac{ik\rho^2}{2}\left(\frac{1}{R-x} - \frac{i\alpha}{1+i\alpha R}\right)\right]$$

$$\times \int_0^\infty dr\, r \exp\left(\frac{ikr^2(1+i\alpha x)}{2(R-x)(1+i\alpha R)}\right)$$

$$\times \int_0^{2\pi} d\phi \exp\left[ir\left(\kappa_r\cos(\phi-\omega) - \frac{k\rho\cos\phi}{R-x}\right)\right]$$

In this expression we have combined exponentials where it is helpful. The azimuth integrations can be done using the following result:

$$\int_0^{2\pi} d\phi \exp[i(a\cos\phi + b\sin\phi)] = 2\pi J_0\left(\sqrt{a^2 + b^2}\right)$$

The radial integral is of the type encountered previously and the result is a relatively simple expression:

$$\Lambda(\kappa_r, \omega, \kappa_x) = \frac{ik}{2} \int_0^R dx \, \exp(ix\kappa_x) \exp\left[i\gamma(x)\left(\rho\kappa_r \cos\omega - \frac{\kappa_r^2(R-x)}{2k}\right)\right]$$

(3.87)

In writing this we have introduced a new symbol for the combination

$$\gamma(x) = \gamma_1(x) - i\gamma_2(x) = \frac{1 + i\alpha x}{1 + i\alpha R}$$

(3.88)

This complex parameter depends on the downrange distance and on the beam parameters contained in α. We will need its real and imaginary parts:

$$\gamma_1(x) = \frac{1 - \alpha_2 R + x[R(\alpha_1^2 + \alpha_2^2) - \alpha_2]}{(1 - \alpha_2 R)^2 + (\alpha_1 R)^2}$$

$$\gamma_2(x) = \frac{\alpha_1(R - x)}{(1 - \alpha_2 R)^2 + (\alpha_1 R)^2}$$

(3.89)

With the result (3.87) we can construct the amplitude weighting function from the defining relationship (2.43):

$$D(\kappa) = \frac{k}{4} \int_0^R dx \, \exp(ix\kappa_x) \left\{ \exp\left[i\gamma(x)\left(\rho\kappa_r \cos\omega - \frac{\kappa_r^2(R-x)}{2k}\right)\right] \right.$$
$$\left. + \exp\left[i\gamma^*(x)\left(\rho\kappa_r \cos\omega + \frac{\kappa_r^2(R-x)}{2k}\right)\right] \right\}$$

(3.90)

The product that determines the amplitude variance is thus given by the double integral

$$D(\kappa)D(-\kappa) = \frac{k^2}{16} \int_0^R dx \int_0^R dx' \, f(x)g(x') \exp[i\kappa_x(x - x')]$$

where $f(x)$ and $g(x')$ are defined by (3.90). The major contribution comes from the region near the diagonal line $x = x'$ and the standard approximation yields

$$D(\kappa)D(-\kappa) = \frac{\pi}{8}\delta(\kappa_x)k^2 \int_0^R dx \, f(x)g(x)$$

or

$$D(\kappa)D(-\kappa) = \frac{\pi}{4}\delta(\kappa_x)k^2 \int_0^R dx \left\{ \exp\left[-\gamma_2(x)\left(2\kappa\rho\cos\omega + \frac{\kappa_r^2(R-x)}{k} \right) \right] \right.$$

$$\left. - \Re\left[\exp\left(-i\frac{\gamma(x)\kappa_r^2(R-x)}{k} \right) \right] \right\} \quad (3.91)$$

We are now at the midpoint of a long calculation. To proceed further one must combine this result with the expression for the amplitude variance (2.37). We assume that isotropy applies and cast the wavenumber integration in spherical coordinates:

$$\langle \chi^2(\rho, R) \rangle = \pi k^2 \int_0^\infty d\kappa\, \kappa^2 \Phi_n(\kappa) \int_0^\pi d\psi\, \sin\psi \int_0^{2\pi} d\omega\, \delta(\kappa\cos\psi)$$

$$\times \int_0^R dx \left\{ \exp\left[-\gamma_2(x)\left(2\kappa\rho\cos\omega + \frac{\kappa_r^2(R-x)}{k} \right) \right] \right.$$

$$\left. - \Re\left[\exp\left(-i\frac{\gamma(x)\kappa_r^2(R-x)}{k} \right) \right] \right\}$$

The angular integrations can be performed:

$$\langle \chi^2(\rho, R) \rangle = 2\pi^2 k^2 \int_0^R dx \int_0^\infty d\kappa\, \kappa\Phi_n(\kappa)\{ J_0(2i\kappa\rho\gamma_2) \exp\left(-\frac{\kappa^2\gamma_2(x)(R-x)}{k} \right)$$

$$- \Re\left[\exp\left(-i\frac{\gamma(x)\kappa^2(R-x)}{k} \right) \right] \right\} \quad (3.92)$$

In using this result we must remember that the real and imaginary parts of γ depend on the downrange distance through the definition (3.89).

For receivers located on the centerline of the beam Ishimaru noted that the path integration in (3.92) can be expressed in terms of error functions [52]. He used this result to plot the wavenumber weighting function in

$$\langle \chi^2(0, R) \rangle = 2\pi^2 k^2 \int_0^\infty d\kappa\, \kappa\Phi_n(\kappa)F_\chi(\kappa, R, w_0, \mathcal{R}_0) \quad (3.93)$$

for seven types of beam waves. The weighting function for a collimated beam rises slightly faster than that for a spherical wave, reaches a maximum value less than 0.5 at the Fresnel wavenumber $\kappa_F = (\lambda R)^{-\frac{1}{2}}$ and then declines steadily. A focused beam whose radius of curvature is greater than the path length has a similar behavior with a peak value five times smaller. Three examples of convergent beams with curvature radii less than the transmission distance all give weighting functions that oscillate beyond the Fresnel wavenumber. A perfectly focused beam has the smallest weighting function of all. These examples illustrate the complex interaction of beam-wave diffraction and the mechanisms that generate amplitude scintillation.

When the receiver is not on the beam axis, one must do the path integration in (3.92) numerically.

Even though one cannot find a simple expression for the wavenumber weighting function, one can calculate the amplitude variances precisely with the Kolmogorov model. To do so we rearrange the terms in (3.92) and perform the wavenumber integration first:

$$\langle \chi^2(\rho, R) \rangle = 2\pi^2 k^2 (0.033 C_n^2) \Re\left(\int_0^R dx \, [I_1(x) - I_2(x) - I_3(\rho, x)] \right)$$

where

$$I_1(x) = \int_0^\infty \frac{d\kappa}{\kappa^{\frac{8}{3}}} \left[1 - \exp\left(-i \frac{\kappa^2 \gamma (R - x)}{k} \right) \right]$$

$$I_2(x) = \int_0^\infty \frac{d\kappa}{\kappa^{\frac{8}{3}}} \left[1 - \exp\left(-\frac{\kappa^2 \gamma_2 (R - x)}{k} \right) \right]$$

$$I_3(\rho, x) = \int_0^\infty \frac{d\kappa}{\kappa^{\frac{8}{3}}} [1 - J_0(2i\kappa\rho\gamma_2)] \exp\left(-\frac{\kappa^2 \gamma_2 (R - x)}{k} \right)$$

The first two integrals are done using the following result from Appendix C:

$$\int_0^\infty \frac{d\kappa}{\kappa^{\frac{8}{3}}} [1 - \exp(-a\kappa^2)] = \frac{3}{5} \Gamma\left(\frac{1}{6} \right) a^{\frac{5}{6}}$$

The third integral is expressed in terms of a Kummer function by noting that

$$\mathcal{J} = \int_0^\infty \frac{d\kappa}{\kappa^{\frac{8}{3}}} [1 - J_0(i\kappa b)] \exp(-a\kappa^2)$$

$$= \int_0^1 dv \, ib \int_0^\infty \frac{d\kappa}{\kappa^{\frac{5}{3}}} J_1(i\kappa v b) \exp(-a\kappa^2)$$

$$= \frac{1}{4} \Gamma\left(\frac{1}{6} \right) b^2 a^{-\frac{1}{6}} \int_0^1 dv \, v M\left(\frac{1}{6}, 2; \frac{v^2 b^2}{4a} \right)$$

where we have used a result from Appendix D. The Kummer function can be integrated using an expression for its derivative found in Appendix G:

$$\mathcal{J} = \frac{3}{5} \Gamma\left(\frac{1}{6} \right) a^{\frac{5}{6}} \left[M\left(-\frac{5}{6}, 1; \frac{b^2}{4a} \right) - 1 \right]$$

When we assemble these results and substitute the appropriate definitions for a and b into the three integrations we find

$$\langle \chi^2(\rho, R) \rangle = \frac{3}{5} \Gamma\left(\frac{1}{6} \right) \pi^2 k^2 (0.033 C_n^2) Q$$

where

$$Q = \Re \left[\int_0^R dx \left(i \frac{\gamma(x)\kappa^2(R-x)}{k} \right)^{\frac{5}{6}} \right] - \int_0^R dx \, M\left(-\frac{5}{6}, 1; \frac{b^2}{4a} \right) \left(\frac{\gamma_2(x)(R-x)}{k} \right)^{\frac{5}{6}}$$

The first term can be expressed as a hypergeometric function using its basic integral definition given in Appendix H. The second integral is easy to do because the argument of the Kummer function does not depend on x:

$$\frac{b^2}{4a} = \frac{4\rho^2\gamma_2^2}{4\gamma_2(R-x)/k} = \frac{\rho^2 k\alpha_1}{(1-\alpha_2 R)^2 + (\alpha_1 R)^2} = \frac{2\rho^2}{w^2(R)}$$

Here we have used the local beam radius defined by (3.83) to simplify the result. The following general solution was first found by Ishimaru [51]:

$$\langle \chi^2(\rho, R) \rangle = 2.176 R^{\frac{11}{6}} k^{\frac{7}{6}} C_n^2 \left\{ \frac{6}{11} \Re \left[i^{\frac{5}{6}} {}_2F_1\left(-\frac{5}{6}, \frac{11}{6}; \frac{17}{6}; \frac{i\alpha R}{1+i\alpha R} \right) \right] \right.$$
$$\left. -\frac{3}{8} \left(\frac{w_0}{w(R)} \right)^{\frac{5}{3}} M\left(-\frac{5}{6}, 1; \frac{2\rho^2}{w^2(R)} \right) \right\} \quad (3.94)$$

3.2.5.3 Graphical Descriptions

The analytical solution (3.94) was investigated by Miller, Ricklin and Andrews in a wide-ranging computational and analytical program to establish the dependence of $\langle \chi^2 \rangle$ on radial distance and path length [53]. The normalized amplitude variance

$$\upsilon(\rho, R) = \frac{\langle \chi^2(\rho, R) \rangle}{R^{\frac{11}{6}} k^{\frac{7}{6}} C_n^2} \quad (3.95)$$

was employed to remove C_n^2 and the customary powers of R and k. What then remains is a function that depends on the combinations αR and $\rho/w(R)$. With the definitions (3.81) and (3.83) we see that the result depends on two independent, dimensionless parameters that completely define the beam wave, namely

$$\frac{\lambda R}{\pi w_0^2} \quad \text{and} \quad \frac{R}{\mathcal{R}_0} \quad (3.96)$$

The first is the Fresnel length squared divided by the area of the transmitter aperture. The second is simply the path length divided by the initial radius of curvature. It is best to examine the different types of wave pictured in Figure 3.13 separately.

The amplitude variance for a collimated beam is reproduced in Figure 3.15. In this type of beam wave the radius of curvature is infinite and the result depends only on $\lambda R/(\pi w_0^2)$. These curves suggest that the scintillation is always smallest at the

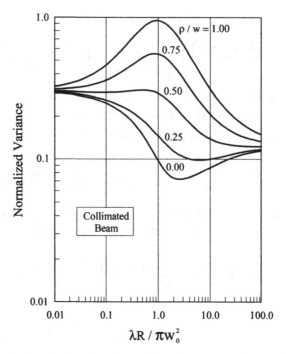

Figure 3.15: The normalized amplitude variance for a collimated beam wave. These curves were computed from (3.94) by Miller, Ricklin and Andrews [53]. The normalization is defined by (3.95) and the radial distance from the centerline is divided by the beam radius at the receiver.

centerline but should increase significantly near the beam edge. This enhancement depends strongly on the independent variable and is maximum when the Fresnel length is approximately equal to the aperture radius. It is reassuring that these curves approach the plane-wave value (3.35) when the aperture is much larger than the Fresnel length. For very small apertures one would expect the spherical-wave result (3.53) and that is confirmed for large values of $\lambda R/(\pi w_0^2)$.

A rather different picture emerges if one considers a convergent beam wave that is perfectly focused at the receiver: $R = \mathcal{R}_0$. The normalized variance for this case is presented in Figure 3.16 for radial distances that range from zero to the beam edge. These predictions are surprising because they suggest that very little scintillation should occur on the centerline of the beam when $\lambda R < \pi w_0^2$. That occurs because the first and second terms can cancel out in the centerline version:

$$v(0, R) = 2.176 \left\{ \frac{6}{11} \Re \left[i^{\frac{5}{6}} {}_2F_1 \left(-\frac{5}{6}, \frac{11}{6}, \frac{17}{6}; \frac{i\alpha R}{1+i\alpha R} \right) \right] - \frac{3}{8} \left(\frac{w_0}{w(R)} \right)^{\frac{5}{3}} \right\}$$

$$(3.97)$$

This cancelation requires that the beam be precisely focused on the receiver. That

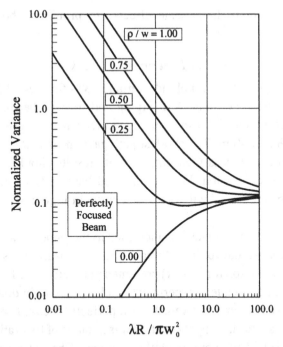

Figure 3.16: The normalized amplitude variance for a convergent beam wave that is perfectly focused at the receiver. These curves were computed from (3.94) by Miller, Ricklin and Andrews [53]. The normalization is defined by (3.95) and the radial distance from the centerline is divided by the beam radius at the receiver.

is difficult to achieve in practice because of beam wander [54], which is also in-duced by refractive-index irregularities.[11] By contrast, large levels of scintillation are expected near the beam edge for this configuration. These forecasts require experimental confirmation before they are adopted.

The program described above provided a second very useful type of description. A family of analytical approximations was developed for the solutions of (3.94) that are relatively easy to use [53]. Those readers who are interested primarily in practical applications should consult this reference before attempting computations based on the formal solution.

The same group investigated the impact of other spectral models on the ampli-tude variance in a second publication [55]. The influence of a finite outer scale length was examined by using the von Karman model with (3.92). They found that the outer scale length has a minimal effect on the scintillation level at the center-line but has a pronounced effect near the beam edge. The role of the inner scale length was examined with the Hill bump model and the Gaussian representation

---

[11] The wander of a beam wave is examined in Section 7.2.4 of Volume 1.

of the dissipation region. Inner-scale effects are prominent both on-axis and off-axis.

3.2.5.4 Experimental Checks

The predicted behavior of the amplitude variance was first tested in the USSR in laser experiments using collimated, diverging and focused beams [56][57]. Path lengths of 500, 2400 and 7000 m over rugged terrain were used, with average heights between 8 and 40 m. The shortest path gave measured rms values for the logarithmic amplitude of less than unity and its results should be described by the weak-scattering expression. For a collimated beam, values measured on the centerline of the beam agreed with the $\rho = 0$ curve in Figure 3.15 as the aperture diameter was varied.

A beam focused at the receiver was then tried at the same distance. In this case, centerline data were similar to the collimated-beam measurements and drastically different than the prediction for $\upsilon(0, R)$ presented in Figure 3.16. This contradiction has been verified in subsequent experiments and there is little doubt that the on-axis amplitude variance does not vanish for a perfectly focused beam wave. This failure of the weak-scattering approximation was a source of frustration and concern for two decades [58]. Using path-integral techniques, Charnotskii and others have shown that double scattering is uniquely important for the on-axis focused-beam case and provides a clear explanation for the measured results [59][60].

3.2.6 The Influence of the Sample Length

In examining phase-fluctuation data in Section 4.1.7 of Volume 1, we found that phase trends and sample-length limitations are important considerations for interpreting such experiments. We should ask whether they also play an influential role in measurements of amplitude fluctuations. In considering this question, we assume that trends in the amplitude time history have been removed by detrending techniques. That involves fitting the running average with straight lines or polynomials, and then subtracting the fitted curve from the raw data. When this process has been completed, one can be reasonably confident that the residual amplitude record is a stationary process.

Consider next the ensemble-averaging process implied by the brackets in our amplitude expressions. In an ideal world, that averaging should extend over all possible configurations of the turbulent atmosphere. In practice, one must approximate the ensemble average by a time average and the mean-square value is invariably approximated by

$$\langle \chi^2 \rangle \simeq \frac{1}{T} \int_0^T dt \, \chi^2(t) \qquad (3.98)$$

We believe that the two estimates will agree if the sample is infinitely long. On the other hand, we usually cannot wait long enough to realize that ideal. We expect that the measured variance will be quite different for sample lengths of a minute, an hour, a day and a month. It is therefore important to identify the sample size. A finite data sample necessarily discriminates against low frequencies in the temporal spectrum and large eddies in the turbulence spectrum

The first step in data processing is to establish a mean value from which the amplitude fluctuations can be measured:

$$\overline{\chi(T)} = \frac{1}{T} \int_0^T dt \, \chi(t) \qquad (3.99)$$

This expression should approach zero as the sample length T increases if the amplitude fluctuations are a stationary random process. However, the measured average need not be zero when a finite sample length is used. In that case, the mean-square amplitude and the amplitude variance are different. The variance for a single data sample is written

$$\overline{|\chi(T) - \overline{\chi(T)}|^2} = \frac{1}{T} \int_0^T dt \left| \chi(t) - \frac{1}{T} \int_0^T dt \, \chi(t) \right|^2$$

$$= \frac{1}{T} \int_0^T dt \, \chi^2(t) - \frac{1}{T^2} \int_0^T dt \int_0^T dt' \, \chi(t)\chi(t')$$

One usually averages this expression over a family of samples. The result should approach the ensemble average if there are enough samples to establish a statistically significant estimate:

$$\overline{\langle |\chi(T) - \overline{\chi(T)}|^2 \rangle} = \frac{1}{T} \int_0^T dt \, \langle \chi^2(t) \rangle - \frac{1}{T^2} \int_0^T dt \int_0^T dt' \, \langle \chi(t)\chi(t') \rangle \quad (3.100)$$

We call this quantity the *sample variance* to indicate that it is derived from a finite portion of the signal history.

The amplitude autocorrelation is needed to complete this calculation. The usual assumption is that a constant wind bears a frozen random medium across the line of sight at a uniform speed v. This is Taylor's hypothesis and is found to be valid in many situations.[12] When it is a valid description and the signal is a plane wave we shall find in Chapter 5 that

$$\langle \chi(t)\chi(t') \rangle = 2\pi^2 R \, k^2 \int_0^\infty d\kappa \, \kappa \, \Phi_n(\kappa) J_0(\kappa v|t - t'|)\left(1 - \frac{\sin(R\kappa^2/k)}{R\kappa^2/k} \right)$$

[12] This matter is discussed in Section 6.1.2 of Volume 1.

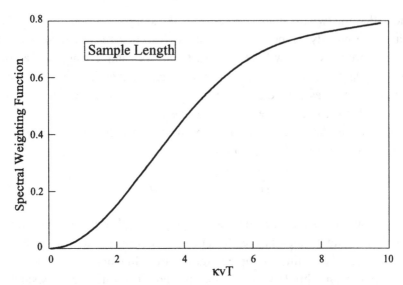

Figure 3.17: The finite-sample-length wavenumber weighting function $S(\kappa v T)$ defined by (3.102). Here v is the wind speed and T is the sample length.

Combining these expressions and taking the time averages,

$$\overline{\langle |\chi(T) - \overline{\chi(T)}|^2 \rangle} = 2\pi^2 R k^2 \int_0^\infty d\kappa \, \kappa \Phi_n(\kappa) S(\kappa v T) \left(1 - \frac{\sin(R\kappa^2/k)}{R\kappa^2/k} \right)$$

$$(3.101)$$

where the *finite-sample-length weighting function*

$$S(x) = 1 + 2\frac{J_1(x)}{x} - \frac{2}{x} \int_0^x du \, J_0(u) \tag{3.102}$$

describes how different wavenumbers influence the result. This function is plotted in Figure 3.17 where one sees that it begins at zero for small values and rises slowly to an asymptotic value of unity for large values.

The influence of large eddies is therefore suppressed both by the diffraction and by the sample-length weighting functions. On the other hand, they differ widely in the rate at which they rise and the effect they have on the variance for the sample. To make this comparison it is helpful to deal with actual values for the parameters that are encountered in practical situations. The sample-length weighting function is appreciable when its argument is greater than five. This suggests that it can be replaced by unity for wavenumbers

$$\kappa > \frac{5}{vT}$$

A typical wind speed is $v = 5$ m s^{-1} and the sample length is usually greater than

20 s. This means that the sample-length term can be ignored when

$$\kappa > 10^{-3} \text{ cm}^{-1}$$

By contrast, the diffraction weighting function reaches its maximum value for

$$\frac{R\kappa^2}{k} > 3 \quad \text{or} \quad \kappa > \frac{4}{\sqrt{\lambda R}}$$

The Fresnel length is a few centimeters for optical links that can be described by weak scattering and the diffraction term is increasing until

$$\text{Optical:} \quad \kappa > 2 \text{ cm}^{-1}$$

The sample-length term has long since reached its asymptotic value when this wavenumber is approached. The amplitude variance for optical links is thus controlled entirely by the diffraction term.

The situation is considerably different for microwave transmissions, for which the Fresnel length varies from 3 to 50 m. This means that the outer scale length and sample length can play a similar role to the diffraction term in weighting the turbulence spectrum.

3.2.7 The Influence of Anisotropic Irregularities

All of our descriptions of amplitude fluctuations developed thus far have assumed that the refractive-index irregularities are isotropic. We know from surface temperature and refractive-index measurements that the eddies are anisotropic as suggested in Figure 3.18.[13] The degree of anisotropy is modest at the surface but the eddies become greatly elongated in the horizontal direction as one ascends in the troposphere.[14] It is important to see whether this reality changes the conclusions we have been drawing.

We can make a reasonable guess about their influence by noting that the amplitude fluctuations are created primarily by irregularities that are comparable in extent to the Fresnel length. This parameter is only a few centimeters for optical paths that are described by weak scattering. In the hierarchy of subdividing eddies described by turbulence theory[15] these eddies are far removed from the large eddies that initiate the process. Rather, they are close to the small dissipation region where diffusion and viscosity remove energy. We know that the decaying eddies become progressively more symmetrical as they divide and redivide in the inertial range. The characteristic anisotropy of large structures is not conveyed to the small eddies that determine amplitude fluctuations of light waves. The amplitude variance ought

[13] See Sections 2.2.7 and 2.3.5 in Volume 1.
[14] As shown in Figure 2.23 of Volume 1.
[15] See Figure 2.5 in Volume 1.

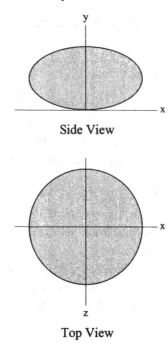

Side View

Top View

Figure 3.18: A description of anisotropic irregularities near the surface suggested by image motion measured in orthogonal directions with collimated laser beams by Lukin [61].

not to be sensitive to the anisotropy of large eddies that is reflected in Figure 3.18. The situation can be quite different for microwave signals because the Fresnel length varies from 3 to 50 m for typical terrestrial links and satellite signals. These values are comparable to the outer scale length and we would expect that eddy asymmetry plays an influential role.

The turbulence spectrum depends on all three components of the wavenumber vector when the random medium is not isotropic. One can describe unsymmetrical irregularities using the ellipsoidal model for the turbulence spectrum that was introduced in Section 2.2.7 of Volume 1:

$$\Phi_n(\kappa) = abc\Phi_n\left(\sqrt{a^2\kappa_x^2 + b^2\kappa_y^2 + c^2\kappa_z^2}\right) \qquad (3.103)$$

This formulation seems to catch the essence of anisotropy and is well suited to analytical operations. It is not unique but it is consistently employed to analyze anisotropic media. The dimensionless *scaling parameters* a, b and c are related to the outer scale length that would be measured in each direction:

$$L_x = aL_0 \qquad L_y = bL_0 \qquad L_z = cL_0$$

The horizontal symmetry suggested in Figure 3.18 is quite typical and we can take $a = b$ without compromising the description. With this model of the irregularities we can express the amplitude variance as follows:

$$\langle \chi^2 \rangle_{\text{anisotropic}} = 4 \int_{-\infty}^{\infty} d\kappa_x \int_{-\infty}^{\infty} d\kappa_y \int_{-\infty}^{\infty} d\kappa_z \, D(\kappa)D(-\kappa)abc$$

$$\times \, \Phi_n\left(\sqrt{a^2\kappa_x^2 + a^2\kappa_y^2 + c^2\kappa_z^2}\right) \tag{3.104}$$

with

$$D(\kappa) = k^2 \int d^3r \, A(\mathbf{R}, \mathbf{r}) \, \exp[i(x\kappa_x + y\kappa_y + z\kappa_z)] \tag{3.105}$$

In these expression we have written out the volume element $d^3\kappa$ and the vector product $\kappa \cdot r$ to suggest the route to follow. On introducing the *stretched wavenumber coordinates*

$$\kappa_x = q_x/a \qquad \kappa_y = q_y/b \qquad \kappa_z = q_z/c$$

we find that the turbulence spectrum becomes isotropic in the new system.

$$\langle \chi^2 \rangle_{\text{anisotropic}} = 4 \int d^3q \, \Phi_n(|\mathbf{q}|)D(\mathbf{q})D(-\mathbf{q})$$

The restated amplitude weighting function is simply

$$D(\mathbf{q}) = k^2 \int_0^R dx \int_{-\infty}^{\infty} dy \int_{-\infty}^{\infty} dz \, A(\mathbf{R}, \mathbf{r}) \exp\left[i\left(\frac{xq_x}{a} + \frac{yq_y}{a} + \frac{zq_z}{c}\right)\right]$$

and can be derived from the complex weighting function expressed in stretched coordinates:

$$\Lambda(\mathbf{q}) = k^2 \int_0^R dx \int_{-\infty}^{\infty} dy \int_{-\infty}^{\infty} dz \, \frac{G(\mathbf{R}, \mathbf{r})E_0(\mathbf{r})}{E_0(\mathbf{R})} \exp\left[i\left(\frac{xq_x}{a} + \frac{yq_y}{a} + \frac{zq_z}{c}\right)\right]$$

$$\tag{3.106}$$

This expression can be evaluated for a plane wave if we make the paraxial approximation for Green's function, as we did in establishing the isotropic result:

$$\Lambda(\mathbf{q}) = \frac{k^2}{4\pi} \int_0^R dx \, \frac{\exp(ixq_x/a)}{R - x} \int_{-\infty}^{\infty} dy \int_{-\infty}^{\infty} dz \exp\left(\frac{ik(y^2 + z^2)}{2(R - x)}\right)$$

$$\times \exp\left[i\left(\frac{yq_y}{a} + \frac{zq_z}{c}\right)\right]$$

We shift to polar coordinates and use two integral results from Appendix D that were introduced previously:

$$
\begin{aligned}
\Lambda(\mathbf{q}) &= \frac{k^2}{4\pi} \int_0^R dx\, \frac{\exp(ixq_x/a)}{R-x} \int_0^\infty dr\, r \exp\left(\frac{ikr^2}{2(R-x)}\right) \\
&\quad \times \int_0^{2\pi} d\phi \exp\left[ir\left(\frac{q_y \cos\phi}{a} + \frac{q_z \sin\phi}{c}\right)\right] \\
&= \frac{k^2}{2} \int_0^R dx\, \frac{\exp(ixq_x/a)}{R-x} \int_0^\infty dr\, r \exp\left(\frac{ikr^2}{2(R-x)}\right) \\
&\quad \times J_0\left(r\sqrt{\left(\frac{q_y}{a}\right)^2 + \left(\frac{q_z}{c}\right)^2}\right) \\
&= \frac{ik}{2} \int_0^R dx \exp\left(\frac{ixq_x}{a}\right) \exp\left\{\frac{R-x}{2ik}\left[\left(\frac{q_y}{a}\right)^2 + \left(\frac{q_z}{c}\right)^2\right]\right\}
\end{aligned}
$$

The corresponding expression for $D(\mathbf{q})$ is given by

$$
D(\mathbf{q}) = \frac{k}{2} \int_0^R dx \exp\left(\frac{ixq_x}{a}\right) \sin\left[\frac{R-x}{2ik}\left(\frac{q_y^2}{a^2} + \frac{q_z^2}{c^2}\right)\right]
$$

The technique used to evaluate the product $D(\mathbf{q})D(-\mathbf{q})$ described following (3.22) is used here with $x' = R - x$ to write

$$
\langle\chi^2\rangle_{\text{anisotropic}} = 2\pi k^2 \int d^3q\, \Phi_n(q)\delta\left(\frac{q_x}{a}\right) \int_0^R dx' \sin^2\left[\frac{x'}{2k}\left(\frac{q_y^2}{a^2} + \frac{q_z^2}{c^2}\right)\right]
$$

On changing to spherical stretched wavenumber coordinates this becomes

$$
\begin{aligned}
\langle\chi^2\rangle_{\text{anisotropic}} &= 2\pi k^2 \int_0^\infty dq\, q^2 \Phi_n(q) \int_0^\pi d\theta \sin\theta \int_0^{2\pi} d\omega\, \delta\left(\frac{q \cos\theta}{a}\right) \\
&\quad \times \int_0^R dx' \sin^2\left[\frac{x'q^2 \sin^2\theta}{2k}\left(\frac{\cos^2\omega}{a^2} + \frac{\sin^2\omega}{c^2}\right)\right] \\
&= 2\pi^2 Rk^2 a \int_0^{2\pi} d\omega \int_0^\infty dq\, q\Phi_n(q)\left(1 - \frac{\sin(\beta q^2)}{\beta q^2}\right)
\end{aligned}
$$

$$(3.107)$$

with

$$
\beta(\omega) = \frac{R}{2k}\left(\frac{\cos^2\omega}{a^2} + \frac{\sin^2\omega}{c^2}\right)
\tag{3.108}
$$

From estimates made for the isotropic case we know that the inertial range of the spectrum is the most important. With the Kolmogorov model we find that

$$\langle \chi^2 \rangle_{\text{anisotropic}} = 2\pi^2 0.033 R C_n^2 k^2 a \int_0^{2\pi} d\omega \int_0^\infty dq\, q^{-\frac{8}{3}} \left(1 - \frac{\sin(\beta q^2)}{\beta q^2}\right)$$

$$= 0.301 R^{\frac{11}{6}} C_n^2 k^{\frac{7}{6}} a \frac{1}{2\pi} \int_0^{2\pi} d\omega \left(\frac{\cos^2 \omega}{a^2} + \frac{\sin^2 \omega}{c^2}\right)^{\frac{5}{6}}$$

This can be written as the isotropic variance multiplied by a function that depends only on the scaling parameters:

$$\langle \chi^2 \rangle_{\text{anisotropic}} = \langle \chi^2 \rangle_{\text{isotropic}} \left\{\frac{a}{c^{\frac{5}{3}}} \frac{1}{2\pi} \int_0^{2\pi} d\omega \left[1 - \left(1 - \frac{c^2}{a^2}\right)\cos^2 \omega\right]^{\frac{5}{6}}\right\} \quad (3.109)$$

The data reflected in Figure 3.18 suggests that $L_z < L_y$, which means that $c < a$. The *axial ratio* therefore falls in the range

$$0 < 1 - c^2/a^2 < 1$$

The integral term in (3.109) varies from 1.0 to 0.54 over this range. The anisotropic and isotropic descriptions of amplitude fluctuations are roughly the same when a and c are not much different. That is the actual situation for links near the surface.

3.3 Optical Astronomy

Scattering of starlight in the atmosphere imposes a fundamental limitation on earth-based astronomy. It has several aspects. The first is the random change of apparent position or *quivering* of a star which was considered in Volume 1. In addition, the intensity of a stellar signal changes randomly with time. This second form of scintillation is called *twinkling* and is observed even on still nights. The level of scintillation varies from night to night and changes with location. It varies with the elevation angle of a star, as indicated in the actual records reproduced in Figure 3.19. Twinkling also depends on the wavelength used and on the diameter of the telescope opening. This is the phenomenon we now want to discuss.

3.3.1 The Transmission Expression

To describe astronomical scintillation we must establish a transmission expression for the arriving wave.[16] It must include two features that were not required for horizontal propagation paths. It must describe inclined paths through the atmosphere

---

[16] Tatarskii developed the description of stellar scintillation in his first book [63] and refined it in his second [64].

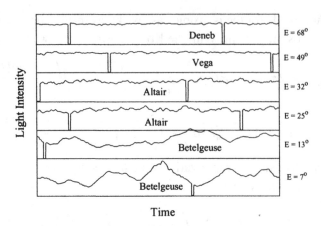

Figure 3.19: Stellar scintillation records measured at various zenith angles over a 90-min period with a 36-inch telescope by Ellison and Seddon [62]. Notice that the time scale for the last two histories is different than that for the first four, reading from the top.

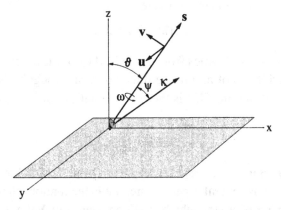

Figure 3.20: The geometry used to describe the scintillation imposed on starlight as it travels through the atmosphere. The line-of-sight vector **s** identifies the source location and makes a zenith angle ϑ with the local vertical. The wavenumber vector κ describes the turbulent irregularities in spherical coordinates referenced to the line of sight.

because stars are observed over a range of elevation angles. It must also recognize the wide range of turbulent conditions that signals encounter during their passage through the atmosphere.

Signals from these distant sources are properly characterized as plane waves. The star's position is described by the zenith angle ϑ or alternatively by its elevation angle $E = \pi/2 - \vartheta$. Unless the source is close to the horizon, we can ignore the spherical nature of the atmosphere and use the flat-earth approximation suggested in Figure 3.20. In the same situations we can omit consideration of refractive bending and identify the path variable s with the distance along the nominal line of sight. These assumptions imply that the *slant range* measured along the line-of-sight vector is related to altitude by

$$s = z \sec \vartheta = z \operatorname{cosec} E \qquad (3.110)$$

The next step is to identify the atmospheric regions that play important roles in generating scintillation. The ionosphere plays no role because we are considering optical and infrared signals whose frequencies are enormous relative to the plasma frequency of the ionized layers above 100 km. This means that refractive-index irregularities in the troposphere are the only source of the amplitude scintillation measured by terrestrial telescopes. They are determined almost completely by temperature fluctuations. We learned in Section 2.3.3 of Volume 1 that tropospheric turbulence changes dramatically as one ascends from ground level. This means that one cannot assume that the random medium is homogeneous – in contrast to terrestrial-links, for which we could avoid that complication.

Two approaches for describing inhomogeneous random media were presented in Chapter 2. In the first method position variability was consolidated in the turbulence spectrum itself and this led to (2.46). In the second approach it was assumed that the spatial correlation of $\delta\varepsilon$ is proportional to the usual wavenumber integral times the profiles taken at the displaced points as in (2.49). The two methods lead to the same result in most cases and we choose the second method,

$$\langle \chi^2 \rangle = 4 \int d^3\kappa \, \Phi_n(\kappa) D(\kappa) D(-\kappa) \qquad (3.111)$$

in which the position variability is contained in the amplitude weighting functions:

$$D(\kappa) = k^2 \int d^3r \, \wp(\mathbf{r}) A(\mathbf{R}, \mathbf{r}) \exp(i\kappa \cdot \mathbf{r}) \qquad (3.112)$$

To establish the amplitude weighting function $D(\kappa)$ we use the paraxial approximation. We can do so because the optical and infrared wavelengths are much smaller than the tropospheric eddies since ℓ_0 is a few millimeters. This means that the forward-scattering angle λ/ℓ in Figure 3.1 is very small and one can use the approximate expression for $D(\kappa)$ developed in (3.11):

$$D(\kappa) = \frac{k}{2} \int_0^\infty ds \, \wp(s) \sin\left(\frac{s\kappa_r^2}{2k}\right) \exp(is\kappa_s) \qquad (3.113)$$

The wavenumber components are related to the angle between κ and s as follows:

$$\kappa_s = \kappa \cos\psi \qquad \text{and} \qquad \kappa_r = \kappa \sin\psi$$

We are now ready to estimate the amplitude variance. We cast (3.111) in terms of the wavenumber coordinates identified in Figure 3.20. The profile of turbulent irregularities is usually a function only of height and we change from slant range to altitude using (3.110):

$$\langle \chi^2 \rangle = k^2 \sec^2 \vartheta \int_0^\infty d\kappa \, \kappa^2 \int_0^\pi d\psi \sin \psi \int_0^{2\pi} d\omega \, \Phi_n(\kappa, \psi, \omega)$$

$$\times \int_0^\infty dz_1 \int_0^\infty dz_2 \exp[i\kappa \cos \psi \sec \vartheta \, (z_1 - z_2)] \, \wp(z_1)\wp(z_2)$$

$$\times \sin \left(\frac{\kappa^2 z_1 \sec \vartheta \sin^2 \psi}{2k} \right) \sin \left(\frac{\kappa^2 z_2 \sec \vartheta \sin^2 \psi}{2k} \right)$$

The double path integration is similar to that considered above (3.91). Its main contribution comes from the region near the diagonal $z_1 = z_2$ because the exponential oscillates rapidly everywhere else:

$$\langle \chi^2 \rangle = 2\pi k^2 \sec^2 \vartheta \int_0^\infty d\kappa \, \kappa^2 \int_0^\pi d\psi \sin \psi \int_0^{2\pi} d\omega \, \Phi_n(\kappa, \psi, \omega)$$

$$\times \delta(\kappa \cos \psi \sec \vartheta) \int_0^\infty dz \, \wp^2(z) \sin^2 \left(\frac{\kappa^2 z \sec \vartheta \sin^2 \psi}{2k} \right)$$

In most studies of astronomical scintillation it is assumed that the irregularities are isotropic. We will make the same simplification for this preliminary estimate. The delta function simplifies the expression above to yield

$$\langle \chi^2 \rangle = 4\pi^2 k^2 \sec \vartheta \int_0^\infty d\kappa \, \kappa \, \Phi_n(\kappa) \int_0^\infty dz \, \wp^2(z) \sin^2 \left(\frac{\kappa^2 z \sec \vartheta}{2k} \right) \qquad (3.114)$$

We can combine the spectrum and the profile function to write an equivalent expression:

$$\langle \chi^2 \rangle = 4\pi^2 k^2 \sec \vartheta \int_0^\infty d\kappa \, \kappa \int_0^\infty dz \, \Phi_n(\kappa, z) \sin^2 \left(\frac{\kappa^2 z \sec \vartheta}{2k} \right) \qquad (3.115)$$

which is Tatarskii's original result for small telescopes [65]. To complete the calculation one must specify the spectrum and how it depends on altitude. There is a good deal of uncertainty and variability associated with the vertical profile of turbulent activity. On the other hand, one is reasonably confident that the Kolmogorov model describes the local behavior of temperature variations along the path. That state of affairs suggests that we should do the wavenumber integration first.

3.3.1.1 The Variance for the Kolmogorov Model

Since the Kolmogorov description of temperature fluctuations is a power-law model, the Fresnel length will play a deciding role. For transmission through the atmosphere the Fresnel length depends on the zenith angle and on the effective height of the turbulent medium:

$$\text{Fresnel length} = \sqrt{\lambda H \sec \vartheta} \tag{3.116}$$

Ten kilometers is a reasonable estimate for H and the Fresnel length is about 10 cm. This is substantially greater than the inner scale length and the dissipation term $\mathcal{F}(\kappa \ell_0)$ can be omitted. The outer scale length is much larger than the Fresnel length everywhere along the path, so the following inequality describes astronomical scintillation:

$$\ell_0 \ll \sqrt{\lambda H \sec \vartheta} \ll L_0 \tag{3.117}$$

We assume that the variation with height occurs only in C_n^2 and the turbulence spectrum can be written in factored form:

$$\Phi_n(\kappa, z) = \frac{0.033 C_n^2(z)}{\kappa^{\frac{11}{3}}} \qquad 0 < \kappa < \infty \tag{3.118}$$

The corresponding amplitude variance is

$$\langle \chi^2(\vartheta, k) \rangle = 0.651 k^2 \sec \vartheta \int_0^\infty dz \, C_n^2(z) \int_0^\infty \frac{d\kappa}{\kappa^{\frac{8}{3}}} \left[1 - \cos\left(\frac{\kappa^2 z \sec \vartheta}{k} \right) \right]$$

The wavenumber integral can be converted to one found in Appendix B:

$$\text{Point Receiver:} \qquad \langle \chi^2(\vartheta, k) \rangle = 0.563 k^{\frac{7}{6}} (\sec \vartheta)^{\frac{11}{6}} \int_0^\infty dz \, z^{\frac{5}{6}} C_n^2(z) \tag{3.119}$$

Experimental data shows that C_n^2 decreases with height as $z^{-\frac{4}{3}}$ near the surface for unstable meteorological conditions, but as $z^{-\frac{2}{3}}$ if they are stable or neutral. This means that the weighted integral in (3.119) can be either sensitive or insensitive to surface values of C_n^2, depending on the meteorology. On the other hand, the altitude weighting $z^{\frac{5}{6}}$ means that turbulence well above the surface may exert a strong influence on the level of scintillation.

3.3.1.2 Aperture Averaging

Astronomical telescopes employ large reflectors or lenses to increase their light-gathering power and angular resolution. Aperture averaging is therefore an important consideration for understanding the scintillation experienced during astronomical observations. One can include this effect by using the geometry of Figure 3.20. The telescope is always aimed at the source and the (u, v) coordinates are useful for describing the averaging indicated by (3.62):

$$\overline{\langle \chi^2 \rangle} = \frac{1}{A^2} \iint_A du_1 \, dv_1 \iint_A du_2 \, dv_2 \, \langle \chi(u_1, v_1) \chi(u_2, v_2) \rangle$$

For a circular aperture we can make this explicit:

$$\overline{\langle\chi^2\rangle} = \frac{1}{\pi^2 a_r^4}\int_0^{a_r} dr_1\, r_1 \int_0^{2\pi} d\phi_1 \int_0^{a_r} dr_2\, r_2 \int_0^{2\pi} d\phi_2\, \langle\chi(r_1,\phi_1)\chi(r_2,\phi_2)\rangle$$

Further progress depends on specifying the spatial covariance of logarithmic amplitude for atmospheric transmission. This is an obvious generalization of (3.115) and the aperture-averaged level of scintillation is given by

$$\overline{\langle\chi^2\rangle} = 4\pi^2 k^2 \sec\vartheta \int_0^\infty d\kappa\,\kappa \int_0^\infty dz\,\Phi_n(\kappa,z)\sin^2\!\left(\frac{\kappa^2 z\sec\vartheta}{2k}\right)\frac{1}{\pi^2 a_r^4}\int_0^{a_r} dr_1\, r_1$$

$$\times \int_0^{2\pi} d\phi_1 \int_0^{a_r} dr_2\, r_2 \int_0^{2\pi} d\phi_2\, J_0\!\left(\kappa\sqrt{r_1^2+r_2^2-2r_1 r_2\cos(\phi_1-\phi_2)}\right)$$

The surface integrals were evaluated in finding (3.64) and give the familiar wavenumber weighting function

$$\overline{\langle\chi^2\rangle} = 2\pi^2 k^2 \sec\vartheta \int_0^\infty dz \int_0^\infty d\kappa\,\kappa\,\Phi_n(\kappa,z)\left(\frac{2J_1(\kappa a_r)}{\kappa a_r}\right)^2\left[1-\cos\!\left(\frac{\kappa^2 z\sec\vartheta}{k}\right)\right]$$

$$(3.120)$$

For the factored spectrum (3.118) this becomes

$$\overline{\langle\chi^2\rangle} = 0.651 k^2 \sec\vartheta \int_0^\infty dz\,C_n^2(z)\int_0^\infty \frac{d\kappa}{\kappa^{\frac{8}{3}}}\left(\frac{2J_1(\kappa a_r)}{\kappa a_r}\right)^2\left[1-\cos\!\left(\frac{\kappa^2 z\sec\vartheta}{k}\right)\right]$$

$$(3.121)$$

The preferred strategy is to do the wavenumber integration first, as we did in establishing (3.119) for small telescopes. That allows one to postpone the definition of the C_n^2 profile until the end. Unfortunately, the resulting wavenumber integral cannot be expressed in terms of familiar functions.[17] There are two options. One can do it numerically or one can reverse the strategy. We adopt the second approach and do the height integration first:

$$\overline{\langle\chi^2\rangle} = 0.651 k^2 \sec\vartheta \int_0^\infty \frac{d\kappa}{\kappa^{\frac{8}{3}}}\left(\frac{2J_1(\kappa a_r)}{\kappa a_r}\right)^2\int_0^\infty dz\,C_n^2(z)\left[1-\cos\!\left(\frac{\kappa^2 z\sec\vartheta}{k}\right)\right]$$

$$(3.122)$$

---

[17] It can be described in terms of the generalized hypergeometric function using a result from A. P. Prudnikov, Yu. A. Brychkov and O. I. Marichev, *Integrals and Series, Volume 2: Special Functions* (Gordon and Breach, New York and London, 1986), page 223:

$$\Im\left[\int_0^1 du\; {}_3F_3\!\left(\frac{3}{2},2,\frac{1}{6};2,2,3;\frac{a_r^2}{i\eta u}\right)i^{\frac{1}{6}}\right]$$

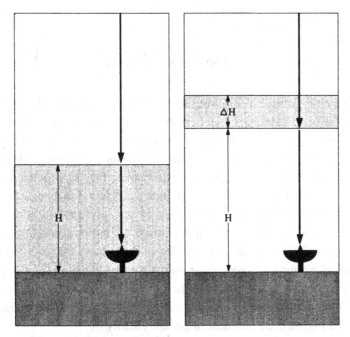

Figure 3.21: Two profile models employed to describe the level of turbulent activity in the atmosphere. On the left, the waves pass through a homogeneous turbulent medium of depth H. The model on the right describes scattering by an elevated turbulent layer of thickness ΔH at height H above the surface.

In modeling the profile of C_n^2 we should be influenced primarily by actual measurements. A secondary consideration is analytical simplicity. Two profile models used in the literature are illustrated for an overhead source in Figure 3.21.

The first panel in Figure 3.21 represents a *slab model* of the lower atmosphere. It assumes that C_n^2 is constant up to a height H and vanishes above. This model was used in Section 5.2 of Volume 1 to interpret phase-difference measurements taken by microwave interferometers. This situation is identical to the terrestrial propagation of a plane wave if one replaces the path length by

$$R = H \sec \vartheta$$

so that the amplitude variance becomes

$$\overline{\langle \chi^2 \rangle} = 0.651\, H \sec\vartheta\, k^2 C_n^2 \int_0^\infty \frac{d\kappa}{\kappa^{\frac{8}{3}}} \left(\frac{2J_1(\kappa a_r)}{\kappa a_r} \right)^2 \left(1 - \frac{\sin\left(\dfrac{\kappa^2 H \sec\vartheta}{k} \right)}{\dfrac{\kappa^2 H \sec\vartheta}{k}} \right)$$

This can be expressed in terms of the receiver gain function for plane waves defined by (3.66) which is plotted in Figure 3.9:

$$\text{Slab:} \qquad \left\langle \chi^2 \right\rangle = 0.307 (H \sec \vartheta)^{\frac{11}{6}} k^{\frac{7}{6}} C_n^2 G\left(a_r \sqrt{\frac{2\pi}{\lambda H \sec \vartheta}} \right) \qquad (3.123)$$

This result would provide a complete description of the scintillation experienced by astronomical observations if the slab model were an accurate description of tropospheric irregularities. Unfortunately the measured profiles do not support that assumption. This did not matter for phase-difference measurements, which are proportional to the simple height integral of C_n^2 for which the structure is unimportant. The altitude structure does matter for amplitude measurements because of the emphasis on elevated turbulence indicated by the $z^{\frac{5}{6}}$ weighting function.

Extensive profiles of C_n^2 have been measured by balloon-borne temperature-difference sensors. Typical profiles for different seasons and times were provided in Figure 2.18 of Volume 1. The data shows that there is usually a strong layer at an altitude between 3 and 4 km where the C_n^2 values are comparable to those at the surface. These measurements suggest the elevated-layer model illustrated in the second panel of Figure 3.21 and described analytically by

$$C_n^2(z) = C_n^2 \begin{cases} 1 & H < z < H + \Delta H \\ 0 & \text{otherwise} \end{cases} \qquad (3.124)$$

In combination with the variance expression (3.122) this model generates the following result:

$$\text{Layer:} \qquad \left\langle \chi^2 \right\rangle = 0.651 C_n^2 k^2 \, \Delta H \sec \vartheta \int_0^\infty \frac{d\kappa}{\kappa^{\frac{8}{3}}} \left(\frac{2 J_1(\kappa a_r)}{\kappa a_r} \right)^2 \mathcal{H}(\kappa, \vartheta)$$

$$(3.125)$$

where the *thick-layer weighting factor* is defined by

$$\mathcal{H}(\kappa, \vartheta) = 1 - \frac{H}{\Delta H} \left(\frac{2k}{\kappa^2 H \sec \vartheta} \right) \sin\left(\frac{\kappa^2 \, \Delta H \sec \vartheta}{2k} \right)$$

$$\times \cos\left[\frac{\kappa^2 H \sec \vartheta}{k} \left(1 + \frac{\Delta H}{2H} \right) \right] \qquad (3.126)$$

This function is plotted versus $x = \kappa^2 H \sec \vartheta / k$ for various values of the thickness-to-height ratio in Figure 3.22. We let $\varepsilon = \Delta H / 2H$

The resulting curves are similar to those for the thin-layer case ($\Delta H = 0$) unless x is quite large. This observation provides the basis for the *thin-screen approximation* which is widely used to describe optical and radio-astronomy scintillation. It is also

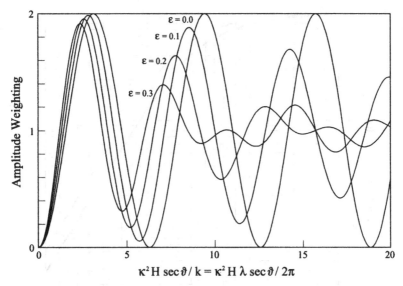

Figure 3.22: The amplitude weighting function for an elevated layer. The parameter ε gives the ratio of the half thickness to the height of the layer.

called the *phase-screen aproximation*:

$$\lim_{\Delta H \ll H} [\mathcal{H}(\kappa, \vartheta)] = 1 - \cos\left(\frac{\kappa^2 H \sec \vartheta}{k}\right) \qquad (3.127)$$

A thin screen imposes phase changes on the downcoming wave but generates no amplitude variations within the layer itself. On the other hand, the irregular wave-front which emerges from a thin layer develops amplitude variations as the wave propagates downward. These fluctuations generate diffraction patterns on the ground, which we will study in the next chapter.

When we combine the phase-screen approximation with (3.125), the aperture-averaged amplitude variance becomes

Thin screen: $\qquad \overline{\langle \chi^2 \rangle} = 0.563 C_n^2 k^{\frac{7}{6}} \Delta H \, H^{\frac{5}{6}} (\sec \vartheta)^{\frac{11}{6}} \mathbb{G}\left(a_r \sqrt{\frac{2\pi}{\lambda H \sec \vartheta}}\right) \qquad (3.128)$

where the *transmission receiver gain factor* for a layered structure is

Thin screen: $\qquad \mathbb{G}(\eta) = 0.578 \int_0^\infty \frac{dx}{x^{\frac{11}{6}}} (1 - \cos x)\left(\frac{2 J_1(\eta \sqrt{x})}{\eta \sqrt{x}}\right)^2 \qquad (3.129)$

This function is plotted in Figure 3.23 together with the receiver gain factor for the slab model defined by (3.124). The phase-screen result is 50% greater than the slab-model version.

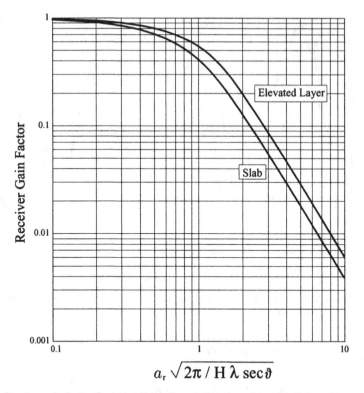

Figure 3.23: Receiver gain factors for transmission through the atmosphere. The slab model corresponds to a homogeneous turbulent medium of height H and is described by (3.66). The receiver gain factor for an elevated layer is defined by (3.129).

The values in Figure 3.23 in combination with (3.128) give a full description for the scaling of scintillation level with wavelength, zenith angle and telescope opening. The ratio of the telescope radius to the Fresnel length is evidently the important parameter. For small telescopes we observe that $\mathbb{G}(0) = 1$ and establish a result similar to (3.119):

$$a_r < \sqrt{\lambda H \sec \vartheta}: \qquad \overline{\langle \chi^2 \rangle} = 0.563 C_n^2 k^{\frac{7}{6}} \Delta H \, H^{\frac{5}{6}} (\sec \vartheta)^{\frac{11}{6}} \qquad (3.130)$$

Most observations are made with large telescopes and the asymptotic expansion

$$\lim_{\eta \to \infty} [\mathbb{G}(\eta)] = 1.293 \eta^{-\frac{7}{3}}$$

shows that the level of scintillation for large openings should be independent of wavelength.

$$a_r > \sqrt{\lambda H \sec \vartheta}: \qquad \overline{\langle \chi^2 \rangle} = 0.728 C_n^2 \Delta H \, H^2 \sec^3 \vartheta \, (a_r)^{-\frac{7}{3}} \qquad (3.131)$$

Aperture smoothing therefore exerts a strong influence on stellar scintillation and the way it scales with wavelength and source position. We need to test these predictions against astronomical measurements.

3.3.2 Scaling with Telescope Diameter

A simple way to do so is to measure the level of intensity fluctuations for different telescope openings using the same source. That can be done by inserting a series of circular masks into the light path of a standard telescope. Since a_r occurs only in the function $T(\eta)$, the scintillation level plotted versus the aperture size should match the curves plotted in Figure 3.23. This experiment was performed by Ellison and Seddon using the 36-inch Cassegrain telescope at the Royal Observatory in the UK [62]. They measured scintillation histories for aperture diameters of 3, 6, 12, 18 and 36 inches, using six stars and Venus at moderate elevations. Their data showed that the scintillation level falls rapidly as the opening is enlarged and agreed with the $-7/3$ relationship predicted by (3.131). Smaller openings could not be used because they did not admit enough light for early detectors and the constant behavior predicted for $a_r < \sqrt{\lambda H \sec \vartheta}$ could not be confirmed.

The subsequent development of sensitive solid-state detectors and photon-counting techniques allowed smaller apertures to be used [66][67]. Measurements made at La Palma with a 60-cm Cassegrain telescope investigated the diameter range 1.2–20 cm using Vega as a source [68]. That data is reproduced in Figure 3.24. These measurements match the predicted curves in Figure 3.23 both in the small- and in the large-aperture range.

One can exploit this agreement to estimate the scattering height. From Figure 3.23 we note that the theoretical curves change from horizontal to $-7/3$ behavior at approximately $\eta = 1$, where

$$\eta = a_r\sqrt{\frac{2\pi}{\lambda H \sec \vartheta}} \tag{3.132}$$

The same transition occurs in the data of Figure 3.24 for diameters of 5 and 7 cm at the two zenith angles. The wavelength was 0.550 μm, which suggests that the layer height was approximately $H = 10$ km, which agrees roughly with measured profiles. A comprehensive survey of aperture averaging for astronomical telescopes with various lens–reflector combinations appeared recently [69].

3.3.3 Wavelength Scaling

It is more difficult to confirm the wavelength dependence of our prediction because it occurs in two places – as the factor $k^{\frac{7}{6}}$ and in the receiver gain function $\mathbb{G}(\eta)$. The

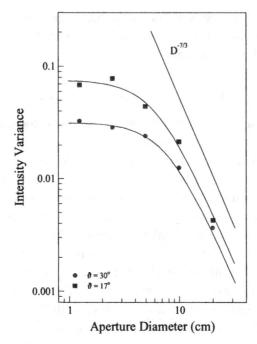

Figure 3.24: Variation of scintillation with aperture diameter measured by Dravins, Lindegren, Mezey and Young [68]. The curves correspond to observations of Vega at the zenith angles indicated. The aperture size was changed rapidly and these data sets represent nearly simultaneous measurements. The straight line corresponds to the scaling law (3.131) predicted for large telescope apertures. © 1997, Astronomical Society of the Pacific; reproduced with permission.

traditional method takes the variance ratio of scintillation measured with filters of different colors in order to avoid the uncertainty in C_n^2:

$$\frac{\left\langle \overline{\chi^2(\lambda_1)} \right\rangle}{\left\langle \overline{\chi^2(\lambda_2)} \right\rangle} = \left(\frac{\lambda_2}{\lambda_1}\right)^{\frac{7}{6}} \frac{\mathbb{G}\left(a_r\sqrt{\dfrac{2\pi}{\lambda_1 H \sec \vartheta}}\right)}{\mathbb{G}\left(a_r\sqrt{\dfrac{2\pi}{\lambda_2 H \sec \vartheta}}\right)} \tag{3.133}$$

The Fresnel length is often comparable to the telescope size and one usually cannot simplify this expression, as we did in Section 3.2.4. The frequency scaling of the ratio can therefore change from $k^{\frac{7}{6}}$ to k^0 depending on the source elevation, aperture diameter and layer height.

The first experiments on the wavelength dependence were done soon after the Second World War using existing astronomical telescopes [70]. They found no wavelength variation, which is not surprising since the large astronomical instruments utilized would have been described by (3.131). Observations with smaller telescopes were undertaken when the nature of aperture averaging became apparent and experiments had confirmed the $k^{\frac{7}{6}}$ scaling for terrestrial paths. Measurements

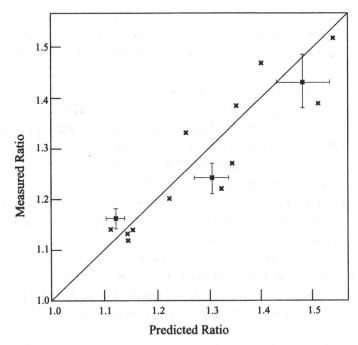

Figure 3.25: Comparisons of measured and theoretical intensity-fluctuation variance ratios for three optical wavelengths made on Mauna Kea by Dainty, Levine, Brames and O'Donnell [72]. The predicted ratio is given by (3.133).

with blue and amber filters were performed with a 3.8-cm aperture for zenith angles $14° < \vartheta < 68°$ [71]. This showed that scintillation does depend on wavelength, but not as strongly as is implied by the limiting form (3.130). The important parameter η would have been in the range $0.6 < \eta < 0.9$ in these experiments and Figure 3.23 indicates that there were significant departures from the $k^{\frac{7}{6}}$ relation.

A more ambitious experiment was conducted later on Mauna Kea in Hawaii (4200 m) with a 61-cm telescope [72]. Filters centered at 0.639, 0.553 and 0.404 μm were used with aperture stops of 2.7 and 4.4 cm. An attempt to estimate the extent of aperture averaging with a 5-mm opening was made. The ratio (3.133) was determined for the three color pairs with this correction. The resulting comparison is reproduced in Figure 3.25 and shows that reasonable agreement between the theoretical and experimental ratios was achieved. The residual mismatch is probably due to imprecision of the profile model and the way its influence was grafted onto the data.

Similar experiments using infrared stellar radiation with a 50-cm telescope were performed at the Observatorio del Tiede (2400 m) in the Canary Islands [73]. Signals from Arcturus at 1.23, 1.65, 2.23 and 3.45 μm were measured for zenith angles in the range $9° < \vartheta < 81°$. The corresponding arguments of the receiver gain function fell in the range $3 < \eta < 12$. This suggests that the wavenumber scaling exponents should be small. The measured values ranged from 0.1 to 0.5, with smaller values relating to longer wavelengths as they should. Nonetheless, the agreement with the prediction is questionable.

Confirmation of a completely different kind comes from measurements made with signals transmitted by geostationary satellites. A research program on propagation was conducted by the Bell Telephone Laboratories using narrow-band beacon signals at 19 and 28 GHz radiated by the Comstar communication satellites [74]. These frequencies are not influenced by the ionosphere and refractive-index irregularities in the troposphere cause any scintillation that is observed. The elevation angle was 40° and the scattering was apparently concentrated in a layer at altitude 4 km. This suggests that the Fresnel lengths were less than the outer scale length, so the optical expressions developed previously ought to describe their data. The 28-GHz signal was measured simultaneously by receivers with diameters of 7 m and 60 cm. The signal ratio confirmed expression (3.129) for the receiver gain factor. This correction was then used to compare the signals at 19 and 28 GHz measured with the same receiver, confirming the basic $k^{\frac{7}{6}}$ scaling law.

3.3.4 Zenith-angle Scaling

We introduced the subject of astronomical intensity fluctuations in Figure 3.19 with time histories taken at various elevation angles. From that data it is apparent that the level of scintillation increases dramatically as the zenith angle of the source increases. We should ask whether our prediction (3.128) accurately describes this behavior.

The scintillation experienced by large telescopes should be described by (3.131) because the parameter η defined by (3.132) for these instruments is invariably greater than unity. This means that traditional astronomical observations should be described by the scaling law

$$\left\langle \overline{\chi^2} \right\rangle = \text{constant} \times (\sec \vartheta)^{\gamma} \tag{3.134}$$

with $\gamma = 3$. This value was confirmed by a number of early observations [70][75]. Measurements using a 22-cm telescope at the La Palma Observatory (2360 m) in the Canary Islands addressed this question anew with improved detectors and data processing [67]. The best fit to a large data set was $\gamma = 2.7$ for sources in the range $1 < \sec \vartheta < 3.4$.

A wider range of zenith angles was investigated using infrared-scintillation measurements made with a 50-cm telescope at the Observatorio del Tiede (2400 m), also in the Canary Islands [73]. Normalized data from this program taken in four

Table 3.2: *Values of the zenith-angle scaling exponent inferred from infrared-scintillation measurements that are reproduced in Figure 3.26; the third line gives the correlation coefficients for the best-fit solutions*

	IR Band			
	J	H	K	L
Wavelength (μm)	1.23	1.65	2.23	3.45
Exponent γ	3.0 ± 0.2	3.0 ± 0.1	3.1 ± 0.1	3.0 ± 0.1
Correlation	0.96	0.99	0.98	0.98

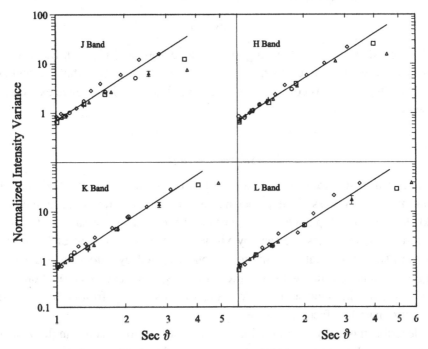

Figure 3.26: Infrared measurements of intensity fluctuations as functions of the zenith angle of the source made by Fuentes, Fuensalida and Sanchez-Magro [73]. These measurements were made with a 50-cm telescope at the Observatorio del Tiede in the Canary Islands.

infrared bands is reproduced in Figure 3.26. This logarithmic data was fitted by straight lines and the best-fit slope values are given in Table 3.2. The coefficients of correlation listed there indicate that excellent fits to the data were obtained. This result provides strong confirmation of (3.131) in the range $1 < \sec \vartheta < 3$.

The experimental data in Figure 3.26 exhibits an important feature of scintillation. The intensity variations begin to saturate for $\sec \vartheta > 3$ and this behavior was noted in early measurements [70] [76]. This represents the onset of strong scattering that is induced by the increasingly long path through the more turbulent region of the troposphere. It is also caused by chromatic dispersion, which can be mitigated with narrow-band filters [77][78].

In view of the previous discussion, one might wonder when the small-aperture version (3.130) is needed. Star trackers have provided navigation fixes for cruise missiles, some aircraft and ballistic missiles. These systems necessarily use small telescopes and aperture smoothing is not available to mitigate the scintillation experienced.

Zenith-angle scaling has also been measured with satellite signals [79]. The 19-GHz beacon signal carried by the Comstar communication satellite was monitored by a 3.7-m ground station. This experiment corresponds to

$$\eta = \sqrt{\cos \vartheta}$$

and the data should fall in the range between (3.130) and (3.131). Scintillation was measured to an accuracy within $1°$ above the horizon and exhibited no sign of saturation. That finding is consistent with the small amplitude variance expected at microwave frequencies.

3.3.5 Source Averaging

If a star and a planet are observed at the same elevation angle, one finds that the fluctuations in brightness of the planet are less than those of the star. Large planets twinkle less than small planets. The scintillation for a planet decreases as the angle subtended by it increases – either by virtue of orbital movement or because of changes in solar illumination. This reduction is caused by *source averaging* and has a good deal in common with aperture smoothing. The observed reduction arises from the way in which the atmosphere scatters waves coming from different points on the illuminated surface of a planet.

To describe this effect we consider a source that subtends an angle α at the telescope. If the lower atmosphere is approximated by a uniform random medium of depth H, the bundle of rays coming from the planet will be separated by a distance $\alpha H \sec \vartheta$ as they enter the troposphere. Waves arriving from opposite sides of the source should be uncorrelated if this distance is greater than the Fresnel length. The critical angular separation is evidently

$$\alpha_0 = \sqrt{\frac{\lambda}{H} \cos \vartheta} \tag{3.135}$$

For optical wavelengths and a height of 10 km, the critical angle is approximately $\alpha_0 \simeq 10^{-5}$ radians or three seconds of arc. The angular sizes of three planets and a

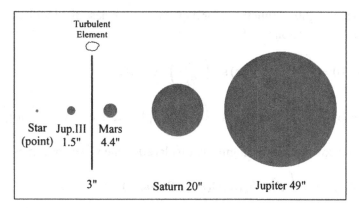

Figure 3.27: A comparison of the angular dimensions for three planets and a moon of Jupiter (Jup. III) suggested by Ellison [80], indicating the critical angular size of random irregularities in the atmosphere.

moon of Jupiter are compared with this value in Figure 3.27. If the planet subtends an angle less than this amount, it should act like a point source and give the same scintillation level as a star. The twinkling will be less than that for a point source if the angle subtended is greater than α_0. This was confirmed in observations comparing scintillation levels for Jupiter, the third moon of Jupiter and a nearby star [62][80]. The brightness of Jupiter was perfectly steady, yet its moon twinkled exactly like the star. Similar comparisons of Jupiter and Mars at the same angle of elevation confirm that the reduction in scintillation begins when the angle subtended by the source exceeds the critical value (3.135).

3.3.5.1 Scintillation Correlation of Adjacent Sources

Tatarskii developed a description of this source averaging based on the Rytov approximation [81][82]. The first step is to calculate the amplitude covariance for plane waves whose propagation vectors are separated by a small angle. We use a coordinate system in which the propagation vector s_1 of the first wave defines the polar axis.[18] The propagation vector s_2 for the second wave makes a small angle δ with this axis and has an azimuthal angle ϕ. The unperturbed plane waves that fall on a scattering eddy at (u, v, s) can then be described as follows:

$$
\begin{aligned}
E_1 &= \mathcal{E}_0 \exp(-iks) \\
E_2 &= \mathcal{E}_0 \exp[-ik(s \cos \delta + u \sin \delta \cos \phi + v \sin \delta \sin \phi)] \\
&\simeq \mathcal{E}_0 \exp\{-ik[s + \delta(u \cos \phi + v \sin \phi)]\}
\end{aligned}
\tag{3.136}
$$

---

[18] Like that illustrated in Figure 7.12 of Volume 1.

The amplitude weighting functions for these waves are described with the paraxial approximation expression (3.11):

$$D(\kappa, 0) = \frac{k}{2} \int_0^\infty ds\, \wp(s) \sin\left(\frac{s\kappa_r^2}{2k}\right) \exp(is\kappa_s)$$

$$D(\kappa, \delta) = \frac{k}{2} \int_0^\infty ds\, \wp(s) \sin\left(\frac{s\kappa_r^2}{2k}\right) \exp\{is[\kappa_s + \delta\kappa_r \cos(\phi - \omega)]\}$$

The amplitude covariance for angularly displaced plane waves becomes

$$\langle \chi(0)\chi(\delta) \rangle = 4 \int d^3\kappa\, \Phi_n(\kappa) D(\kappa, 0) D(-\kappa, \delta)$$

$$= k^2 \int d^3\kappa\, \Phi_n(\kappa) \int_0^\infty ds_1\, \wp(s_1) \int_0^\infty ds_2\, \wp(s_2) \sin\left(\frac{s_1\kappa_r^2}{2k}\right)$$

$$\times \sin\left(\frac{s_2\kappa_r^2}{2k}\right) \exp[-i\delta\kappa_r s_2 \cos(\phi - \omega)] \exp[i\kappa_s(s_1 - s_2)]$$

Transforming to sum and difference coordinates reduces the path integrations to an integral taken along the line-of-sight path:

$$\langle \chi(0)\chi(\delta) \rangle = 2\pi k^2 \int d^3\kappa \int_0^\infty ds\, \Phi_n(\kappa, s) \sin^2\left(\frac{s\kappa_r^2}{2k}\right)$$

$$\times \delta(\kappa_s) \exp[-i\delta\kappa_r s \cos(\phi - \omega)]$$

In writing this expression we have combined the local spectrum and profile functions. We employ spherical wavenumber coordinates if the temperature irregularities are isotropic:

$$\langle \chi(0)\chi(\delta) \rangle = 2\pi k^2 \int_0^\infty d\kappa\, \kappa^2 \int_0^\pi d\psi \sin\psi \int_0^{2\pi} d\omega \int_0^\infty ds\, \Phi_n(\kappa, s)$$

$$\times \sin^2\left(\frac{s\kappa_r^2 \sin^2\psi}{2k}\right) \delta(\kappa \cos\psi) \exp[-i\delta\kappa s \sin\psi \cos(\phi - \omega)]$$

so that finally

$$\langle \chi(0)\chi(\delta) \rangle = 4\pi^2 k^2 \int_0^\infty d\kappa\, \kappa \int_0^\infty ds\, \Phi_n(\kappa, s) \sin^2\left(\frac{s\kappa^2}{2k}\right) J_0(\delta\kappa s) \qquad (3.137)$$

3.3.5.2 Averaging for a Planet or Extended Source

This result can be used to estimate the source-averaged scintillation caused by waves that come from different parts of the planet, like that illustrated in Figure 3.28. It is easy to calculate the average for a fully illuminated planet facing the observer; partially illuminated surfaces can be analyzed numerically in the same way. Solar

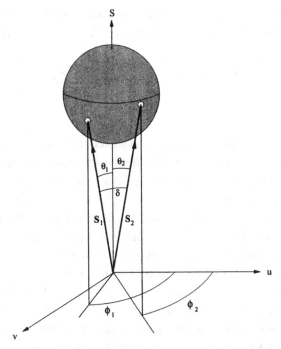

Figure 3.28: The geometry for analyzing the correlation between amplitude fluctuations imposed on plane waves coming from different points on a planet or an extended source.

illumination reflected from different points on the planet is not phase coherent. The combined effect of all waves reaching the earth is therefore the average of (3.137) taken over the half-angle β subtended by the planet or source:

$$\overline{\langle \chi^2 \rangle} = \frac{1}{(\pi\beta^2)^2} \int_0^\beta d\theta_1\, \theta_1 \int_0^\beta d\theta_2\, \theta_2 \int_0^{2\pi} d\phi_1 \int_0^{2\pi} d\phi_2\, \langle \chi(0)\chi(\delta) \rangle \qquad (3.138)$$

The angle between the propagation vectors is related to the bearing angles of the source elements identified in the Figure 3.28:

$$\cos\delta = \cos\theta_1 \cos\theta_2 + \sin\theta_1 \sin\theta_2 \cos(\phi_1 - \phi_2)$$

For the small dimensions which are relevant to such observations

$$\delta = \sqrt{\theta_1^2 + \theta_2^2 - 2\theta_1\theta_2 \cos(\phi_1 - \phi_2)}$$

The source-averaged scintillation is found by combining these expressions:

$$\overline{\langle \chi^2 \rangle} = \frac{4k^2}{\beta^4} \int_0^\infty d\kappa\, \kappa \int_0^\infty ds\, \Phi_n(\kappa, s) \sin^2\left(\frac{s\kappa^2}{2k}\right) \int_0^\beta d\theta_1\, \theta_1 \int_0^\beta d\theta_2\, \theta_2$$

$$\times \int_0^{2\pi} d\phi_1 \int_0^{2\pi} d\phi_2\, J_0\left(s\kappa\sqrt{\theta_1^2 + \theta_2^2 - 2\theta_1\theta_2 \cos(\phi_1 - \phi_2)}\right)$$

The four angular integrations were evaluated in connection with aperture averaging of phase fluctuations.[19] We convert from path length to vertical height and find that

$$\overline{\langle \chi^2 \rangle} = 4\pi^2 k^2 \sec \vartheta \int_0^\infty d\kappa\, \kappa \int_0^\infty dz\, \Phi_n(\kappa, z) \sin^2\left(\frac{\kappa^2 z \sec \vartheta}{2k}\right) \left[\frac{2J_1^2(\kappa\beta z \sec \vartheta)}{\kappa\beta z \sec \vartheta}\right]^2$$

(3.139)

The factor in square brackets indicates that wavenumbers greater than $1/\beta z \sec \vartheta$ are suppressed by source averaging. The filter factor that describes this effect is identical to the term for aperture smoothing. This means that one can use the transmission receiver gain factor plotted in Figure 3.23 to describe source averaging for an elevated thin layer. The predicted reduction of planetary scintillation level agrees with the measurements discussed earlier. This development reminds us that there is likely to be an interaction between aperture and source averaging in astronomical observations. That mixing is observed experimentally [80].

The diffraction theory of weak scattering also provides a powerful way to measure the sizes of radio sources in the distant universe. It was recognized that the level of scintillation of microwave signals from a quasar of finite angular dimension should be less than that of an adjacent point-source quasar. The ratio of these levels gives good estimates of the angular sizes of stars and galaxies [83][84]. These techniques have led to remarkable measurements of the interstellar plasma, in addition to defining the shapes and sizes of microwave sources [85]. Improvements in experiments have stimulated a continuing program of theoretical developments, which have taken the description well beyond the simple model presented here [86][87][88].

3.4 The Ionospheric Influence

Microwave signals that arrive at the earth from distant galaxies exhibit random fluctuations in amplitude.[20] Experiments show that these fluctuations are caused primarily by irregularities in electron density in the ionosphere. Satellite signals are also affected by the ionospheric plasma unless the frequency is above approximately 10 000 MHz. Scintillation can impair the quality of radio-astronomy observations and the performance of communication systems. The way in which the scintillation level varies with radio frequency and zenith angle can also tell one a good deal about the ionosphere.

---

[19] See Section 4.1.8 of Volume 1.
[20] The discovery and exploration of this phenomenon is discussed in Section 2.4.2 of Volume 1.

3.4.1 A General Description

Weak-scattering theory provides a valid description for these amplitude fluctuations unless the source is close to the horizon or the frequency is at the low end of the VHF range. The microwave frequencies used in these observations are therefore much greater than the plasma frequency of the ionosphere. This means that one can use the simplified relationship between variations of the dielectric constant and the electron density:

$$\varepsilon = \varepsilon_0 \left(1 - \frac{Ne^2}{\epsilon_0 m \omega^2} \right) \tag{3.140}$$

This means that fluctuations in these quantities are related by

$$\delta\varepsilon = -r_e \lambda^2 \, \delta N \tag{3.141}$$

where $r_e = 2.818 \times 10^{-15}$ m is the *classical electron radius*. On combining this expression with the single-scattering description (2.31) we see that the logarithmic amplitude fluctuation can be written as

$$\chi(\mathbf{R}) = 4\pi^2 r_e \int d^3r \, A(\mathbf{R}, \mathbf{r}) \, \delta N(\mathbf{r}) \tag{3.142}$$

where the kernel $A(\mathbf{R}, \mathbf{r})$ is defined in terms of the unperturbed field strength and Green's function by (2.30). The amplitude variance therefore depends on the co-variance of electron-density variations:

$$\langle \chi^2 \rangle = \left(4\pi^2 r_e \right)^2 \int d^3r \, A(\mathbf{R}, \mathbf{r}) \int d^3r' \, A(\mathbf{R}, \mathbf{r}') \langle \delta N(\mathbf{r}) \, \delta N(\mathbf{r}') \rangle \tag{3.143}$$

We learned in Section 2.4.3 of Volume 1 that the spatial covariance can be expressed as the Fourier transformation of the *wavenumber spectrum of electron-density fluctuations*:

$$\langle \delta N(\mathbf{r}, t) \, \delta N(\mathbf{r}', t) \rangle = \int d^3\kappa \, \Psi_N(\kappa) \exp[i\kappa \cdot (\mathbf{r} - \mathbf{r}')] \tag{3.144}$$

The spectrum Ψ_N depends on all three components of the wavenumber because the eddies are elongated in the direction of the magnetic field. This approach provides significant advantages for describing the electron-density irregularities, just as it does in the tropospheric case. It is important to remember that $\Psi_N(\kappa)$ is entirely different than the spectrum of refractive-index variations $\Phi_n(\kappa)$. They have completely different dimensions. More to the point, they describe phenomenon controlled by quite different physics.

3 Amplitude Variance

When we combine (3.143) and (3.144) we find that the amplitude variance for transmission of signals through the ionosphere can be written as

$$\langle \chi^2 \rangle = \left(4\pi^2 r_e\right)^2 \int d^3r\, A(\mathbf{R},\mathbf{r}) \int d^3r'\, A(\mathbf{R},\mathbf{r}') \int d^3\kappa\, \Psi_N(\kappa) \exp[i\kappa \cdot (\mathbf{r}-\mathbf{r}')]$$

The volume integrations are now separated and can be expressed in terms of the amplitude weighting function that was introduced in (2.36):

$$\langle \chi^2 \rangle = \left(4\pi^2 r_e\right)^2 \int d^3\kappa\, \Psi_N(\kappa) D(\kappa) D(-\kappa)\, k^{-4} \qquad (3.145)$$

The wavelengths of radio-astronomy signals range from 10 m to a few millimeters. By contrast, the outer scale length of ionospheric irregularities is greater than 1 km. The strongest scattering eddies are thus large relative to the wavelength. The description (3.11) based on the paraxial approximation can therefore be used to express the amplitude weighting function. When we shift to sum and difference path-length coordinates we find that

$$\langle \chi^2 \rangle = (\pi r_e \lambda)^2 \int d^3\kappa \int_0^\infty ds\, \Psi_N(\kappa, s) \sin^2\left(\frac{s\kappa^2}{2k}\right) \delta(\kappa_s)\, 2\pi \qquad (3.146)$$

in the same way as that in which (3.27) was established. In this result we have absorbed the profile function $\wp(s)$ into the turbulence spectrum.

The next task is to estimate the four basic length parameters that define the propagation conditions. To do so we rely on the experimental evidence and resulting models presented in Section 2.4.3.2 of Volume 1. We are confident that the power-law portion of the spectrum is responsible for amplitude scintillation, so the Fresnel length will play a discriminating role. The Fresnel length for ionospheric transmission falls in the range

$$500\ \text{m} < \sqrt{\lambda H} < 2000\ \text{m}$$

for scattering irregularities at altitude 350 km and the wavelengths used in satellite services and radio-astronomy observations. These values are larger than the sizes of antennas of most radio telescopes and one can ignore aperture averaging. The inner scale length is a few meters and thus considerably less than the sizes of the antennas which gather these signals. The outer scale length is best understood as the transition from horizontal traveling waves to the field-aligned irregularities. It is about 1 km and thus comparable to the Fresnel length. The following relationships therefore characterize transmission measurements:

$$\ell_0 \ll a_r \ll \sqrt{\lambda H} \approx L_0 \qquad (3.147)$$

This means that we can ignore the upper wavenumber limit for the spectrum. On the other hand, the outer-scale wavenumber κ_0 may play an important role if the spectrum is sufficiently steep.

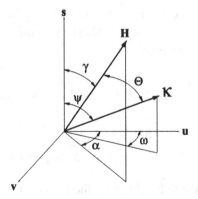

Figure 3.29: Coordinate systems used to describe amplitude scintillations imposed on arriving microwave signals by electron-density irregularities in the ionosphere. The magnetic field **H** and wavenumber vector κ are described in spherical coordinates that use the line-of-sight vector **s** as their polar axis.

To proceed further we must establish an appropriate set of coordinates to describe the wavenumber and line-of-sight vectors that occur in the amplitude-variance expression (3.146). We use spherical coordinates centered on the line-of-sight vector to describe the magnetic field and wavenumber vectors as shown in Figure 3.29:

$$\kappa = \kappa(\mathbf{i}_u \sin \psi \cos \omega + \mathbf{i}_v \sin \psi \sin \omega + \mathbf{i}_s \cos \psi)$$

$$\mathbf{H} = H(\mathbf{i}_u \sin \gamma \cos \alpha + \mathbf{i}_v \sin \gamma \sin \alpha + \mathbf{i}_s \cos \gamma)$$

The angle between the wavenumber vector and the magnetic field can be found by taking the scalar product:

$$\kappa \cdot \mathbf{H} = \kappa H \cos \Theta = \kappa H[\cos \psi \cos \gamma + \sin \psi \sin \gamma \cos(\omega - \alpha)] \qquad (3.148)$$

We will use the power-law spectrum of electron-density fluctuations introduced by (2.135) in Volume 1. The irregularities are symmetrical about the magnetic-field lines and the spectrum can be expressed in terms of the magnitude of the wavenumber and the angle between κ and **H** given by (3.148):

$$\Psi_N(\kappa) = \frac{Q_\nu}{2\pi} \frac{\langle \delta N^2 \rangle \kappa_0^{\nu-3}}{\kappa^\nu [1 + (\mathcal{A}^2 - 1)\cos^2 \Theta]^{\frac{\nu}{2}}} \qquad \kappa_0 < \kappa < \frac{2\pi}{\ell_0} \qquad (3.149)$$

Here $\mathcal{A}$ is the axial ratio of the elongated irregularities and $\langle \delta N^2 \rangle$ is the mean-square variation in electron density. The normalization Q_ν is found by inserting this expression into (3.144) and setting $\mathbf{r} = \mathbf{r}'$. The exponent ν is the important variable. It is left undefined and will be determined by transmission experiments. In view of the relationships noted in (3.147) we can replace the upper limit by infinity when applying this to estimating scintillation levels. This form is consistent with ionospheric measurements and is widely used in ionospheric studies.

The next step is to combine this spectrum model with the expression (3.146) for the amplitude variance. When we express these quantities in the spherical coordinates defined by Figure 3.29 we find

$$\langle \chi^2 \rangle = (\pi \lambda r_e)^2 \int_{\kappa_0}^{\infty} d\kappa \, \kappa^2 \int_0^{\pi} d\psi \, \sin \psi \int_0^{2\pi} d\omega \int_0^{\infty} ds \, \delta(\kappa \cos \psi) \sin^2 \left(\frac{s\kappa^2 \sin^2 \psi}{2k} \right)$$

$$\times \frac{\langle \delta N^2 \rangle Q_\nu \kappa_0^{\nu-3}}{\kappa^\nu \{1 + (A^2 - 1)[\cos \psi \cos \gamma + \sin \psi \sin \gamma \cos(\omega - \alpha)]^2\}^{\frac{1}{2}}}$$

We convert from slant range to height using the flat-earth approximation explained in Section 3.3 and make the replacement $\omega = \zeta + \alpha$ to establish

$$\langle \chi^2 \rangle = (\pi \lambda r_e)^2 Q_\nu \kappa_0^{\nu-3} \sec \vartheta \int_{\kappa_0}^{\infty} \frac{d\kappa \, \kappa}{\kappa^\nu} \int_0^{\infty} dz \, \sin^2 \left(\frac{\kappa^2 z \sec \vartheta}{2k} \right) \langle \delta N^2(z) \rangle$$

$$\times \int_0^{2\pi} d\zeta \, [1 + (A^2 - 1) \sin^2 \gamma \cos^2 \zeta]^{-\frac{\nu}{2}}$$

To proceed further one must describe how the electron-density variance changes with altitude. Rocket and satellite measurements suggest that the thick-layer model portrayed in Figure 3.21 is an appropriate way to do so. The vertical integration can then be expressed in terms of the thick-layer weighting factor defined by (3.124):

$$\langle \chi^2 \rangle = \pi^3 r_e^2 \langle \delta N^2 \rangle Q_\nu \kappa_0^{\nu-3} \lambda^2 \, \Delta H \int_{\kappa_0}^{\infty} \frac{d\kappa \, \kappa}{\kappa^\nu} \mathcal{H}(\kappa, \vartheta)$$

$$\times \sec \vartheta \int_0^{\frac{\pi}{2}} d\zeta \, [1 + (A^2 - 1) \sin^2 \gamma \cos^2 \zeta]^{-\frac{\nu}{2}}$$

The substitution $x = \kappa^2 H \sec \vartheta / k$ brings this into a form that makes clear its dependence on the measurement variables:

$$\langle \chi^2 \rangle = \frac{1}{4} \pi^2 r_e^2 \langle \delta N^2 \rangle Q_\nu \kappa_0^{\nu-3} \, \Delta H \left(\frac{H}{2\pi} \right)^{\frac{\nu}{2}-1} \lambda^{1+\frac{\nu}{2}} \mathcal{L} \left(\zeta_0, \frac{\Delta H}{2H}; \nu \right)$$

$$\times (\sec \vartheta)^{\frac{\nu}{2}} \int_0^{\frac{\pi}{2}} d\zeta \, [1 + (A^2 - 1) \sin^2 \gamma \cos^2 \zeta]^{-\frac{\nu}{2}} \qquad (3.150)$$

where the function generated by the thick-layer model is

$$\mathcal{L}(\zeta_0, \varepsilon; \nu) = \int_{\zeta_0}^{\infty} \frac{dx}{x^{\frac{\nu}{2}}} \left(1 - \frac{\sin(\varepsilon x)}{\varepsilon x} \cos[2x(1 + \varepsilon)] \right) \qquad (3.151)$$

The outer-scale scattering parameter ζ_0 plays a central role in the analysis of ionospheric scintillation. It depends on the wavelength and zenith angle:

$$\zeta_0 = \frac{H\kappa_0^2 \sec \vartheta}{k} = \frac{2\pi H\lambda \sec \vartheta}{L_0^2} \tag{3.152}$$

The outer scale length for plasma irregularities is best understood as the transition from the horizontal traveling ionospheric disturbances to the field-aligned plasma irregularities [89][90]. Horizontal disturbances are characterized by wavelengths of 100 km or more. By contrast, the anisotropic plasma turbulence is well developed only for scale lengths less than 1 km. This means that the Fresnel filtering term $\sin^2[s\kappa^2/(2k)]$ suppresses the influence of large eddies before one reaches the outer-scale wavenumber. One can safely take the limit $\zeta_0 = 0$ and the wavelength and zenith-angle dependences are thereby uncoupled:

$$\langle \chi^2(\lambda, \vartheta)\rangle = B_\nu \lambda^{1+\frac{\nu}{2}}(\sec \vartheta)^{\frac{\nu}{2}} \int_0^{\frac{\pi}{2}} d\zeta \left[1 + (A^2 - 1)\sin^2 \gamma \cos^2 \zeta\right]^{-\frac{\nu}{2}} \tag{3.153}$$

The constant is defined in terms of the various parameters by

$$B_\nu = \frac{1}{4}\pi^2 r_e^2 Q_\nu \langle \delta N^2\rangle \kappa_0^{\nu-3} \Delta H \left(\frac{H}{2\pi}\right)^{\frac{\nu}{2}-1} \mathcal{L}\left(0, \frac{\Delta H}{2H}; \nu\right) \tag{3.154}$$

Now let us turn to the experimental data to see whether we can identify the spectral index ν.

3.4.2 Frequency Scaling

We consider first the frequency scaling of the variance. The scintillation index S_4 has become a measurement standard for such experiments and is equivalent to the rms intensity fluctuations defined earlier in (3.2). In the weak-scattering regime it is linearly related to the rms logarithmic amplitude:

$$S_4 = \left\langle \frac{I^2 - \langle I\rangle^2}{\langle I\rangle^2}\right\rangle^{\frac{1}{2}} = \sqrt{4\langle \chi^2\rangle} \tag{3.155}$$

The wavelength dependence of this quantity implied by (3.153) is

$$S_4 = \text{constant} \times \lambda^{\frac{1}{4}(2+\nu)} \tag{3.156}$$

There is a considerable history of using this relationship to estimate the spectral index [91]. The ratio of scintillation levels taken at two frequencies from the same

Table 3.3: *A summary of results of wavelength-scaling experiments using galactic sources, primarily in Cygnus and Cassiopeia*

Location	Frequencies (MHz)	n	ν	Ref.
Cambridge	38 vs 45	2	6	92
Sydney	40–300	1–2	2–6	93
Jodrell Bank	79 vs 400	1.9	5.6	94
	79 vs 100	2.1	6.4	
	36 vs 100	2.0	6.0	
	36 vs 79	2.1	6.4	
Boulder	53 vs 108	2	6	95
Boston	113 vs 400	2	6	96
	30 vs 113	0–1	≤ 2	
Jodrell Bank	79 vs 1390	1.45–1.55	3.8–4.2	97
Holland	327 vs 609		3.5 ± 0.6	98

source provides a good way to measure the exponent n in the expression:

$$\frac{S_4(\lambda_1)}{S_4(\lambda_2)} = \left(\frac{\lambda_1}{\lambda_2}\right)^n$$

The frequency-scaling exponent n is related to the spectral index ν by

$$\nu = 4n - 2 \tag{3.157}$$

Inferred values of the spectral index ν are sensitive to small changes in n.

3.4.2.1 Radio-astronomy Observations

The first wavelength-scaling experiments used radio-star sources at low VHF frequencies – primarily Cygnus and Cassiopeia. We now know that these signals are influenced both by the terrestrial ionosphere and by the interstellar plasma. Our description captures only the first effect. The findings of radio-astronomy wavelength-scaling experiments are summarized in Table 3.3. Early comparisons suggested that $n = 2$, which implies $\nu = 6$. This is surprisingly large relative to the *in situ* electron-density measurements that were discussed in Section 2.4.2.3 of Volume 1. On the other hand, many of these observations were made with low frequencies at small angles of elevation. It is likely that multiple scattering replaced weak scattering in many of those experiments. Later comparisons of 79- and 1390-MHz signals indicate that $1.45 \leq n \leq 1.55$ [97], which places the spectral index in the more

Table 3.4: *A summary of results of wavelength-scaling experiments using satellite signals*

Satellite	Frequencies (MHz)	n	ν	Ref.
Intelsat	4000 vs 6000	2	6	99
Marisat	1542 vs 3946	1.56–1.95	4.24–5.80	100
Transit	150 vs 400	1.5	4.0	101
ATS-5	138 vs 412	1.31–1.97	3.24–5.9	102
ATS-6	40 vs 140	1.5–1.7	4–4.8	103
	40 vs 140	1.6	4.4	104
	40 vs 140	1.4	3.6	105
DNA	138, 379, 390, 402, 413, 424, 436, 447, 1239, 2891	1.5	4	106

reasonable range $3.8 \leq \nu \leq 4.2$. In a recent experiment levels of scintillation at 327 and 608.5 MHz [98] were compared, giving the value $\nu = 3.5 \pm 0.6$. Determinations using UHF and L-band signals are more likely to be described by weak scattering.

3.4.2.2 Satellite-signal Observations

Earth-orbiting satellites provided an important new way to determine the wavelength scaling of ionospheric scintillation since their signals are not affected by the interstellar plasma. Such experiments began when the first artificial satellite was launched by the USSR in 1957. The results of satellite wavelength-scaling experiments done after 1973 are summarized in Table 3.4. The first experiments used signals of opportunity, primarily from wideband communication satellites. Comparison of signals from an Intelsat spacecraft at 4 and 6 GHz suggested that the scintillation level varies as λ^2. Subsequent observations of Marisat signals at 1541.5 and 3945.5 MHz found wavelength exponents in the range $1.58 < n < 1.94$. Comparisons of 150- and 400-MHz signals from low-altitude navigation satellites gave $\nu = 4$, which provided a good fit to amplitude, Doppler and angle-of-arrival scintillation data [101].

Specially designed synchronous satellites were then built and launched by NASA. These Advanced Technology Satellites radiated narrow-band, phase-coherent signals at several frequencies and were ideal for wavelength-scaling measurements. The fifth spacecraft in this series carried beacons at 137.5 and 412 MHz whose

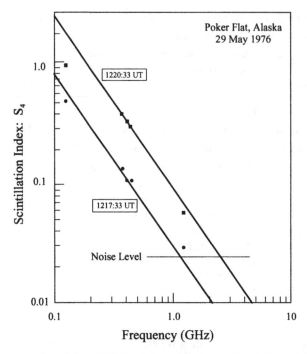

Figure 3.30: Frequency scaling of the scintillation index using phase-coherent signals radiated by the DNA wideband satellite that were measured by Fremouw, Leadabrand, Livingston, Cousins, Rino, Fair and Long in Alaska [106]. The straight lines correspond to $S_4 = \text{constant} \times \lambda^{1.5}$.

comparison gave $1.31 < n < 1.97$. This range was narrowed when the sixth ATS spacecraft, with coherent signals at 40, 140 and 360 MHz, was launched. Propagation data from this satellite was analyzed to give $1.5 < n < 1.7$ under weak-scattering conditions [91][103] [104]. These experiments were repeated at equatorial sites, where a value of $n = 1.4$ was measured [105]. The ATS experiments show that the frequency-scaling exponent decreases as the scintillation level increases, suggesting the onset of multiple scattering.

Even more accurate scintillation measurements became possible with the launching of the DNA wideband satellite into a 1000-km orbit in 1976. This spacecraft was specifically designed for propagation studies and radiated phase-coherent signals at ten frequencies. The wavelength scaling of the scintillation index was measured at equatorial, mid-latitude and auroral ground stations. Typical data from this program is reproduced in Figure 3.30 and indicates that the wavelength exponent is close to $n = 1.5$. This implies that $\nu = 4$, which is consistent with *in situ* measurements. Results of satellite frequency-scaling experiments therefore suggest that a spectral index close to $\nu = 4$ describes the ionospheric irregularities. This value is often used to interpret experimental results [107].

3.4.3 Zenith-angle Scaling

The variation of scintillation level with zenith angle can be established from the basic expression (3.153). In doing so one must remember that the angle γ between **H** and **s** may also depend on the zenith angle:

$$\langle \chi^2(\vartheta) \rangle = \text{constant} \times (\sec \vartheta)^{\frac{\nu}{2}} \frac{2}{\pi} \int_0^{\frac{\pi}{2}} d\zeta \, [1 + (A^2 - 1)\sin^2 \gamma(\vartheta) \cos^2 \zeta]^{-\frac{\nu}{2}}$$

(3.158)

This integral was addressed while analyzing angle-of-arrival fluctuations induced by ionospheric irregularities[21] and we found that its values are surprisingly insensitive to the spectral index. This means that we can use the value $\nu = 4$ suggested above and the resulting integral is found in Appendix B:

$$\langle \chi^2(\vartheta) \rangle = \text{constant} \times (\sec \vartheta)^{\frac{\nu}{2}} \frac{1 + \frac{1}{2}(A^2 - 1)\sin^2 \gamma(\vartheta)}{[1 + (A^2 - 1)\sin^2 \gamma(\vartheta)]^{\frac{3}{2}}}$$

(3.159)

Radio-astronomy measurements of zenith-angle dependence have been made primarily at mid-latitude observing sites. The strongest radio sources lie to the north of these sites and the line of sight makes a steep angle with the downcoming magnetic-field vector. This means that the angle γ changes very little as the source rises and sets. The zenith-angle dependence of the scintillation index can therefore be written in the simple form

$$S_4(\vartheta) = S_4(0)(\sec \vartheta)^{\frac{\nu}{2}}$$

(3.160)

In principle, one can use this expression to determine the index ν in the spectral model. As a practical matter it does not provide a strong indicator. Booker compared early observations made at Cambridge and Manchester with this formula and found that $\nu = 3$ fits the data for angles of elevation greater than $10°$ [108]. The problem is that (3.159) generates shallow curves that are not much different for likely values of ν. The curves for different ν do separate rapidly at small angles of elevation. However, multiple scattering becomes important for small elevations at the low VHF frequencies used in the early experiments and one cannot relate their results to the weak-scattering result above. This problem was addressed briefly in a series of experiments using signals at 1390 MHz received by the 250-ft dish at Jodrell Bank [97]. The higher frequency means that the propagation is more likely to be characterized by weak scattering and the data from these experiments was consistent with a value of $\nu = 4$. Crane measured the scintillation level of satellite signals as a function of the angle of elevation both for 400 MHz and for 7300 MHz [109]. It is unlikely that the X-band signal was influenced by the ionosphere. The rms

---

[21] See Section 7.5.1 of Volume 1.

scintillation at 400 MHz varied approximately as $\sec \vartheta$ in the range $80° < \vartheta < 89°$, which agrees with the value for ν suggested above.

3.5 Problems

Problem 1

We established an exact solution for the amplitude weighting function in (3.19) for a plane wave propagating along a terrestrial link. Evaluate the amplitude-variance expression (3.21) for isotropic turbulence and show that it leads to the familiar expression (3.29). You will need to assume that $\kappa R \gg 1$ and $k \gg \kappa$, as we did implicitly in the short-cut method.

Problem 2

Use the weak-scattering condition (3.1) to estimate the maximum path length that can be described by the Rytov approximation. Calculate this distance versus wavelength from 0.3 μm to 10 m using appropriate values of C_n^2 and L_0 from the discussion of tropospheric properties in Section 2.3 of Volume 1. Plot the maximum transmission distance versus C_n^2 for half a dozen selected frequencies and investigate how it depends on atmospheric conditions.

Problem 3

The *Poynting vector* for the electromagnetic field transmitted by a harmonic source

$$\mathbf{S} = \frac{c}{8\pi k}\Im(E\,\nabla E)$$

describes the flow of energy per unit area per unit time. Confirm the relationship (3.45) between the transmitted power and the constant $\mathcal{E}_0$ for a spherical wave.

Make a similar connection between the transmitted power and the constant $\mathcal{E}_0$ that scales the field strength for a beam wave in (3.80). In this case one must integrate the Poynting vector over the circular exit aperture of the transmitter identified in Figure 3.14. If this seems confusing, wait until we discuss energy flow in Chapter 9 and then return to this question.

Problem 4

Consider propagation along a horizontal path for which C_n^2 is not constant. Assume that the Kolmogorov spectrum describes the local spectrum of irregularities for each volume element along the path and show that

$$\langle \chi^2(R) \rangle = 0.503 k^{\frac{7}{6}} \int_0^R dx\, C_n^2(x)(R-x)^{\frac{5}{6}}$$

where x is the distance measured from the transmitter. This integral equation connects the horizontal profile of C_n^2 and the amplitude variance considered as a function of path length. Can you invert this relationship? Try to discover how sensitive the solution is to measurement errors. Can you devise a reliable experiment that would measure the amplitude variance for different transmission distances and thereby reconstruct the profile of turbulent activity along the path?

Problem 5

Use the expressions for plane and spherical waves derived in Section 3.2 to estimate the amplitude variance for the microwave experiments identified in Table 3.1. Be sure to include aperture averaging. Use the measured values of C_n^2, L_0 and ℓ_0 provided in Section 2.3 of Volume 1 to make these computations. How do your estimates compare with the values measured in references [36][37][38][39] which all represented weak scattering?

Problem 6

Assume that the profile of C_n^2 varies exponentially with altitude. This is a reasonable model if C_n^2 is proportional to the gradient of atmospheric temperature or humidity. It also characterizes the underside of a Chapman layer that is often used to describe ionospheric layers. Include aperture smoothing and show that

$$\langle \chi^2 \rangle = 0.651 C_n^2 h_0 k^2 \int_{\kappa_0}^{\infty} \frac{d\kappa}{\kappa^{\frac{8}{3}}} \left(\frac{2J_1(\kappa a_r)}{\kappa a_r} \right)^2 \left[1 + \left(\frac{k}{\kappa^2 h_0} \right)^2 \right]^{-1}$$

where h_0 is the scale height. Calculate the amplitude variance for a point receiver. Then estimate the influence of receiver gain by setting $\kappa_0 = 0$. Plot the gain factor on logarithmic coordinates and compare it with the curves in Figure 3.23 which describe a homogeneous atmosphere and elevated layer. What is the apparent sensitivity of the amplitude variance to specific profile models?

References

[1] A. M. Obukhov, "On the Influence of Weak Atmospheric Inhomogeneities on the Propagation of Sound and Light," *Izvestiya Akademii Nauk SSSR, Seriya Geofizicheskaya (Bulletin of the Academy of Sciences of the USSR, Geophysical Series)*, No. 2, 155–165 (1953). (English translation by W. C. Hoffman, published by U. S. Air Force Project RAND as Report T-47, Santa Monica, CA, 28 July 1955.)

[2] V. I. Tatarskii, *Wave Propagation in a Turbulent Medium* (translated by R. A. Silverman, Dover, New York, 1967), 127.

[3] V. I. Tatarskii, *The Effects of the Turbulent Atmosphere on Wave Propagation* (translated from the Russian and issued by the National Technical Information Office, U. S. Department of Commerce, Springfield, Virginia 22161, 1971), 218–258.

[4] L. A. Chernov, *Wave Propagation in a Random Medium* (translated by R. A. Silverman, Dover, New York, 1967), 66.

[5] M. Born and E. Wolf, *Principles of Optics, 6th Edition* (Pergamon Press, New York, 1980), 633–664.

[6] See [3], pages 222–223.

[7] S. M. Rytov, Yu. A. Kravtsov and V. I. Tatarskii, *Principles of Statistical Radiophysics 4, Wave Propagation Through Random Media* (Springer-Verlag, Berlin, 1989), 51–53.

[8] See [3], pages 189–193.

[9] See [7], page 24.

[10] See [2], page 153.

[11] M. E. Gracheva, A. S. Gurvich and M. A. Kallistratova, "Dispersion of 'Strong' Atmospheric Fluctuations in the Intensity of Laser Radiation," *Izvestiya Vysshikh Uchebnykh Zavedenii Radiofizika (Radiophysics and Quantum Electronics)*, **13**, No. 1, 40–42 (January 1970).

[12] See [3], pages 65–67 and 232–239.

[13] A. Ishimaru, *Wave Propagation and Scattering in Random Media*, vol. 2 (Academic Press, San Diego, 1978), 368 and 379.

[14] R. J. Hill, "Models of the Scalar Spectrum for Turbulent Advection," *Journal of Fluid Mechanics*, **88**, part 3, 541–562 (13 October 1978).

[15] R. J. Hill and S. F. Clifford, "Modified Spectrum of Atmospheric Temperature Fluctuations and its Application to Optical Propagation," *Journal of the Optical Society of America*, **68**, No. 7, 892–899 (July 1978).

[16] J. H. Churnside, "A Spectrum of Refractive Turbulence in the Turbulent Atmosphere," *Journal of Modern Optics*, **37**, No. 1, 13–16 (January 1990).

[17] L. C. Andrews, "An Analytical Model for the Refractive Index Power Spectrum and its Application to Optical Scintillations in the Atmosphere," *Journal of Modern Optics*, **39**, No. 9, 1849–1853 (September 1992).

[18] R. J. Hill and R. G. Frehlich, "Onset of Strong Scintillation with Application to Remote Sensing of Turbulence Inner Scale," *Applied Optics*, **35**, No. 6, 986–997 (20 February 1996).

[19] R. S. Cole, K. L. Ho and N. D. Mavrokoukoulakis, "The Effect of the Outer Scale of Turbulence and Wavelength on Scintillation Fading at Millimeter Wavelengths," *IEEE Transactions on Antennas and Propagation*, **AP-26**, No. 5, 712–715 (September 1978).

[20] See [2], pages 148–149.

[21] D. L. Fried, "Propagation of a Spherical Wave in a Turbulent Medium," *Journal of the Optical Society of America*, **57**, No. 2, 175–180 (February 1967).

[22] R. T. Aiken, "Propagation from a Point Source in a Randomly Refracting Medium," *The Bell System Technical Journal*, **48**, No. 5, 1129–1165 (May–June 1969).

[23] S. F. Clifford, "Temporal-Frequency Spectra for a Spherical Wave Propagating through Atmospheric Turbulence," *Journal of the Optical Society of America*, **61**, No. 10, 1285–1292 (October 1971).

[24] See [2], pages 187–188.

[25] M. Abramowitz and I. A. Stegun, *Handbook of Mathematical Functions* (Dover Publications, New York, 1972), 295–329.

[26] See [2], pages 229–238.

[27] D. L. Fried, "Aperture Averaging of Scintillation," *Journal of the Optical Society of America*, **57**, No. 2, 169–175 (February 1967).

[28] M. E. Gracheva and A. S. Gurvich, "Averaging Effect of the Receiving Aperture on Fluctuations in Light Intensity," *Izvestiya Vysshikh Uchebnykh Zavedenii Radiofizika (Radiophysics and Quantum Electronics)*, **12**, No. 2, 202–204 (February 1969).

[29] See [3], pages 272–275.

[30] J. H. Churnside, "Aperture Averaging of Optical Scintillations in the Turbulent Atmosphere," *Applied Optics*, **30**, No. 15, 1982–1994 (20 May 1991).

[31] A. I. Kon, "Averaging of Spherical-Wave Fluctuations over a Receiving Aperture," *Izvestiya Vysshikh Uchebnykh Zavedenii Radiofizika (Radiophysics and Quantum Electronics)*, **12**, No. 1, 122–124 (January 1969).

[32] D. H. Höhn, "Effects of Atmospheric Turbulence on the Transmission of a Laser Beam at 6328 Å. 1 – Distribution of Intensity," *Applied Optics*, **5**, No. 9, 1427–1431 (September 1966).

[33] D. L. Fried, G. E. Mevers and M. P. Keister, "Measurements of Laser-Beam Scintillation in the Atmosphere," *Journal of the Optical Society of America*, **57**, No. 6, 787–797 (June 1967).

[34] G. E. Homstad, J. W. Strohbehn, R. H. Berger and J. M. Heneghan, "Aperture-Averaging Effects for Weak Scintillations," *Journal of the Optical Society of America*, **64**, No. 2, 162–165 (February 1974).

[35] J. H. Churnside, "Aperture-Averaging Factor for Optical Propagation Through the Turbulent Atmosphere," NOAA Technical Memorandum ERL WPL-188 (November 1990).

[36] G. M. Babler, "Scintillation Effects at 4 and 6 GHz on a Line-of-Sight Microwave Link," *IEEE Transactions on Antennas and Propagation*, **AP-19**, No. 4, 574–575 (July 1971).

[37] M. C. Thompson, L. E. Wood, H. B. Janes and D. Smith, "Phase and Amplitude Scintillations in the 10 to 40 GHz Band," *IEEE Transactions on Antennas and Propagation*, **AP-23**, No. 6, 792–797 (November 1975).

[38] K. L. Ho, N. D. Mavrokoukoulakis and R. S. Cole, "Wavelength Dependence of Scintillation Fading at 110 and 36 GHz," *Electronic Letters*, **13**, No. 7, 181–183 (31 March 1977).

[39] R. W. McMillan, R. A. Bohlander, G. R. Ochs, R. J. Hill and S. F. Clifford, "Millimeter Wave Atmospheric Turbulence Measurements: Preliminary Results and Instrumentation for Future Measurements," *Optical Engineering*, **22**, No. 1, 032–039 (January–February 1983).

[40] R. J. Hill, S. F. Clifford and R. S. Lawrence, "Refractive-Index and Absorption Fluctuations in the Infrared Caused by Temperature, Humidity, and Pressure Fluctuations," *Journal of the Optical Society of America*, **70**, No. 10, 1192–1205 (October 1980).

[41] N. A. Armand, A. O. Izyumov and A. V. Sokolov, "Fluctuations of Submillimeter Waves in a Turbulent Atmosphere," *Radio Engineering and Electron Physics (USA)*, **16**, No. 8, 1259–1266 (August 1971).

[42] M. W. Fitzmaurice, J. L. Bufton and P. O. Minott, "Wavelength Dependence of Laser-Beam Scintillation," *Journal of the Optical Society of America*, **59**, No. 1, 7–10 (January 1969).

[43] R. H. Kleen and G. R. Ochs, "Measurements of the Wavelength Dependence of Scintillation in Strong Turbulence," *Journal of the Optical Society of America*, **60**, No. 12, 1695–1697 (December 1970).

[44] J. R. Kerr, "Experiments on Turbulence Characteristics and Multiwavelength Scintillation Phenomena," *Journal of the Optical Society of America*, **62**, No. 9, 1040–1049 (September 1972).

[45] W. F. Dabberdt and W. B. Johnson, "Analysis of Multiwavelength Observations of Optical Scintillation," *Applied Optics*, **12**, No. 7, 1544–1548 (July 1973).

[46] L. C. Andrews and R. L. Phillips, *Laser Beam Propagation through Random Media* (SPIE Optical Engineering Press, Bellingham, Washington, 1998), 72–73.

[47] A. I. Kon and V. I. Tatarskii, "Parameter Fluctuations of a Space-Limited Light Beam in a Turbulent Atmosphere," *Izvestiya Vysshikh Uchebnykh Zavedenii Radiofizika (Soviet Radiophysics)*, **8**, No. 5, 617–620 (September–October 1965).

[48] D. L. Fried and J. B. Seidman, "Laser-Beam Scintillation in the Atmosphere," *Journal of the Optical Society of America*, **57**, No. 2, 181–185 (February 1967).

[49] Y. Kinoshita, M. Suzuki and T. Matsumoto, "Fluctuations of a Gaussian Light Beam Propagating Through a Random Medium," *Radio Science*, **3**, No. 3, 287–294 (March 1968).

[50] A. S. Gurvich, "Light-Intensity Fluctuations in Divergent Beams," *Izvestiya Vysshikh Uchebnykh Zavedenii Radiofizika (Radiophysics and Quantum Electronics)*, **12**, No. 1, 119–121 (January 1969).

[51] A. Ishimaru, "Fluctuations of a Beam Wave Propagating Through a Locally Homogeneous Medium," *Radio Science*, **4**, No. 4, 295–305 (April 1969).

[52] A. Ishimaru, "Fluctuations of a Focused Beam Wave for Atmospheric Turbulence Probing," *Proceedings of the IEEE*, **57**, No. 4, 407–414 (April 1969).

[53] W. B. Miller, J. C. Ricklin and L. C. Andrews, "Log-Amplitude Variance and Wave Structure Function: A New Perspective for Gaussian Beams," *Journal of the Optical Society of America A*, **10**, No. 4, 661–672 (April 1993).

[54] J. R. Kerr and J. R. Dunphy, "Experimental Effects of Finite Transmitter Apertures on Scintillations," *Journal of the Optical Society of America*, **63**, No. 1, 1–8 (January 1973).

[55] W. B. Miller, J. C. Ricklin and L. C. Andrews, "Effects of the Refractive Index Spectral Model on the Irradiance Variance of a Gaussian Beam," *Journal of the Optical Society of America A*, **11**, No. 10, 2719–2726 (October 1994).

[56] S. S. Khmelevtsov and R. Sh. Tsvyk, "Intensity Fluctuations of a Laser Beam Propagating in a Turbulent Atmosphere," *Izvestiya Vysshikh Uchebnykh Zavedenii Radiofizika (Radiophysics and Quantum Electronics)*, **13**, No. 1, 111–113 (January 1970).

[57] S. S. Khmelevtsov, "Propagation of Laser Radiation in a Turbulent Atmosphere," *Applied Optics*, **12**, No. 10, 2421–2433 (October 1973).

[58] W. B. Miller, J. C. Ricklin and L. C. Andrews, "Scintillation of Initially Convergent Gaussian Beams in the Vicinity of the Geometric Focus," *Applied Optics*, **34**, No. 30, 7066–7073 (20 October 1995).

[59] M. I. Charnotskii, "Asymptotic Analysis of Finite-Beam Scintillation in a Turbulent Medium," *Waves in Random Media*, **4**, No. 3, 243–273 (July 1994).

[60] V. A. Banakh and I. N. Smalikho, "Statistical Characteristics of the Laser Beam Propagating Along Vertical and Sloping Paths Through a Turbulent Atmosphere," *Proceedings of the SPIE Vol. 1968 Atmospheric Propagation and Remote Sensing II*, 303–311 (1993).

[61] V. P. Lukin, *Atmospheric Adaptive Optics* (SPIE Optical Engineering Press, Bellingham, Washington, 1995; originally published in Russian in 1986), 99–101.

[62] M. A. Ellison and H. Seddon, "Some Experiments on the Scintillation of Stars and Planets," *Monthly Notices of the Royal Astronomical Society* (Blackwell Sciences, Ltd), **112**, No. 1, 73–87 (1952).

[63] See [2], pages 224–257.

[64] See [3], pages 323–334.

[65] See [2], page 239.

[66] R. S. Iyer and J. L. Bufton, "Aperture Averaging Effects in Stellar Scintillation," *Optics Communications*, **22**, No. 3, 377–381 (September 1977).

[67] A. H. Mikesell, A. A. Hoag and J. S. Hall, "The Scintillation of Starlight," *Journal of the Optical Society of America*, **41**, No. 10, 689–695 (October 1951).

[68] D. Dravins, L. Lindegren, E. Mezey and A. T. Young, "Atmospheric Intensity Scintillation of Stars. I. Statistical Distributions and Temporal Properties," *Publications of the Astronomical Society of the Pacific*, **109**, No. 732, 173–207 (February 1997).

[69] D. Dravins, L. Lindegren, E. Mezey and A. T. Young, "Atmospheric Intensity Scintillation of Stars. III. Effects for Different Telescope Apertures," *Publications of the Astronomical Society of the Pacific*, **110**, No. 747, 610–633 (May 1998).

[70] W. M. Protheroe, "Preliminary Report on Stellar Scintillation," Air Force Cambridge Research Center Report No. AFCRC-TR-54-115 (November 1954).

[71] J. J. Burke, "Observations of the Wavelength Dependence of Stellar Scintillation," *Journal of the Optical Society of America*, **60**, No. 9, 1262–1264 (September 1970).

[72] J. C. Dainty, B. M. Levine, B. J. Brames and K. A. O'Donnell, "Measurements of the Wavelength Dependence and Other Properties of Stellar Scintillation at Mauna Kea, Hawaii," *Applied Optics*, **21**, No. 7, 1196–1200 (1 April 1982).

[73] F. J. Fuentes, J. J. Fuensalida and C. Sanchez-Magro, "Measurements of the Near-Infrared Stellar Scintillation Above the Observatorio Del Teide (Tenerife)," *Monthly Notices of the Royal Astronomical Society* (Blackwell Sciences, Ltd), **226**, No. 4, 769–783 (15 June 1987).

[74] D. C. Cox, H. W. Arnold, and H. H. Hoffman, "Observations of Cloud-Produced Amplitude Scintillation on 19- and 28-GHz Earth–Space Paths," *Radio Science*, **16**, No. 5, 885–907 (September–October 1981).

[75] L. N. Zhukova, "The Registration of Stellar Scintillation by the Photoelectric Method," *Izvestiya Glavnoi Astronomicheskoi Observatorii v Pulkove*, **21**, No. 162, 72–82 (1958). (No English translation is available.)

[76] See [2], page 228.

[77] A. T. Young, "Aperture Filtering and Saturation of Scintillation," *Journal of the Optical Society of America*, **60**, No. 2, 248–250 (February 1970).

[78] A. T. Young, "Saturation of Scintillation," *Journal of the Optical Society of America*, **60**, No. 11, 1495–1500 (November 1970).

[79] J. M. Titus and H. W. Arnold, "Low-Elevation-Angle Propagation Effects on COMSTAR Satellite Signals," *The Bell System Technical Journal*, **61**, No. 7, 1567–1572 (September 1982).

[80] M. A. Ellison, "Location, Size and Speed of Refractional Irregularities Causing Scintillation," *Quarterly Journal of the Royal Meteorological Society*, **80**, No. 344, 246–248 (April 1954).

[81] See [2], pages 252–257.

[82] See [3], pages 278–284.

[83] R. Hanbury Brown and R. Q. Twiss, "A New Type of Interferometer for Use in Radio Astronomy," *Philosophical Magazine*, Series 7, **45**, No. 366, 663–682 (July 1954).

[84] R. C. Jennison and M. K. Das Gupta, "The Measurement of the Angular Diameter of Two Intense Radio Sources – II: Diameter and Structural Measurements of the Radio Stars Cygnus A and Cassiopeia A," *Philosophical Magazine*, Series 8, **1**, No. 1, 65–75 (January 1956).

[85] I. N. Zhuk, "Scintillation Studies of Cosmic Source Angular Structure (Review)," *Izvestiya Vysshikh Uchebnykh Zavedenii Radiofizika (Radiophysics and Quantum Electronics)*, **23**, No. 8, 597–615 (August 1980).

[86] V. I. Shishov and T. D. Shishova, "Influence of Source Sizes on the Spectra of Interplanetary Scintillations Theory," *Astronomicheskii Zhurnal (Soviet Astronomy)*, **22**, No. 2, 235–239 (March–April 1978).

[87] J. L. Codona and R. G. Frehlich, "Scintillation from Extended Incoherent Sources," *Radio Science*, **22**, No. 4, 469–480 (July–August 1987).

[88] M. I. Charnotskii, "Asymptotic Analysis of the Flux Fluctuations Averaging and Finite-size Source Scintillations in Random Media," *Waves in Random Media*, **1**, No. 4, 223–243 (October 1991).

[89] H. G. Booker, "The Role of Acoustic Gravity Waves in the Generation of Spread-*F* and Ionospheric Scintillation," *Journal of Atmospheric and Terrestrial Physics*, **41**, No. 5, 501–515 (May 1979).

[90] C. L. Rino, "A Power Law Phase Screen Model for Ionospheric Scintillation; 1. Weak Scatter," *Radio Science*, **14**, No. 6, 1135–1145 (November–December 1979).

[91] K. C. Yeh and C. H. Liu, "Radio Wave Scintillations in the Ionosphere," *Proceedings of the IEEE*, **70**, No. 4, 324–360 (April 1982).

[92] A. Hewish, "The Diffraction of Galactic Radio Waves as a Method of Investigating the Irregular Structure of the Ionosphere," *Proceedings of the Royal Society A*, **214**, No. 1119, 494–514 (9 October 1952).

[93] J. G. Bolton, O. B. Slee and G. J. Stanley, "Galactic Radiation at Radio Frequencies," *Australian Journal of Physics*, **6**, 434–451 (December 1953).

[94] H. J. A. Chivers, "The Simultaneous Observation of Radio Star Scintillations on Different Radio-Frequencies," *Journal of Atmospheric and Terrestrial Physics*, **17**, No. 3, 181–187 (1960).

[95] R. S. Lawrence, J. L. Jesperson and R C. Lamb, "Amplitude and Angular Scintillations of the Radio Source Cygnus-A Observed at Boulder, Colorado," *Journal of Research of the NBS – D. Radio Propagation*, **65**, No. 4, 333–350 (July–August 1961).

[96] J. Aarons, R. S. Allen and T. J. Elkins, "Frequency Dependence of Radio Star Scintillations," *Journal of Geophysical Research*, **72**, No. 11, 2891–2902 (1 June 1967).

[97] H. J. A. Chivers and R. D. Davies, "A Comparison of Radio Star Scintillations at 1390 and 79 Mc/s at Low Angles of Elevation," *Journal of Atmospheric and Terrestrial Physics*, **24**, 573–584 (July 1962).

[98] T. A. T. Spoelstra, "Effects of Amplitude and Phase Scintillations on Decimeter Wavelength Observations at Mid-latitudes," *Astronomy and Astrophysics*, **148**, No. 1, 21–28 (January 1985).

[99] R. R. Taur, "Ionospheric Scintillation at 4 and 6 GHz," *Comsat Technical Review*, **3**, No. 1, 145–163 (Spring 1973).

[100] S. J. Franke, C. H. Liu and D. J. Fang, "Multifrequency Study of Ionospheric Scintillation at Ascension Island," *Radio Science*, **19**, No. 3, 695–706 (May–June 1984).

[101] R. K. Crane, "Spectra of Ionospheric Scintillation," *Journal of Geophysical Research*, **81**, No. 13, 2041–2050 (1 May 1976).

[102] C. L. Rino, R. C. Livingston and H. E. Whitney, "Some New Results on the Statistics of Radio Wave Scintillation; 1. Empirical Evidence for Gaussian Statistics," *Journal of Geophysical Research*, **81**, No. 13, 2051–2057 (1 May 1976).

[103] R. Umeki, C. H. Liu, and K. C. Yeh, "Multifrequency Studies of Ionospheric Scintillations," *Radio Science*, **12**, No. 2, 311–317 (March–April 1977).

[104] See [21], page 343, where the authors reproduced wavelength-scaling data from an earlier conference paper that is difficult to obtain.

[105] A. Bhattacharyya, R. G. Rastogi and K. C. Yeh, "Signal Frequency Dependence of Ionospheric Amplitude Scintillations," *Radio Science*, **25**, No. 4, 289–297 (July–August 1990).

[106] E. J. Fremouw, R. L. Leadabrand, R. C. Livingston, M. D. Cousins, C. L. Rino, B. C. Fair and R. A. Long, "Early Results from the DNA Wideband Satellite Experiment – Complex-signal Scintillation," *Radio Science*, **13**, No. 1, 167–187 (January–February 1978).

[107] R. K. Crane, "Ionospheric Scintillation," *Proceedings of the IEEE*, **65**, No. 2, 180–1199 (February 1977).

[108] H. G. Booker, "The Use of Radio Stars to Study Irregular Refraction of Radio Waves in the Ionosphere," *Proceedings of the IRE*, **46**, No. 1, 298–314 (January 1958).

[109] R. K. Crane, "Low Elevation Angle Measurement Limitations Imposed by the Troposphere: An Analysis of Scintillation Observations made at Haystack and Millstone," MIT Lincoln Laboratory Technical Report No. 518 (18 May 1976).

4

Spatial Covariance

The similarity of signals measured at adjacent receivers is an important feature of electromagnetic scintillation. The way in which their correlation decreases with their separation provides a powerful insight into such phenomena. We discussed the correlation of phase fluctuations measured at nearby receivers in Volume 1. We found there that the phase difference defines the angular accuracy of direction-finding systems and the resolution of interferometric imaging techniques.

In this chapter we will investigate the correlation of signal amplitudes and intensities measured at adjacent receivers. This type of experiment played a crucial role in the development of scintillation physics. Optical measurements of intensity correlation on short paths first validated the description based on Rytov's approximation and Kolmogorov's model of atmospheric turbulence. These experiments were later replicated with microwave and millimeter-wave signals on longer paths.

Diffraction patterns are created on the ground when starlight is scattered by tropospheric irregularities. These patterns are correlated over tens of centimeters and are sometimes visually apparent at the onset of solar eclipses. Astronomical signals are thus spatially correlated over distances that are smaller than most telescope openings. This means that the arriving field is coherent over only a modest portion of the reflector surface. The light-gathering power of astronomical telescopes is limited by this effect unless modern speckle-interferometry techniques are used to reconstruct the original wave-front.

By contrast, variations in electron density in the ionosphere generate diffraction patterns for radio-astronomy signals that are correlated over several kilometers. This scintillation is generated in the ionosphere, in the solar wind and in the interstellar plasma.

4.1 Terrestrial Links

The spatial correlation depends primarily on the inertial range of the turbulence spectrum and is similar to amplitude variance in this regard. Aperture averaging often plays an important role in suppressing the effect of small eddies. When it does not, the dissipation region controls the spatial correlation for small inter-receiver separations. By contrast, the outer-scale region of large eddies is seldom influential. In studying optical and microwave propagation along paths near the surface, we will examine plane and spherical waves separately because their spatial correlations are surprisingly different.

4.1.1 Plane Waves

One can measure the correlation between signals received at two receivers placed along the propagation path or normal to it. The lateral deployment is usually chosen because it better represents practical applications and because it provides more useful data. We shall assume that the receivers are deployed horizontally as shown in Figure 4.1. This is the most convenient experimental arrangement for microwave experiments, although some measurements have been made with vertically deployed receivers.

4.1.1.1 The Covariance Expression

It is a straightforward task to describe the spatial covariance of weak amplitude fluctuations for plane waves that travel along horizontal paths. The logarithmic

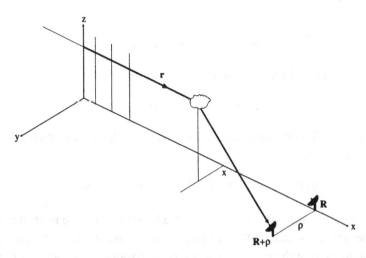

Figure 4.1: The coordinate system used to describe the spatial covariance of a plane wave for receivers deployed normal to the line of sight.

amplitude fluctuations measured at separated receivers are described by expression (2.31) taken at $\mathbf{R}$ and $\mathbf{R} + \rho$:

$$\chi(\mathbf{R}) = -k^2 \int d^3r \, A(\mathbf{R}, \mathbf{r}) \, \delta\varepsilon(\mathbf{r}) \tag{4.1}$$

$$\chi(\mathbf{R} + \rho) = -k^2 \int d^3r \, A(\mathbf{R} + \rho, \mathbf{r}) \, \delta\varepsilon(\mathbf{r}) \tag{4.2}$$

We translate the coordinate system in the second integral by setting $\mathbf{r} = \mathbf{r}' + \rho$. The inter-receiver separation disappears at the infinite limits if the arriving signal is an unbounded plane wave:

$$\chi(\mathbf{R} + \rho) = -k^2 \int d^3r' \, A(\mathbf{R} + \rho, \mathbf{r}' + \rho) \, \delta\varepsilon(\mathbf{r}' + \rho) \tag{4.3}$$

The kernel is defined by (2.30) in terms of the unperturbed field strength and Green's function:

$$A(\mathbf{R} + \rho, \mathbf{r}' + \rho) = \Re\left(\frac{G(\mathbf{R} + \rho, \mathbf{r}' + \rho)E_0(\mathbf{r}' + \rho)}{E_0(\mathbf{R} + \rho)} \right)$$

The field strength for an infinite plane wave is unchanged by lateral translations and depends only on the downrange distance:

$$E_0(\mathbf{r}' + \rho) = E_0(x') \quad \text{and} \quad E_0(\mathbf{R} + \rho) = E_0(R)$$

As explained in Appendix L, Green's function depends on the scalar distance between the two points in question and is invariant with respect to the same shift of both points:

$$G(\mathbf{R} + \rho, \mathbf{r}' + \rho) = G(\mathbf{R}, \mathbf{r}')$$

The kernel is therefore independent of the separation vector:

$$A(\mathbf{R} + \rho, \mathbf{r}' + \rho) = A(\mathbf{R}, \mathbf{r}')$$

One can rewrite (4.3) in the same form as (4.1) if the dielectric variation is measured at the shifted location:

$$\chi(\mathbf{R} + \rho) = -k^2 \int d^3r' \, A(\mathbf{R}, \mathbf{r}') \, \delta\varepsilon(\mathbf{r}' + \rho) \tag{4.4}$$

Let us pause to examine the assumptions made so far. We have made no approximation if the arriving signal really is an unbounded plane wave. That is the situation for starlight arriving at the earth. However, on horizontal paths one must approximate a plane wave by a collimated beam. One can measure the spatial covariance for such signals only if the inter-receiver separation is small relative to the beam radius w_0. This ensures that the unperturbed field E_0 is not changed appreciably by the lateral shift $\rho = \rho \mathbf{i}_y$ and allows one to avoid redefinition of the integration boundaries.

We can construct the spatial covariance with the expressions (4.1) and (4.4) as follows:

$$\langle \chi(\mathbf{R})\chi(\mathbf{R}+\rho)\rangle = k^4 \int d^3r\, A(\mathbf{R}, \mathbf{r}) \int d^3r'\, A(\mathbf{R}, \mathbf{r}')\langle \delta\varepsilon(\mathbf{r})\, \delta\varepsilon(\mathbf{r}'+\rho)\rangle \quad (4.5)$$

The wavenumber-integral description (2.35) for the ensemble average of dielectric variations leads to the following expression:

$$\langle \chi(\mathbf{R})\chi(\mathbf{R}+\rho)\rangle = \int d^3\kappa\, \Phi_\varepsilon(\kappa, x)\exp(i\kappa \cdot \rho)$$
$$\times \left| k^2 \int d^3r\, A(\mathbf{R}, \mathbf{r})\exp(i\kappa \cdot \mathbf{r}) \right|^2$$

The volume integral in absolute brackets is simply the amplitude weighting function defined by (2.36). When one is dealing with propagation near the surface it is appropriate to shift to the spectrum of refractive-index irregularities:

$$\langle \chi(\mathbf{R})\chi(\mathbf{R}+\rho)\rangle = 4 \int d^3\kappa\, \Phi_n(\kappa, x)\exp(i\kappa \cdot \rho)\, D(\kappa, \mathbf{R})D(-\kappa, \mathbf{R}) \quad (4.6)$$

We shall rely here on the paraxial (small-scattering angle) approximation because it describes a wide range of optical, microwave and radio propagation situations. One can use the expression for $D(\kappa)D(-\kappa)$ given by (3.27) in this case:

$$\langle \chi(\mathbf{R})\chi(\mathbf{R}+\rho)\rangle = 2\pi k^2 \int_0^R dx \int d^3\kappa\, \Phi_n(\kappa, x)\delta(\kappa_x)\sin^2\left(\frac{x\kappa_r^2}{2k}\right)\exp(i\rho\kappa_y)$$
$$(4.7)$$

Near the surface the irregularities are approximately isotropic. Moreover, we found in Section 3.2.7 that anisotropy has little influence on amplitude fluctuations. The wavenumber integration is naturally evaluated in spherical coordinates:

$$\langle \chi(\mathbf{R})\chi(\mathbf{R}+\rho)\rangle = 2\pi k^2 \int_0^R dx \int_0^\infty d\kappa\, \kappa^2 \Phi_n(\kappa, x) \int_0^\pi d\psi\, \sin\psi\, \delta(\kappa\cos\psi)$$
$$\times \sin^2\left(\frac{x\kappa^2\sin^2\psi}{2k}\right) \int_0^{2\pi} d\omega \exp(i\kappa\rho\sin\psi\cos\omega)$$

The azimuth integration gives the zeroth-order Bessel function and the following general description first found by Tatarskii emerges [1]:

$$\langle \chi(\mathbf{R})\chi(\mathbf{R}+\rho)\rangle = 4\pi^2 k^2 \int_0^R dx \int_0^\infty d\kappa\, \kappa\, \Phi_n(\kappa, x) J_0(\kappa\rho) \sin^2\left(\frac{x\kappa^2}{2k}\right) \quad (4.8)$$

4.1.1.2 The First Moment of Covariance

Let us pause to note an interesting consequence of (4.8). The first moment of the amplitude covariance taken over all possible separations must vanish. This conclusion is independent of the spectrum of irregularities. It derives instead from the orthogonality of Bessel functions noted in Appendix D:

$$\int_0^\infty d\rho\, \rho\langle \chi(\mathbf{R})\chi(\mathbf{R}+\rho)\rangle = 4\pi^2 k^2 \int_0^R dx \int_0^\infty d\kappa\, \kappa\, \Phi_n(\kappa, x)$$

$$\times \sin^2\left(\frac{x\kappa^2}{2k}\right) \int_0^\infty d\rho\, \rho J_0(\kappa\rho)$$

$$= 4\pi^2 k^2 \int_0^R dx \int_0^\infty d\kappa\, \kappa\, \Phi_n(\kappa, x)$$

$$\times \sin^2\left(\frac{x\kappa^2}{2k}\right)\frac{\delta(\kappa)}{\kappa} = 0 \quad (4.9)$$

This implies that the correlation must assume negative as well as positive values. We shall find that it is true both for the theoretical prediction and for measurements made with plane waves. Tatarskii found an interesting connection between conservation of energy and this property [2].

4.1.1.3 The Covariance for the Kolmogorov Model

For a signal propagating near the earth's surface the spectrum is nearly constant and the path integration in (4.8) can be performed:

$$\langle \chi(\mathbf{R})\chi(\mathbf{R}+\rho)\rangle = 2\pi^2 k^2 R \int_0^\infty d\kappa\, \kappa\, \Phi_n(\kappa) J_0(\kappa\rho)\left(1 - \frac{\sin(\kappa^2 R/k)}{\kappa^2 R/k}\right) \quad (4.10)$$

To test this result we must specify the spectrum of irregularities. The Fresnel length is a few centimeters for optical signals propagating less than 1 km. That is larger than the inner scale length and substantially smaller than the outer scale length for temperature fluctuations:

$$\ell_0 < \sqrt{\lambda R} \ll L_0$$

The Fresnel length is much larger for microwave and millimeter waves, but the same relationship holds in most cases of interest. We first analyze the situation by ignoring inner-scale effects. The inertial range of the wavenumber spectrum is all-important in this situation and the Kolmogorov model can be used to evaluate the covariance:

$$\langle \chi(\mathbf{R})\chi(\mathbf{R}+\rho)\rangle = 2\pi^2 \times 0.033 C_n^2 k^2 R \int_0^\infty \frac{d\kappa}{\kappa^{\frac{8}{3}}} J_0(\kappa\rho)\left(1 - \frac{\sin(\kappa^2 R/k)}{\kappa^2 R/k}\right)$$

The *spatial correlation* is formed by dividing this expression by the amplitude variance. It is often the preferred measurement because it removes the familiar combination of path and turbulence parameters. Using the substitution $x = \kappa^2 R/k$ the following expression emerges:

$$C_\chi(q) = 1.060 \int_0^\infty \frac{dx}{x^{\frac{11}{6}}} J_0(q\sqrt{x})\left(1 - \frac{\sin x}{x}\right) \tag{4.11}$$

This function depends uniquely on the ratio of the inter-receiver separation and the Fresnel length:

$$q = \rho\sqrt{k/R} \quad \text{or} \quad q = 2.507\frac{\rho}{\sqrt{\lambda R}} \tag{4.12}$$

and the integral can be expressed in terms of a Kummer function [3]:

$$C_\chi(q) = 3.8643\left[\exp\left(i\pi\frac{11}{12}\right) M\left(-\frac{11}{6}, 1, \frac{iq^2}{4}\right)\right] - 7.532\left(\frac{q^2}{4}\right)^{\frac{5}{6}} \tag{4.13}$$

but it is a simple matter to compute the integral directly. The results are plotted in Figure 4.2.

It is significant that the spatial-correlation expression (4.11) contains no adjustable parameters. One can measure ρ, λ and R with great precision for each experiment. The theory is therefore subject to a stern test. The spatial-correlation curve in Figure 4.2 either agrees with plane-wave measurements – or it does not. We shall find that it does.

One might ask why the combination (4.12) plays a unique role in defining the spatial correlation. Since a power-law spectrum has no inherent length scale, the diffraction term in (4.10) must supply the needed metric. The only scale available there is the Fresnel length. The dimensionless ratio (4.12) emerges as a unique and inevitable combination, prompting Tatarskii [4] to observe the following: "*Thus all attempts to determine the mean size of inhomogeneities from the correlation radius*

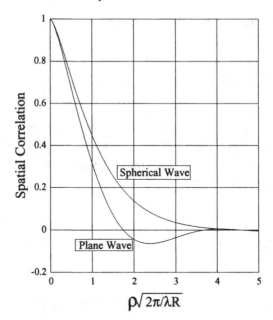

Figure 4.2: Spatial correlation of intensity fluctuations for plane and spherical waves predicted by the Kolmogorov model. These curves do not include aperture averaging or inner-scale effects.

of the intensity fluctuations are doomed to failure since these observations will give only the parameter $\sqrt{\lambda R}$."

We should recall that a very different result is generated by geometrical optics [5]:

$$\text{Ray theory:} \qquad \langle \chi(\mathbf{R})\chi(\mathbf{R}+\rho)\rangle = \frac{\pi^2}{3}R^3 \int_0^\infty d\kappa\, \kappa^5 \Phi_n(\kappa) J_0(\kappa\rho) \quad (4.14)$$

The only metrics available in this expression are the inner and outer scale lengths of the turbulence spectrum itself. One is too large and the other too small to explain the experimental results.

4.1.2 Spherical Waves

Microwave and optical transmitters are often quite small and the waves they radiate are essentially spherical for the path lengths employed in terrestrial experiments. We cannot use the translation that simplified the plane-wave calculation because spherical waves vary with radial distance from the line of sight. Instead, we must return to the basic description for weak scattering provided by (2.42).

The natural enlargement of the definition (2.37) represents the covariance in terms of the amplitude weighting functions evaluated at the centerline location and

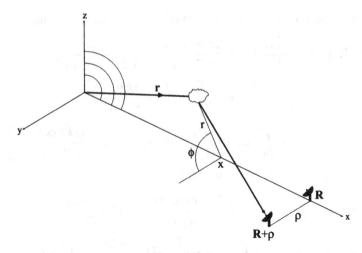

Figure 4.3: The geometry used to describe the spatial covariance of a spherical wave for receivers deployed normal to the line of sight.

the displaced receiver location:

$$\langle \chi(\mathbf{R})\chi(\mathbf{R}+\rho)\rangle = 4 \int d^3\kappa \ \Phi_n(\kappa)D(\kappa,\mathbf{R})D(-\kappa,\mathbf{R}+\rho) \qquad (4.15)$$

The amplitude weighting function for scattered waves that reach the on-axis receiver was established in (3.48) using the coordinates identified in Figure 4.3:

$$D(\kappa,\mathbf{R}) = \frac{k}{2}\int_0^R dx \exp(ix\kappa_x)\sin\left(\frac{\kappa_r^2 x(R-x)}{2kR}\right) \qquad (4.16)$$

We can therefore concentrate on the displaced receiver, whose location is

$$\mathbf{R}+\rho = R\mathbf{i}_x + \rho\mathbf{i}_y$$

Using the cylindrical coordinates defined in Figure 4.3 the complex weighting function becomes

$$\Lambda(\kappa,\mathbf{R}+\rho) = k^2 \int_0^R dx \int_0^\infty dr\, r \int_0^{2\pi} d\phi \frac{\sqrt{R^2+\rho^2}}{\sqrt{x^2+r^2}}$$

$$\times \frac{\exp\left(ik\sqrt{(R-x)^2+r^2+\rho^2-2r\rho\cos\phi}\right)}{4\pi\sqrt{(R-x)^2+r^2+\rho^2-2r\rho\cos\phi}}$$

$$\times \exp\left[ik\left(\sqrt{x^2+r^2}-\sqrt{R^2+\rho^2}\right)\right]$$

$$\times \exp\{i[x\kappa_x + r\kappa_r\cos(\phi-\omega)]\}$$

We expand the square roots using the paraxial approximation and find that

$$\Lambda(\kappa, \mathbf{R} + \rho) = \frac{k^2 R}{4\pi} \int_0^R dx \, \frac{\exp(ix\kappa_x)}{x(R-x)} \exp\left(\frac{ikx\rho^2}{2R(R-x)}\right)$$

$$\times \int_0^\infty dr \, r \exp\left[-r^2\left(\frac{-ikR}{2x(R-x)}\right)\right]$$

$$\times \int_0^{2\pi} d\phi \exp\left[i\left(r\kappa_r \cos(\phi - \omega) - \frac{ir k\rho \cos\phi}{R-x}\right)\right]$$

The azimuth integration gives a Bessel function:

$$\Lambda(\kappa, \mathbf{R} + \rho) = \frac{k^2 R}{2} \int_0^R dx \, \frac{\exp(ix\kappa_x)}{x(R-x)} \exp\left(\frac{ikx\rho^2}{2R(R-x)}\right)$$

$$\times \int_0^\infty dr \, r \exp\left[-r^2\left(\frac{-ikR}{2x(R-x)}\right)\right] J_0(nr)$$

where

$$n = \sqrt{\kappa_r^2 + \left(\frac{k\rho}{R-x}\right)^2 - 2\cos\omega \, \frac{\kappa_r k\rho}{R-x}}$$

The radial integral is to be found in Appendix D and gives

$$\Lambda(\kappa, \mathbf{R} + \rho) = \frac{ik}{2} \int_0^R dx \exp\left[i\left(x\kappa_x + \frac{\kappa_r x\rho}{R}\cos\omega\right)\right] \exp\left(-\frac{i\kappa_r^2 x(R-x)}{2kR}\right)$$

$$(4.17)$$

The amplitude weighting function for the displaced receiver is calculated from this result using (2.43):

$$D(\kappa, \mathbf{R} + \rho) = \frac{k}{2} \int_0^R dx \exp\left[i\left(x\kappa_x + \frac{\kappa_r x\rho}{R}\cos\omega\right)\right] \sin\left(\frac{\kappa_r^2 x(R-x)}{2kR}\right) \quad (4.18)$$

With this result we can estimate the spatial covariance from (4.15) using the expressions (4.16) and (4.18):

$$\langle \chi(\mathbf{R})\chi(\mathbf{R}+\rho) \rangle = k^2 \int d^3\kappa \, \Phi_n(\kappa) \int_0^R dx \int_0^R dx' \exp[ix\kappa_x(x-x')]$$

$$\times \sin\left(\frac{\kappa_r^2 x(R-x)}{2kR}\right)$$

$$\times \sin\left(\frac{\kappa_r^2 x'(R-x')}{2kR}\right)$$

$$\times \exp\left(i\frac{\kappa_r x\rho}{R}\cos\omega\right)$$

The double path integrations collapse to one integral and emphasize very small values of κ_x:

$$\langle \chi(\mathbf{R})\chi(\mathbf{R}+\rho)\rangle = 2\pi k^2 \int_0^R dx \int d^3\kappa \, \Phi_n(\kappa)\delta(\kappa_x)\exp\left(i\frac{\kappa_r x \rho}{R}\cos\omega\right)$$

$$\times \sin^2\left(\frac{\kappa_r^2 x(R-x)}{2kR}\right)$$

We assume that the refractive-index irregularities are isotropic because small eddies dominate the formation of amplitude fluctuations. It is then natural to express the result in terms of spherical wavenumber coordinates and rescale the downrange coordinate using $x = Ru$:

$$\langle \chi(\mathbf{R})\chi(\mathbf{R}+\rho)\rangle = 2\pi k^2 R \int_0^1 du \int_0^\infty d\kappa \, \kappa \Phi_n(\kappa)$$

$$\times \int_0^\pi d\psi \sin\psi \, \delta(\kappa \cos\psi) \sin^2\left(\frac{\kappa^2 \sin^2 \psi \, Ru(1-u)}{2k}\right)$$

$$\times \int_0^{2\pi} d\omega \exp(i\kappa \sin\psi \, u\rho \cos\omega)$$

The angular integrations are easily done to give a description that is valid for any model of the turbulence spectrum [6]:

$$\langle \chi(\mathbf{R})\chi(\mathbf{R}+\rho)\rangle = 4\pi^2 k^2 R \int_0^1 du \int_0^\infty d\kappa \, \kappa \Phi_n(\kappa) J_0(\kappa u\rho)$$

$$\times \sin^2\left(\frac{\kappa^2 Ru(1-u)}{2k}\right) \tag{4.19}$$

This expression was first evaluated with a Gaussian spectrum model for terrestrial links [7] and for propagation through the ionosphere [8].

As the importance of the Kolmogorov model became apparent, it was clear that (4.19) should be recalculated with the new spectrum.[1] The spatial correlation for this model is plotted in Figure 4.2, where it can be compared with the plane-wave result.

---

[1] Fried showed that one can express the spatial correlation in terms of Kummer and hypergeometric functions [6]:

$$C_\chi(\rho) = \Re\left[\frac{i^{\frac{5}{6}}}{\cos(5\pi/12)} M\left(\frac{11}{6}, 1; -\frac{ik\rho^2}{4R}\right) - 1.808\left(\frac{k\rho^2}{4R}\right)^{\frac{5}{6}}\right.$$

$$\left. + \frac{1.434i}{\cos(5\pi/12)}\left(\frac{k\rho^2}{4R}\right)^{\frac{11}{6}} {}_2F_2\left(\frac{11}{3}, 1, \frac{17}{6}, \frac{17}{6}; -\frac{ik\rho^2}{4R}\right)\right]$$

Using series approximations for them, he was able to define the spherical-wave correlation and was apparently the first to recognize the considerable difference between plane and spherical waves.

One sees there that the two results are substantially different. The plane-wave curve assumes negative values over a considerable range. By contrast, the spherical-wave correlation is positive almost everywhere and reaches only small negative values. Evidently one must identify the type of wave that is actually transmitted and use the appropriate curve from Figure 4.2 for comparisons with data.

4.1.3 Experimental Confirmations

Diffraction theory tells us that the spatial correlation should depend uniquely on the dimensionless ratio $\rho/\sqrt{\lambda R}$. This prediction was confirmed in a historic experiment performed in the USSR [9]. The output of an arc lamp was passed through a pin hole and then partially collimated by a 10-cm lens. The spatial correlation was measured at three distances; namely 500, 1000 and 1500 m. The inter-receiver separations were adjusted at each distance so that data was recorded for the same six values of inter-receiver spacing divided by the Fresnel length:

$$\frac{\rho}{\sqrt{\lambda R}} = 0.25,\ 0.50,\ 1.00,\ 2.00,\ 4.00,\ 8.00$$

The original data is reproduced in Figure 4.4 and demonstrates convincingly that the spatial correlation depends on the ratio of the inter-receiver separation and the Fresnel length. This data does not precisely match either the plane-wave or the spherical-wave curve in Figure 4.2. On the other hand, the incoherent optical signal had a divergence of 1/500 and the wave type was somewhere between plane and spherical [10].

The availability of laser light sources made a more precise test of the theory possible. The first experiment was done in the USSR using coherent light from a

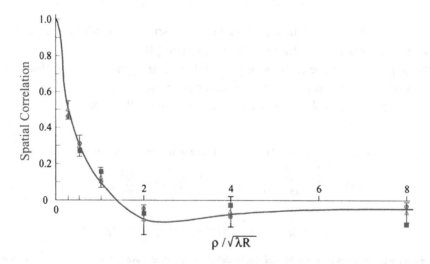

Figure 4.4: Spatial correlation of intensity fluctuations measured under weak-scattering conditions by Gurvich, Tatarskii and Tsvang using an arc-lamp source and a partially collimated beam [9].

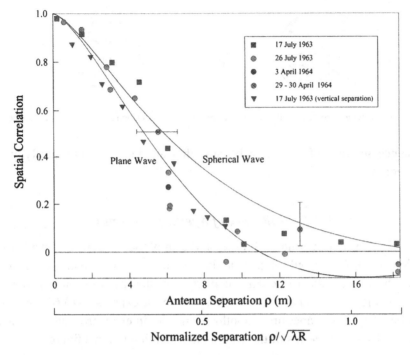

Figure 4.5: Spatial correlation of amplitude fluctuations measured at 35 GHz on a 28-km path by Lee and Waterman [12].

He–Ne laser [11]. The source output was collimated with a 50-cm parabolic reflector and transmitted over a 650-m path. The receiver used openings of 0.3 mm with separations ranging from 0.5 to 200 mm. The data for weak-scattering conditions gave remarkably close agreement with the plane-wave result of Figure 4.2. This experiment and others like it have confirmed the predictions of the Rytov approximation.

Similar experiments were performed using microwave transmissions on a 28-km path [12]. A 35-GHz signal was monitored by eight identical receivers connected to a main receiver. This array provided simultaneous measurements of phase and amplitude fluctuations from which the spatial correlation could be established for spacings in the range $0 < \rho < 24$ m. Amplitude correlations measured on five days are reproduced in Figure 4.5, where they are compared with the plane- and spherical-wave predictions. The transmitted signal was nearly spherical with a beam width of $0.4°$, yet the data is better fitted by the plane-wave curve. Notice that there is a good deal of scatter in the data and that the error bars are quite wide.

The data in Figure 4.5 includes measurements both for vertical and for horizontal deployments of the receivers. There is virtually no difference in the results. This confirms the conclusion drawn in Section 3.2.7, namely that anisotropy should

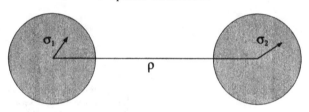

Figure 4.6: Coordinates used to describe the aperture-averaged spatial correlation for separated circular receivers.

not influence amplitude fluctuations because they are determined primarily by the inertial region.

4.1.4 The Influence of Aperture Averaging

Aperture size has an important influence on the amplitude variance, as we learned in Section 3.2.3. One would expect a similar effect for the spatial covariance. Consider the two identical circular receivers separated by a distance ρ that are illustrated in Figure 4.6. A natural extension of the basic expression (3.62) describes the combined effect of aperture smoothing and the inter-receiver separation for parabolic reflectors[2] in terms of the coordinates identified in this figure:

$$\overline{\langle \chi_1 \chi_2 \rangle} = \frac{1}{A^2} \iint_A d^2\sigma_1 \iint_A d^2\sigma_2 \, \langle \chi(\sigma_1)\chi(\rho + \sigma_2)\rangle \tag{4.20}$$

The general expression (4.6) describes the integrand for plane waves:

$$\overline{\langle \chi_1 \chi_2 \rangle} = \frac{4}{A^2} \iint_A d^2\sigma_1 \iint_A d^2\sigma_2 \int d^3\kappa \, \Phi_n(\kappa)$$
$$\times D(\kappa)D(-\kappa)\exp[i\kappa \cdot (\rho + \sigma_2 - \sigma_1)]$$

The aperture coordinates occur only in the exponential so that

$$\overline{\langle \chi_1 \chi_2 \rangle} = 4 \int d^3\kappa \, \Phi_n(\kappa)D(\kappa)D(-\kappa)\exp(i\kappa \cdot \rho)\,|Q(\kappa)|^2 \tag{4.21}$$

where

$$Q(\kappa) = \frac{1}{A} \iint_A d^2\sigma \, \exp(i\kappa \cdot \sigma)$$
$$= \frac{1}{\pi a_r^2} \int_0^{a_r} dr\, r \int_0^{2\pi} d\phi \exp(i\kappa r \cos\phi)$$
$$= \frac{2J_1(\kappa a_r)}{\kappa a_r} \tag{4.22}$$

---

[2] See Figure 4.10 in Volume 1 and the surrounding discussion.

The *aperture-averaged covariance for a plane wave* therefore becomes

$$\overline{\langle \chi_1 \chi_2 \rangle} = 4 \int d^3\kappa \; \Phi_n(\kappa) D(\kappa) D(-\kappa) \exp(i\kappa \cdot \rho) \left(\frac{2J_1(\kappa a_r)}{\kappa a_r} \right)^2 \tag{4.23}$$

For terrestrial links we can assume that isotropy and homogeneity pertain. We exploit the paraxial expression for $D(\kappa)D(-\kappa)$ given in (3.25). To establish the spatial correlation we must divide (4.23) by the variance (3.64) which also includes aperture smoothing. For the Kolmogorov model,

$$C_\chi(\rho, a_r) = \frac{1}{\mathcal{N}} \int_0^\infty \frac{du}{u^{\frac{11}{6}}} \left(1 - \frac{\sin u}{u} \right) J_0(q\sqrt{u}) \left(\frac{2J_1(\eta\sqrt{u})}{\eta\sqrt{u}} \right)^2 \tag{4.24}$$

where

$$q = \rho\sqrt{\frac{2\pi}{\lambda R}} \qquad \text{and} \qquad \eta = a_r\sqrt{\frac{2\pi}{\lambda R}} \tag{4.25}$$

and $\mathcal{N}$ is the normalization that corresponds to $q = 0$.

The expression (4.24) was evaluated numerically and is plotted in Figure 4.7 for several values of η. The curves are superficially similar because both the numerator and the denominator are modified in the same way by aperture smoothing. On closer inspection, it is apparent that significant differences in correlation occur in the range $0 < q < 2.5$. One is advised to include this correction in making careful comparisons with measurements.

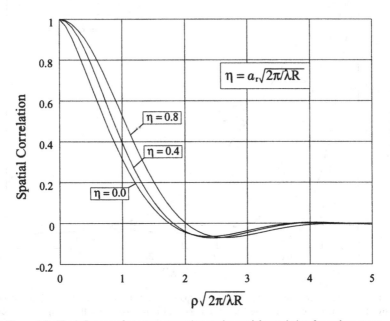

Figure 4.7: The influence of aperture averaging on the spatial correlation for a plane wave.

4.1.5 The Inner-scale Influence

We learned in Section 3.2.1 that the dissipation region of the spectrum plays an important role for the amplitude variance measured on short paths. It should also affect measurements of spatial covariance. We can describe this effect using the spectrum model (3.36) in the plane-wave covariance expression (4.8):

$$\langle \chi(\mathbf{R})\chi(\mathbf{R}+\rho)\rangle = 4\pi^2 0.033 C_n^2 k^2 \int_0^R dx \int_0^\infty \frac{d\kappa}{\kappa^{\frac{8}{3}}} \mathcal{F}(\kappa \ell_0) J_0(\kappa \rho)$$
$$\times \sin^2\left(\frac{x\kappa^2}{2k}\right) \qquad (4.26)$$

The substitution $u = R\kappa^2/k$ shows that the Fresnel length scales both the inter-receiver separation and the inner scale length. The normalized spatial correlation becomes

$$C_\chi(\rho, \ell_0) = \frac{1}{\mathcal{N}} \int_0^\infty \frac{du}{u^{\frac{11}{6}}} \mathcal{F}\left(\ell_0 \sqrt{\frac{u 2\pi}{R\lambda}}\right) J_0\left(\rho \sqrt{\frac{u 2\pi}{R\lambda}}\right)\left(1 - \frac{\sin u}{u}\right) \qquad (4.27)$$

where $\mathcal{N}$ is the same expression for zero separation. This result was computed numerically for the Hill bump model of the dissipation function [13] and the results are reproduced in Figure 4.8 for three values of $\ell_0/\sqrt{\lambda R}$. Notice that the

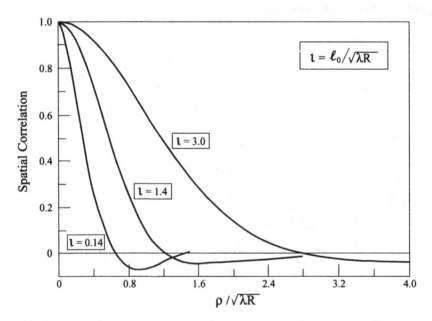

Figure 4.8: The influence of the inner scale length on the spatial correlation for a plane wave. These curves were calculated by Hill and Clifford for a plane wave using the Hill bump model for the dissipation region of the turbulence spectrum [13].

inner-scale effect is similar to aperture smoothing in that both corrections raise the spatial correlation above the reference curve of Figure 4.2. This makes sense because they both provide wavenumber cutoffs that suppress the influence of small eddies.

The spatial correlation is altered very substantially when the inner scale length is comparable to the Fresnel length. Similar curves were prepared for spherical waves and yield the same conclusion [14]. A good rule of thumb is that one can ignore dissipation effects when

$$\sqrt{\lambda R} > 7\ell_0. \qquad (4.28)$$

Since ℓ_0 is less than 1 cm near the surface, this distortion of Figure 4.8 should occur only for optical links over distances less than 50 m.[3] It should be completely absent for microwave links, which have much larger Fresnel lengths.

In view of this discussion, one might wonder why the laser experiment described following Figure 4.4 should have provided close agreement with a prediction that ignored the dissipation region. The weak-scattering data for that experiment was measured on the 650-m path corresponding to $\sqrt{\lambda R} \simeq 20$ mm. The condition (4.28) suggests that $\ell_0 < 3$ mm and this is quite reasonable for the nighttime measurements that were employed.

While the inner scale length is not important for most applications, it provides an important opportunity for atmospheric scientists. Frehlich showed that one can make accurate measurements of the inner scale length in this way [15]. He used a diverging laser signal on a 20-m path and an array of photodiodes with $0 < \rho < 3.7$ cm. That arrangement provided very accurate spatial correlations, which he compared with calculations of $C_\chi(\rho, \ell_0)$ for various values of ℓ_0. The inner scale lengths measured in this way were accurate to within 1 mm.

4.1.6 Beam Waves

The description of beam-wave fluctuations presented in Section 3.2.5 provides the basis for estimating their spatial covariance. Using the field-strength expression (3.80) for the unperturbed beam wave and the coordinates defined in Figure 3.14, Ishimaru established the following expression [16]:

$$\langle \chi(R, \rho_1)\chi(R, \rho_2) \rangle = 2\pi^2 k^2 \int_0^R dx \int_0^\infty d\kappa \, \kappa \, \Phi_n(\kappa, x) \mathcal{J}(x, \kappa) \qquad (4.29)$$

---

[3] See Section 2.3.4 of Volume 1 and notice especially Figure 2.19.

where

$$\mathcal{J}(x, \kappa) = \Re\left[J_0(\kappa P) \exp\left(-\frac{\kappa^2 \gamma_2 (R - x)}{k} \right) \right.$$
$$\left. - J_0(\kappa Q) \exp\left(-\frac{\kappa^2 i \gamma (R - x)}{k} \right) \right] \tag{4.30}$$

The receiver locations (y_1, z_1) and (y_2, z_2) enter this expression through the Bessel-function arguments in the following combinations:

$$P = \sqrt{(\gamma y_1 - \gamma^* y_2)^2 + (\gamma z_1 - \gamma^* z_2)^2} \tag{4.31}$$

$$Q = \gamma \sqrt{(y_1 - y_2)^2 + (z_1 - z_2)^2} \tag{4.32}$$

For the Kolmogorov model the wavenumber integration can be performed analytically in terms of Kummer functions [17]. The important step is to rearrange the terms as follows:

$$e^{-\kappa^2 a} J_0(\kappa c) - e^{-\kappa^2 (a+b)} J_0(\kappa d) = \left(1 - e^{-\kappa^2 (a+b)} \right) - \left(1 - e^{-\kappa^2 a} \right)$$
$$+ [1 - J_0(\kappa d)] e^{-\kappa^2 (a+b)}$$
$$- [1 - J_0(\kappa c)] e^{-\kappa^2 a}$$

so that the resulting integrals become familiar. The spatial correlation is thus

$$C_\chi(\rho_1, \rho_2) = \frac{\Re}{\mathcal{N}} \int_0^R dx\, C_n^2(x) \left[\left(\frac{i\gamma(R-x)}{k} \right)^{\frac{5}{6}} M\left(-\frac{5}{6}, 1; -\frac{kQ^2}{4i\gamma(R-x)} \right) \right.$$
$$\left. - \left(\frac{\gamma_2(R-x)}{k} \right)^{\frac{5}{6}} M\left(-\frac{5}{6}, 1; -\frac{kP^2}{4\gamma_2(R-x)} \right) \right] \tag{4.33}$$

where $\mathcal{N}$ is the same integral evaluated for $\rho_1 = \rho_2$. This result still requires an integration along the centerline coordinate x. Recall that the quantities γ and γ_2 depend on the downrange distance through the relations (3.88) and (3.89). This presents a formidable challenge and little progress has been made in reducing (4.33) to a form that can be compared with experiments.

4.2 Optical Astronomy

Starlight falls on the earth as a plane wave and is scattered in the lower atmosphere. Weak-scattering theory often provides a good description of astronomical observations for two reasons. Telescopes are usually placed on high mountains to reduce blurring of images. Observations are seldom made close to the horizon for the same reason.

The alternating positive and negative portions of the spatial correlation explain the phenomenon of *shadow bands*. If two points are separated by a Fresnel length, one point will be in a region that is brighter than the mean irradiance while the other will lie in a region that is darker. Ancient civilizations observed alternating light and dark bands on white walls during solar eclipses. The bands were separated by approximately 20 cm. They occur because the sun acts like a point source just before and just after total eclipse [18][19]. The sun's vanishing image is modulated by the plane-wave correlation function plotted in Figure 4.2 when this occurs.

Similar *speckle bands* are routinely observed in telescopes trained on bright stars [18]. The nonuniform patterns reproduced in Figure 4.9 can be observed with the naked eye if the eyepiece of a telescope is removed. They are another manifestation of scintillation phenomenon and can be explained by the theory of weak scattering we have been developing. The fine-scale features of these shadow patterns change in times as short as a few milliseconds. They can be recorded with photomultiplier tubes, photodiode arrays or fast film. Such techniques provide a simple way to measure the spatial correlation of light that has passed completely through the atmosphere.

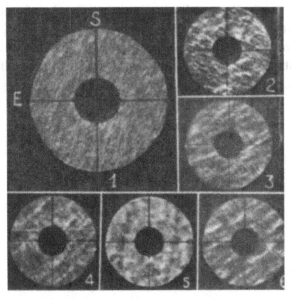

Figure 4.9: Shadow bands photographed by Gaviola at the focal plane of a 60-inch reflecting Cassegrain telescope [18].

4.2.1 Describing the Spatial Covariance

We encounter two complications in describing the covariance of intensity fluctuations. Starlight passes through the entire atmosphere and thus encounters a wide variety of turbulent conditions. This means that one must recognize the vertical profile of turbulent irregularities that was discussed in Section 2.3.3 of Volume 1. The second complication occurs because most stars are not overhead at the time of observation and hence one must describe oblique signal paths through the atmosphere. These features were easily included in the calculations of the amplitude variance. A more difficult problem occurs here because the bearing to the star and the inter-receiver separation vector both represent preferred directions. One can use (4.11) to describe the spatial correlation if the stellar line of sight is perpendicular to the line connecting the two receivers. The opposite extreme occurs when the baseline vector and the propagation vector lie in the same plane. The correlation depends on the foreshortened baseline $\rho \cos \vartheta$ in that case. Most observations are made with receiver orientations that lie between these two extremes.

To construct a general description we employ the geometry indicated in Figure 4.10. The observatory ground plane is defined by x and y, while z is the zenith axis. We choose the x direction so that the light arriving at the first receiver travels in the (x, z) plane. The two receivers are described in terms of their separation ρ and its orientation β relative to the source. To describe the downcoming waves, we erect a rectangular coordinate system (u_1, v_1, s_1) on the first bearing line and a similar system (u_2, v_2, s_2) on the second. We choose the u_1 and u_2 directions to be perpendicular to the propagation planes so that they coincide with the y axis. The logarithmic amplitude fluctuations of the signals received can be written

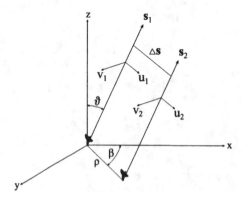

Figure 4.10: Coordinate systems employed to analyze the spatial covariance of a plane wave arriving from a stellar source.

in terms of these coordinates as

$$\chi_1 = -k^2 \int_0^\infty ds_1 \int_{-\infty}^\infty du_1 \int_{-\infty}^\infty dv_1 \, A(u_1, v_1, s_1; 0, 0) \, \delta\varepsilon(u_1, v_1, s_1)$$

$$\chi_2 = -k^2 \int_0^\infty ds_2 \int_{-\infty}^\infty du_2 \int_{-\infty}^\infty dv_2 \, A(u_2, v_2, s_2; \rho, \beta) \, \delta\varepsilon(u_2, v_2, s_2) \quad (4.34)$$

We would like to express amplitude fluctuations of the second system in terms of the coordinate system used to describe the first. We can do so by making the translation

$$\mathbf{s}_2 = \mathbf{s}_1 + \boldsymbol{\Delta}\mathbf{s}$$

where

$$\boldsymbol{\Delta}\mathbf{s} = \rho(\mathbf{i}_u \sin\beta + \mathbf{i}_v \cos\vartheta \cos\beta - \mathbf{i}_s \sin\vartheta \cos\beta) \quad (4.35)$$

With this shift of coordinates the kernels become the same,

$$A(u_2, v_2, s_2; \rho, \beta) = A(u_1, v_1, s_1; 0, 0)$$

so that

$$\langle \chi_1 \chi_2 \rangle = k^4 \int d^3 s_1 \int d^3 s_1' \, A(\mathbf{s}_1; 0) A(\mathbf{s}_1'; 0) \langle \delta\varepsilon(\mathbf{s}_1) \, \delta\varepsilon(\mathbf{s}_1' + \boldsymbol{\Delta}\mathbf{s}) \rangle$$

We use the wavenumber description (2.49) to write this in terms of the refractive-index fluctuations:

$$\langle \chi_1 \chi_2 \rangle = 4 \int d^3\kappa \, \Omega(\kappa) \exp(i\boldsymbol{\kappa} \cdot \boldsymbol{\Delta}\mathbf{s}) \left| k^2 \int d^3 s \, C_n(s) A(\mathbf{s}_1; 0) \exp(i\boldsymbol{\kappa} \cdot \mathbf{s}) \right|^2$$

The volume integral is just the amplitude weighting function for inclined paths introduced in (3.112):

$$\langle \chi_1 \chi_2 \rangle = 4 \int d^3\kappa \, \Omega(\kappa) \exp(i\boldsymbol{\kappa} \cdot \boldsymbol{\Delta}\mathbf{s}) \, D(\kappa) D(-\kappa) \quad (4.36)$$

This is identical to (4.6) in a formal sense but the functions $D(\kappa)$ are now quite different. They must reflect the inhomogeneity of the troposphere along the path. Fortunately, they were evaluated in (3.113) *en route* to calculating the amplitude variance for such paths and we can immediately write the spatial covariance as

$$\langle \chi_1 \chi_2 \rangle = 2\pi k^2 \int d^3\kappa \, \Omega(\kappa) \exp(i\boldsymbol{\kappa} \cdot \boldsymbol{\Delta}\mathbf{s}) \, \delta(\kappa_s) \int_0^\infty ds \, C_n^2(s) \sin^2\left(\frac{s\kappa_r^2}{2k}\right) \quad (4.37)$$

We assume that the tropospheric irregularities are isotropic because we believe that amplitude fluctuations are determined by small eddies. It is then natural to use the bearing line to the star from the first receiver as the polar axis for spherical wavenumber coordinates:

$$\langle \chi(\mathbf{R})\chi(\mathbf{R}+\rho)\rangle = 2\pi \int_0^\infty d\kappa\,\kappa^2\Omega(\kappa) \int_0^\pi d\psi\,\sin\psi \int_0^{2\pi} d\omega\exp(i\kappa\rho\cos\xi)$$
$$\times \int_0^\infty ds\,C_n^2(s)\sin^2\left(\frac{s\kappa^2\sin^2\psi}{2k}\right)\delta(\kappa\cos\psi)$$

where

$$\cos\xi = \sin\beta\sin\psi\sin\omega + \cos\beta\cos\vartheta\sin\psi\cos\omega - \cos\beta\sin\vartheta\cos\psi$$

The delta function collapses the polar integration and the azimuth integration can be expressed in terms of the zeroth Bessel function. When we change from distance along the oblique path to height above the observatory, the following expression emerges:

$$\langle \chi(\mathbf{R})\chi(\mathbf{R}+\rho)\rangle = 4\pi^2 k^2\sec\vartheta \int_0^\infty d\kappa\,\kappa\Omega(\kappa)J_0(\kappa\rho_e)$$
$$\times \int_0^\infty dz\,C_n^2(z)\sin^2\left(\frac{\kappa^2 z\sec\vartheta}{2k}\right) \qquad (4.38)$$

which accommodates any deployment of the receivers. The *effective baseline* is defined by the zenith angle and its orientation relative to the line-of-sight plane:

$$\rho_e = \rho\sqrt{1 - \cos^2\beta\sin^2\vartheta} \qquad (4.39)$$

There are two ways to proceed. One can select a profile model for C_n^2 and use it to calculate the spatial covariance. Alternatively, one can regard (4.38) as an integral equation whose solution determines the profile from spaced-receiver measurements. We shall consider the first approach in this section and deal with the second in the next section.

4.2.2 Atmospheric Modeling

There is a good deal of uncertainty and variability associated with the profile of C_n^2 that occurs in (4.38). On the other hand, one is reasonably confident that the Kolmogorov model describes the local behavior of the spectrum along the line of sight. The important question is that of which portions of the spectrum are needed. The Fresnel length for transmission through the atmosphere is defined by (3.116) and varies between 10 and 20 cm. That is considerably larger than the inner scale length of the spectrum and significantly less that the outer scale length:

$$\ell_0 \ll \sqrt{\lambda H\sec\vartheta} \ll L_0 \qquad (4.40)$$

We can therefore rely on the inertial portion of the spectrum. Most astronomical observations are made with telescopes whose openings are larger than the Fresnel length and one should include the aperture-averaging factor in those cases. On the other hand, the original verification of the prediction was made with small telescopes, for which aperture averaging may be ignored.

We can calculate the influence of different profile models with (4.38) and compare their predictions with astronomical observations. The simplest is the slab model illustrated in Figure 3.21, which assumes that the turbulent atmosphere is homogeneous up to a fixed height H and vanishes above that height. This model is equivalent to horizontal propagation if the path length R is identified with the effective height H. The results of Figure 4.2 can then be used to describe the spatial correlation, provided that the inter-receiver separation is replaced by $\rho_e \cos \vartheta$ to represent the oblique passage.

A second profile is suggested by meteorological observations, which indicate that strong turbulence occurs in a region near the bottom of the tropopause at an altitude of about 10 km. The elevated-layer model illustrated in Figure 3.21 describes that situation:

$$\langle \chi(\mathbf{R})\chi(\mathbf{R}+\rho) \rangle = 0.651 C_n^2 k^2 \, \Delta H \, \sec \vartheta \int_0^\infty \frac{d\kappa}{\kappa^{\frac{8}{3}}} J_0(\kappa\rho_e) \mathcal{H}(\kappa, H, \Delta H, \vartheta)$$

$$(4.41)$$

where the thick-layer weighting function $\mathcal{H}(\kappa, H, \Delta H, \vartheta)$ is defined by (3.126). The thin-phase-screen approximation for this function provides a reasonable description for scattering by turbulent layers in the troposphere and is equivalent to setting $\Delta H = 0$ in the weighting function:

$$\langle \chi(\mathbf{R})\chi(\mathbf{R}+\rho) \rangle = 0.651 C_n^2 k^2 \, \Delta H \, \sec \vartheta \int_0^\infty \frac{d\kappa}{\kappa^{\frac{8}{3}}} J_0(\kappa\rho_e)$$

$$\times \left[1 - \cos\left(\frac{\kappa^2 H \sec \vartheta}{k} \right) \right]$$

$$(4.42)$$

The substitution $x = \kappa^2 H \sec \vartheta / k$ presents the normalized spatial correlation in terms of a new function:

Thin screen: $$\mathbb{T}(q) = 0.578 \int_0^\infty \frac{dx}{x^{\frac{11}{6}}} (1 - \cos x) J_0(q\sqrt{x}) \qquad (4.43)$$

The independent variable is defined in terms of the effective baseline (4.39) and the various angles by

$$q = 2.51 \rho \sqrt{1 - \cos^2 \beta \sin^2 \vartheta} \, \sqrt{\frac{\cos \vartheta}{\lambda H}} \qquad (4.44)$$

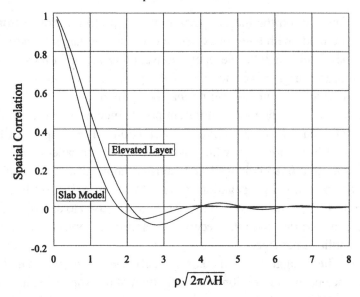

Figure 4.11: Predicted spatial correlations for a plane wave falling vertically on tropospheric irregularities described by a homogenous slab of depth H, and an elevated thin layer at height H above the receiver.

The function defined by (4.43) was first expressed in terms of a Kummer function [20]:

$$\mathbb{T}(q) = 3.864 \Re\left[\exp\left(i\pi\frac{5}{12}\right) M\left(-\frac{5}{6}, 1, \frac{iq^2}{4}\right)\right] - 4.108\left(\frac{q^2}{4}\right)^{\frac{5}{6}} \qquad (4.45)$$

but it is now a simple matter to compute it directly. The numerical results are plotted in Figure 4.11, together with the correlation for a slab model. There is little difference between the two curves even if the layer height is the same as the slab depth.

A family of tapered profile models was subsequently proposed in order to describe the variation of C_n^2 with height in the lower atmosphere [21][22]. It is a simple matter to generate the spatial correlation for virtually any profile model with modern computers. This allows one to experiment numerically with different profiles. The correlations generated in this way are characterized by a rapid drop to zero at approximately the Fresnel length. The negative portion extends for some distance and is followed by small oscillations at large separations. These predictions are similar to the models plotted in Figure 4.11.

4.2.3 Experimental Confirmations

Two early astronomical experiments confirmed the description given above. The first involved measuring the spatial correlation of intensity fluctuations with two telescopes trained on the same bright star [23]. The telescopes had apertures of

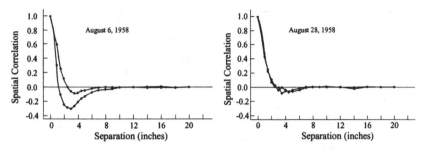

Figure 4.12: Spatial correlation of shadow bands measured by Barnhart, Keller and Mitchell using two telescopes sighted on Vega [23]. The separate curves for each date correspond to rotation of the telescope baseline by 90°.

2.5 inches and focal lengths of 8 inches. A Fabry lens in each telescope focused the image from the objective plane onto a photomultiplier tube. Signals from the photomultipliers were fed to an analog computer, which computed their cross correlation. The separation between the two telescopes was varied along one direction and the correlation measured for various separations. The apparatus was then rotated by 90° and the correlation measured along the orthogonal direction. Typical data from this experiment is reproduced in Figure 4.12. It exhibits both positive and negative values as required by (4.9). These results agree with the predicted curves in Figure 4.11 if the Fresnel length is 10 cm, which corresponds to a layer height of 11 km. We note that the correlations measured along orthogonal directions are sometimes quite different, reflecting the directional influence of the effective baseline.

In the second method for measuring the covariance a spatial filter was placed in front of a single telescope [24][25]. The first spatial filters were made by cutting two holes in an opaque mask, like that illustrated in Figure 4.13. A photomultiplier near the focus first measured the mean-square variation in intensity with both holes open. The measurement was repeated with first one hole closed – and then the other. The normalized difference in intensity

$$\Pi = \frac{\langle \delta I_{both}^2 \rangle - \langle \delta I_1^2 \rangle \; \langle \delta I_2^2 \rangle}{\langle \delta I_1^2 \rangle + \langle \delta I_2^2 \rangle} \tag{4.46}$$

is proportional to the spatial correlation evaluated at the separation ρ.

To establish this connection we note that scintillations at the focal plane are proportional to those which arrive at the objective plane. Since variations in intensity are linearly related to χ for weak scattering,

$$\Pi = \frac{\langle 4(\chi_1 + \chi_2)^2 \rangle - \langle 4\chi_1^2 \rangle - \langle 4\chi_2^2 \rangle}{\langle 4\chi_1^2 \rangle + \langle 4\chi_2^2 \rangle} = \frac{\langle \chi_1 \chi_2 \rangle}{\langle \chi^2 \rangle} = C_\chi(\rho) \tag{4.47}$$

It is thus possible to measure the spatial correlation with a single telescope using masks with different hole spacings. One must include aperture-averaging effects in

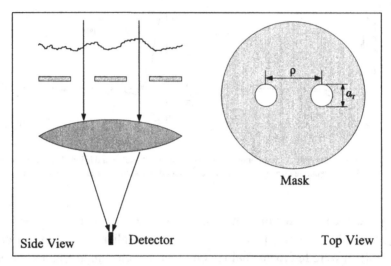

Figure 4.13: A schematic diagram of the spatial filter arrangement used by Protheroe to measure the correlation of intensity variations [25]. The output of a photodetector at the focus of the telescope is proportional to the spatial correlation function of the shadow pattern for the separation ρ. Masks with different separations were used to measure the complete correlation function.

doing so, since the openings are usually comparable to the separation in order to pass sufficient light. Protheroe used this technique to study atmospheric conditions with a 12.5-inch refracting telescope [25]. The mask holes were one inch in diameter and separated by distances ranging from one to eleven inches. He obtained results similar to the curves shown in Figure 4.12.

Microwave signals from earth-orbiting satellites provide an independent check on this description. Communication satellites in synchronous orbit are ideal for this purpose because they appear stationary to ground stations. The waves they transmit become plane before reaching the atmosphere. These satellites operate between 4 and 30 GHz so their signals are influenced primarily by the troposphere. The curves in Figure 4.11 should also describe the spatial correlation of these microwave signals. An experiment was conducted at 28 GHz using two receivers separated by 7.3 m [26]. The antennas were fixed to a large tracking mount that was boresighted on the source so the effective separation ρ_e and the actual spacing were identical. The satellite elevation was 40° and the dimensionless parameter defined by (4.44) became

$$q = 2.51\rho\sqrt{\frac{\cos\vartheta}{\lambda H}} = \frac{3.8}{\sqrt{H_{km}}}$$

From meteorological data the height of the scattering layer was estimated to be 10 km, suggesting that $q = 1.2$. Figure 4.11 indicates that the correlation should be 0.2 for an elevated layer at this height. The measured values ranged from 0.20 to

0.25. We judge that (4.43) and (4.44) provide a reliable system for estimating the spatial correlation of satellite signals above 4 GHz.

4.3 Inversion of Shadow Patterns

Now let us turn to the second approach and regard (4.38) as an integral equation connecting the unknown profile of C_n^2 with intensity measurements made at spaced receivers. If one reverses the order of integration in that relationship, one has

$$\langle \chi(\mathbf{R})\chi(\mathbf{R}+\rho)\rangle = 0.651k^2 \sec \vartheta \int_0^\infty dz\, C_n^2(z) \int_0^\infty \frac{d\kappa}{\kappa^{\frac{8}{3}}} J_0(\kappa\rho_e)$$

$$\times \left(\frac{2J_1(\kappa a_r)}{\kappa a_r}\right)^2 \left[1 - \cos\left(\frac{\kappa^2 z \sec \vartheta}{k}\right)\right] \qquad (4.48)$$

With the substitution $x = \kappa^2 z \sec \vartheta / k$ the κ integration can be performed:

$$\langle \chi(\mathbf{R})\chi(\mathbf{R}+\rho)\rangle = 0.563k^{\frac{7}{6}} \sec^{\frac{11}{6}} \vartheta \int_0^\infty dz\, C_n^2(z) L(\rho_e, z) \qquad (4.49)$$

If one ignores aperture averaging, the kernel can be expressed in terms of the phase-screen spatial correlation function defined by (4.43):

$$L(\rho_e, z) = z^{\frac{5}{6}}\, \mathbb{T}(\rho_e \cos \vartheta \sqrt{k/z}) \qquad (4.50)$$

This weighting function indicates how irregularities at each level contribute to the spatial correlation for a given receiver configuration. It is plotted as a three-dimensional surface in Figure 4.14 and it can be seen that the profile at each level is determined by a relatively small range of inter-receiver separations. This gives advance warning of mathematical problems that will be encountered

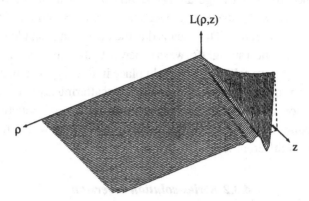

Figure 4.14: A three-dimensional plot of the single-star weighting function generated by Vernin [27]. This surface describes how irregularities at height z contribute to the spatial correlation for a separation ρ. It corresponds to the expression (4.50) for $L(\rho_e, z)$.

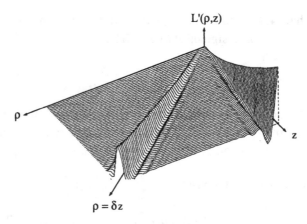

Figure 4.15: A three-dimensional plot of the weighting function for double stars, indicating how irregularities at height z contribute to the spatial correlation for receivers separated by ρ. The angular separation of the double stars is denoted by δ. This surface was generated by Vernin [27].

in trying to reconstruct the profile from the shadow patterns cast by a single star.

4.3.1 Inversion with Double Stars

By contrast, one can use double stars to establish the profile of C_n^2 with considerable precision. The concept involves triangulating on regions of different heights using light received from the two nearby sources [28][29]. If the light is scattered by an elevated layer, two speckle patterns are produced in the telescope. The patterns are displaced from one another by a distance proportional to the height of the scattering layer. The weighting function which describes the contribution of each height level to the spatial correlation is plotted in Figure 4.15 as a three-dimensional surface [27]. The prominent ridge along the line $\rho = z\delta$ indicates that there is a strong connection between spaced-receiver measurements and the C_n^2 profile in a relatively narrow height range. This means that the inversion of shadow patterns cast by double stars should be a straightforward process. Unfortunately, there are too few pairs of stars and they are seldom at the right place in the sky when they are needed.

Multiple patterns occur if there is more than one scattering layer. For a continuous vertical distribution of scatterers one must solve an integral equation similar to (4.48). The angular separation provides a strong lever with which to identify the contributions from different levels.

4.3.2 Series-solution Inversion

This brings one back to the single-star technique and the need for a method to determine the profile from shadow-pattern measurements. The traditional way to do so is to assume that the profile can be represented as a linear combination of known functions:

$$C_n^2(z) = A_1 f_1(z) + A_2 f_2(z) + \cdots A_N f_N(z) \qquad (4.51)$$

The functions $f_n(z)$ can be any set of basis functions.[4] The prospects for a successful inversion are improved substantially if they are each chosen to be plausible models of the turbulence profile. A linear combination of known terms emerges to represent the covariance when this series is substituted into (4.48) and the integrations are performed:

$$\langle \chi(\mathbf{R})\chi(\mathbf{R}+\rho) \rangle = A_1 F_1(\rho) + A_2 F_2(\rho) + \cdots A_N F_N(\rho) \qquad (4.52)$$

The calculated functions of separation are established from

$$F_n(\rho) = 0.651 k^2 \sec \vartheta \int_0^\infty dz\, f_n(z) \int_0^\infty \frac{d\kappa}{\kappa^{\frac{8}{3}}} J_0(\kappa \rho_e)$$
$$\times \left(\frac{2 J_1(\kappa a_r)}{\kappa a_r} \right)^2 \left[1 - \cos\left(\frac{\kappa^2 z \sec \vartheta}{k} \right) \right] \qquad (4.53)$$

If one measures the spatial correlation for various inter-receiver separations, this approach generates a series of algebraic equations that can be summarized by the following matrix equation:

$$[\langle \chi(\mathbf{R})\chi(\mathbf{R}+\rho_m) \rangle] = [F_n(\rho_m)][A_n] \qquad (4.54)$$

One must invert this matrix equation to obtain the coefficients A_n. The number of measurements needed in order to do so may exceed the number of trial functions in the original series. If the difference between the largest and smallest eigenvalue of the matrix is large, this is an *ill-posed problem* and it is then difficult or impossible to find a solution. The mathematical problems are made easier if (a) good choices are made for the trial functions, (b) the inter-receiver spacings are numerous and wisely chosen, and (c) the correlation measurements are relatively free of noise.

---

[4] These basis functions are often *assumed* to be orthonormal so that

$$\int_0^\infty dz\, f_n(z) f_m(z) = \delta_{nm}$$

With this relationship the coefficients can be estimated from a typical profile or one measured by other means as follows:

$$A_n = \int_0^\infty dz\, f_n(z) C_n^2(z)$$

However, this assumption is not required for the inversion of shadow patterns.

4.3.3 The Exact Solution for Point Receivers

Peskoff showed that the basic integral equation for shadow patterns can be solved exactly for very small receivers [30][31]. His solution relates the profile of C_n^2 to a weighted average of amplitude measurements made with different inter-receiver spacings. It is convenient to assume that we have an overhead source in explaining this solution. The first step is to take the Fourier–Bessel transformation [32] of both sides of (4.48) with the aperture radius set equal to zero:

$$
\int_0^\infty d\rho\, \rho J_0(\upsilon\rho)\langle\chi(\mathbf{R})\chi(\mathbf{R}+\rho)\rangle = 0.651 k^2 \int_0^\infty \frac{d\kappa}{\kappa^{\frac{8}{3}}} \int_0^\infty dz\, C_n^2(z)
$$

$$
\times \left[1 - \cos\left(\frac{\kappa^2 z}{k}\right)\right] \int_0^\infty d\rho\, \rho J_0(\kappa\rho) J_0(\upsilon\rho)
$$

Since the Bessel functions are orthogonal,

$$
\int_0^\infty d\rho\, \rho J_0(\kappa\rho) J_0(\upsilon\rho) = \frac{\delta(\kappa - \upsilon)}{\kappa}
$$

and the wavenumber integration is collapsed:

$$
\upsilon^{\frac{11}{3}} \int_0^\infty d\rho\, \rho J_0(\upsilon\rho)\langle\chi(\mathbf{R})\chi(\mathbf{R}+\rho)\rangle = 0.651 k^2 \int_0^\infty dz\, C_n^2(z)\left[1 - \cos\left(\frac{\upsilon^2 z}{k}\right)\right]
$$

We multiply both sides by $\upsilon^{-1}\sin(\upsilon^2 y/k)$ and integrate over all values of υ:

$$
0.651 k^2 \int_0^\infty dz\, C_n^2(z) \int_0^\infty \frac{d\upsilon}{\upsilon} \sin\left(\frac{\upsilon^2 y}{k}\right)\left[1 - \cos\left(\frac{\upsilon^2 z}{k}\right)\right]
$$

$$
= \int_0^\infty d\upsilon\, \upsilon^{\frac{8}{3}} \sin\left(\frac{\upsilon^2 y}{k}\right) \int_0^\infty d\rho\, \rho J_0(\rho\upsilon)\langle\chi(\mathbf{R})\chi(\mathbf{R}+\rho)\rangle
$$

The left-hand side can be evaluated by setting $\zeta = \upsilon^2$ and using the discontinuous integral

$$
\frac{2}{\pi}\int_0^\infty \frac{d\zeta}{\zeta} \sin(a\zeta) = \begin{cases} +1, & a > 0 \\ -1, & a < 0 \end{cases}
$$

An exact solution emerges when one reverses the order of integration on the right-hand side:

$$
\int_z^\infty dy\, C_n^2(y) = \frac{0.920}{k^{\frac{1}{6}} z^{\frac{11}{6}}} \int_0^\infty d\rho\, \rho\langle\chi(\mathbf{R})\chi(\mathbf{R}+\rho)\rangle K(\rho, z) \qquad (4.55)
$$

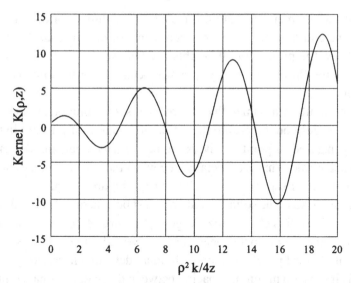

Figure 4.16: A plot of the kernel defined by (4.56), which weights measurements of the spatial corrleation to yield the profile of C_n^2 in the Peskoff solution.

The kernel is defined by

$$K(\rho, z) = \lim_{\epsilon \to 0} \left[\int_0^\infty dv\, v^{\frac{8}{3}} J_0(v\rho) \sin\left(\frac{v^2 z}{k}\right) \exp(-\epsilon v^2) \right]$$

and can be expressed in terms of a Kummer function:

$$K(\rho, z) = \Im\left[\exp\left(i\frac{\pi}{12}\right) M\left(\frac{11}{6}, 1, \frac{i\rho^2 k}{4z}\right) \right] \tag{4.56}$$

This kernel weights the contribution of each spaced-receiver measurement in a way that depends on the altitude. It is plotted in Figure 4.16 and is characterized by a series of oscillations whose amplitudes increase almost linearly with the argument. That behavior is also apparent from the asymptotic expression for Kummer functions found in Appendix G:

$$\lim_{\rho^2 k/(4z) \to \infty} [K(\rho, z)] = 1.063 \left(\frac{\rho^2 k}{4z}\right)^{\frac{5}{6}} \cos\left(\frac{\rho^2 k}{4z}\right)$$

The Peskoff solution is characterized by a delicate balance between the rising kernel and the declining correlation measurements. A rising weighting function means that measurements made with large separations have an unusually strong influence on profile determinations. On the other hand, the oscillations tend to average out the contributions from large separations – just as the sinusoidal terms do in Fourier analysis.

It is important to remember that (4.55) is an exact solution for the basic integral equation. If one could make precise measurements of the correlation for all separations and knew the profile for all heights, we believe that (4.55) would connect the measured data. Problems occur in practice because one seldom has noise-free measurements.

The Peskoff solution assumes that the receivers are omnidirectional. A linear array of photodiodes could measure the diffraction pattern of starlight in this way. The signals generated in point receivers are necessarily weak and detector noise plays a significant role. This is especially true when the correlation is small, as it is for large inter-receiver separations. We know from Figure 4.16 that large spacings play an important role in the solution and one must be concerned about the fact that low signal-to-noise ratios characterize large separations. A second problem is that the array has a finite length so the measurements needed for very large spacings are entirely absent. A third problem is that individual detectors necessarily have finite areas and this imposes a minimum spacing between detectors. A final complication is that one must differentiate the solution (4.55) to recover the profile and that further exaggerates the influence of noise sources. Practical limitations thus limit the utility of Peskoff's ingenious solution.

A sensitivity analysis of the analytical solution that relates variations in the profile to errors in the measurement of the spatial correlation is needed. Strohbehn examined this question by calculating the spatial correlation for a family of profile models [33]. He found that the results were surprisingly sensitive to small changes in the profile. They were even more sensitive to departures from the Kolmogorov model.[5] This unwelcome conclusion is supported by a careful examination of the integral equation [34]. From a mathematical perspective (4.48) presents an ill-posed problem and one should not be surprised that it is difficult to solve it.

4.3.4 Inversion Using Spatial Filters

This problem is partly due to the computational strategy we have adopted. Recall that we first used shadow-pattern measurements to synthesize the spatial correlation. We then used that summation to initiate the solution of a sensitive integral equation. In essence, we first condensed and then expanded the data using the spatial correlation as a "switching center."

There is a way to bypass the spatial correlation function when one is inverting shadow patterns. The concept is to work directly with spatial components of the

---

[5] In this connection it is signficiant that Peskoff's solution is valid for a general power law spectrum

$$\Phi(\kappa) = \text{constant} \times \kappa^{-\nu}$$

subject to certain restrictions on the index ν that are described in [31].

arriving wave-front. These are generated if one performs a Fourier analysis of them. The relative strength of these components defines a *spatial spectrum* for the shadow pattern $S(\mu)$ – which must not be confused with the turbulence spectrum $\Phi_n(\kappa)$ of atmospheric irregularities. The spatial spectrum and amplitude covariance provide equivalent descriptions of the wave-front if all the data is present. They are formally linked to one another by cosine transforms:

$$S(\mu) = 2 \int_0^\infty d\rho \cos(\rho\mu) \langle \chi(\mathbf{R})\chi(\mathbf{R}+\rho)\rangle \qquad (4.57)$$

$$\langle \chi(\mathbf{R})\chi(\mathbf{R}+\rho)\rangle = \frac{1}{\pi} \int_0^\infty d\mu \cos(\mu\rho) S(\mu) \qquad (4.58)$$

There is a substantial advantage to working with the spatial spectrum $S(\mu)$. It can be measured directly by passing the incoming light through a spatial filter placed at the focus or entry plane of a telescope. The mask with two holes illustrated in Figure 4.13 is such a spatial filter and was the first configuration employed for this purpose [25]. It was soon realized that more light could be gathered if the spatial filter were constructed with alternating opaque and transparent stripes. This concept was first exploited using a diffraction grating placed at the coudé focus of the 152-cm reflector at the Haut Provence Observatory in France [28]. Light reflected in opposite directions by the grating was sensed by two photomultipliers, whose sum and difference signals were then used to measure the spatial spectrum for the selected wavelength. Different wavelengths were investigated by changing the angle of reflection. By measuring also the power spectrum of the scintillations Roddier and Vernin were able to establish rough profiles for the wind velocity above this observatory.

A less expensive experimental arrangement is provided by the spatial filter illustrated in Figure 4.17. The mask is constructed by laying opaque stripes on an otherwise transparent diaphragm and placed at the entry plane of a telescope. This filter passes only those spatial components which are about the same size as the stripe spacing. A detector placed at the focus of the telescope should be especially sensitive to the relative strength of wavenumber components near $\mu = 2\pi/d$. One can establish $S(\mu)$ by repeating the experiment with different stripe spacings. This possibility provides the basis for the *optical scintillometer* whose profile measurements were described in Section 2.3.3 of Volume 1. We need to describe how such a system operates in more detail.

Let us begin with its theory of operation. One cannot ignore aperture averaging since the spatial filter blocks a significant fraction of the arriving light. As a practical matter, this means that one must use numerical techniques to establish the profile from spatially filtered telescope observations. The aperture-averaged result for the

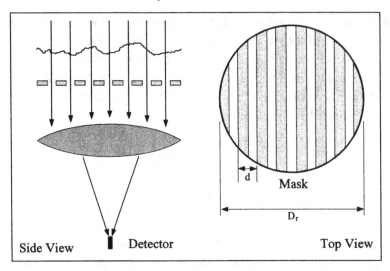

Figure 4.17: A spatial filter created by a series of striped masks placed at the entry plane of a telescope operating as an optical scintillometer.

spatial filter shown in Figure 4.17 is directly related to the spatial correlation for the unfiltered wave-front [35]:

$$
\begin{aligned}
S(\mu) &= \frac{1}{D_r^2} \int_{-\frac{D_r}{2}}^{\frac{D_r}{2}} dx \int_{-\frac{D_r}{2}}^{\frac{D_r}{2}} dx' \cos(\mu x) \cos(\mu x') \, C_\chi(|x - x'|) \\
&= \frac{1}{D_r} \int_0^{D_r} d\xi \, C_\chi(\xi) \left[\frac{\sin[\mu(D_r - \xi)]}{\mu D_r} - \left(1 - \frac{\xi}{D_r}\right) \cos(\mu \xi) \right] \quad (4.59)
\end{aligned}
$$

On substituting the thin-phase-screen approximation for the spatial correlation from (4.43) and rearranging terms, the result can be expressed as a weighted integral of the profile:

$$
S(\mu) = 0.264\pi k^2 \int_0^\infty dz \, C_n^2(z) N(z, \mu) \quad (4.60)
$$

where the kernel

$$
\begin{aligned}
N(z, \mu) &= \int_0^\infty dw \left(\frac{\sin\left(\frac{D_r}{2}(\mu - w)\right)}{\frac{D_r}{2}(\mu - w)} + \frac{\sin\left(\frac{D_r}{2}(\mu + w)\right)}{\frac{D_r}{2}(\mu + w)} \right)^2 \\
&\quad \times \int_w^\infty \frac{d\kappa}{\kappa^{\frac{8}{3}} \sqrt{\kappa^2 - w^2}} \sin^2\left(\frac{\kappa^2 z}{2k}\right) \quad (4.61)
\end{aligned}
$$

indicates how each height region influences the wave-front spectrum for a given spacing $\mu = 2\pi/d$.

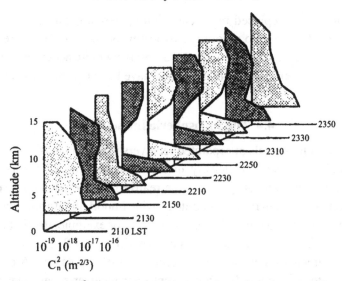

Figure 4.18: Vertical profiles of C_n^2 measured during a single evening by Ochs, Lawrence, Wang and Zieske [36] that indicate that there is significant diurnal variability.

The function (4.61) was evaluated numerically in [35]. It is characterized by seven broad humps centered at altitudes ranging from 2.25 km to more than 14.5 km. These weighting functions therefore select values of C_n^2 that are *representative* of a broad height range. They cannot yield precise profiles, as radar back scattering and thermosonde balloon flights do. The virtue of this technique is that it provides a general description of the profile and can be updated frequently at modest cost.

Now let us turn to the implementation of this concept. The first experiment to measure the vertical profile in this way used a striped spatial filter between the entry plane and the focus of a 35.6-inch lens [35][36]. It was realized that the spatial filter need not be changed if one moves the filter along the optical axis, thereby changing the scale of the shadow pattern projected onto the filter. The stripes were aluminized so that one photomultiplier tube can collect the reflected light while a second receives light passing between the stripes. The difference of the two outputs is sensitive to the spatial spectrum $S(\mu)$. The profiles reproduced in Figure 4.18 were constructed in this way using measurements taken during 20-min observation periods. They indicate that significantly different profiles are experienced on separate days and even within a 2-h period. They emphasize the importance of measuring the profile rather than assuming it when accuracy and confidence are required. Similar profile determinations were made in Hawaii and compared favorably with airborne measurements of C_n^2 near the site [36].

The spatial-filtering technique has been developed and refined in France [37][38]. Astronomers there have replaced the transmission masks by diffraction gratings and experimented with various arrangements of detectors. In the most recent version, the spatial filters were replaced by a television camera combined with an image intensifier. This allows one to record the two-dimensional shadow pattern in real time. These developments have provided progressively more accurate and versatile ways to measure the vertical profile of C_n^2 and have been reviewed by Vernin [27].

4.4 The Ionospheric Influence

Comparisons of radio signals at separated receivers played a decisive role in identifying fluctuations in electron density in the ionosphere as a primary cause of radio-astronomy scintillations.[6] It was soon realized that detailed comparisons of signal variations at adjacent receivers could provide important information about the ionosphere that could be obtained in no other way. For example, the first indication of significant anisotropy emerged from such experiments [39].

Early measurements relied on radio sources in distant galaxies. A comparison of amplitude histories recorded with different inter-receiver separations using the source in Cassiopeia is reproduced in Figure 4.19. This indicates that there is good correlation between *radiowave shadow patterns* if the equivalent baseline is 4 km but little resemblance if the separation is twice as great. Artificial satellites of the earth presented important new opportunities to investigate scintillations with controlled transmissions after 1957. Signals transmitted by earth-orbiting satellites are not influenced by the solar wind and interstellar plasma and thus provide a way to focus attention on ionospheric irregularities.

4.4.1 The Covariance Expression

The first task is to calculate the spatial covariance of amplitude fluctuations. In doing so we build on the analysis of optical transmissions developed in Section 4.2.1 and on the techniques used in Section 3.4.1 to estimate $\langle \chi^2 \rangle$ for transionospheric propagation. We use the weak-scattering expression (3.142) for the amplitude of a plane wave that has passed through an ionized random medium. The covariance for signals at separated receivers is described by (4.36) with the tropospheric refractivity spectrum replaced by the scaled spectrum of electron-density fluctuations:

$$\langle \chi(\mathbf{R})\chi(\mathbf{R} + \rho) \rangle = (r_e\lambda^2)^2 \int d^3\kappa \, \Psi_N(\kappa) \exp(i\kappa \cdot \Delta\mathbf{s}) \, D(\kappa)D(-\kappa) \qquad (4.62)$$

Here $\Delta\mathbf{s}$ is the vector separation (4.35) between the line-of-sight paths that lead to the different receivers. We noted earlier that the small-scattering-angle approximation provides a good description both for radio-astronomy and for

---

[6] As discussed in Section 2.4.2 of Volume 1.

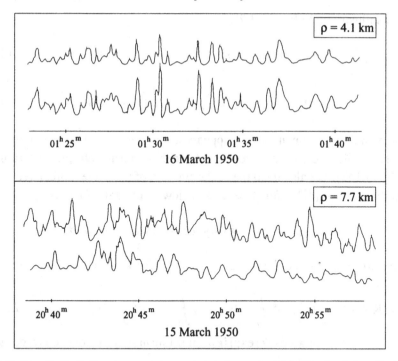

Figure 4.19: Intensity fluctuations measured at separated receivers using 81.5-MHz signals from Cassiopeia by Little and Maxwell [40]. The first pair of records was obtained with an effective baseline of 4.1 km and exhibits a correlation of approximately 90%. For the second pair a separation of 7.7 km was used and no meaningful correlation was found.

satellite signals transiting the ionosphere. The covariance is thus given by (4.37) with the same replacements as those noted above:

$$\langle \chi(\mathbf{R}) \chi(\mathbf{R} + \rho) \rangle = 2\pi (\pi r_e \lambda)^2 \int d^3\kappa \, \Psi_N(\kappa) \exp(i\kappa \cdot \Delta s)$$

$$\times \, \delta(\kappa_s) \int_0^\infty ds \, \wp^2(s) \sin^2\left(\frac{s\kappa_r^2}{2k}\right) \qquad (4.63)$$

We have omitted the aperture-averaging term because the Fresnel length for ionospheric heights and VHF wavelengths is larger than most receivers. We are confident that the power-law portion of the spectrum $\Psi_N(\kappa)$ is responsible for amplitude scintillations.

Anisotropy of ionospheric electron-density irregularities is the complicating element in this analysis. The first task is to choose a coordinate system in which one can evaluate the integrations in (4.63) for all orientations of the propagation direction s, magnetic field **H** and receiver baseline ρ. In this situation it is convenient to refer both the magnetic field and the wavenumber vector to the propagation direction as illustrated in Figure 3.29:

$$\boldsymbol{\kappa} = \kappa(\mathbf{i}_s \cos\psi + \mathbf{i}_v \sin\psi \cos\omega + \mathbf{i}_u \sin\psi \sin\omega) \tag{4.64}$$

$$\mathbf{H} = H(\mathbf{i}_s \cos\gamma + \mathbf{i}_v \sin\gamma \cos\alpha + \mathbf{i}_u \sin\gamma \sin\alpha) \tag{4.65}$$

The wavenumber component which appears in the delta function is simply $\kappa_s = \kappa \cos\psi$. The angle Θ between these vectors occurs in the anisotropic spectrum and is given by (3.148). We also need the scalar product of the wavenumber vector and the separation vector (4.35) for $\psi = \pi/2$ in view of the delta function:

$$\boldsymbol{\kappa} \cdot \boldsymbol{\Delta s}|_{\psi=\pi/2} = \kappa\rho(\sin\omega \sin\beta + \cos\omega \cos\beta \cos\vartheta)$$

With these conventions and the power-law model (3.149) the covariance becomes

$$\langle\chi(\mathbf{R})\chi(\mathbf{R}+\rho)\rangle = (\pi r_e\lambda)^2 \int_0^\infty ds\,\wp^2(s) \int_0^\infty d\kappa\,\kappa^2 \int_0^\pi d\psi\,\sin\psi \int_0^{2\pi} d\omega$$

$$\delta(\kappa\cos\psi)\exp[i\kappa\rho\sin\psi\,(\sin\omega\sin\beta + \cos\omega\cos\beta\cos\vartheta)]$$

$$\times \sin^2\left(\frac{s\kappa^2\sin^2\psi}{2k}\right) \frac{Q_\nu\langle\delta N^2\rangle\kappa_0^{\nu-3}}{\kappa^\nu[1+(A^2-1)\cos^2\Theta]^{\frac{\nu}{2}}}$$

A lower limit on the wavenumber spectrum has been omitted in accordance with the discussion of ionospheric conditions that followed (3.152). The delta function drives ψ to $\pi/2$ so that $\cos\Theta = \sin\gamma\,\cos(\omega-\alpha)$ and we obtain:

$$\langle\chi(\mathbf{R})\chi(\mathbf{R}+\rho)\rangle = (\pi r_e\lambda)^2 \sec\vartheta \int_0^\infty dz\,\langle\delta N^2(z)\rangle Q_\nu\kappa_0^{\nu-3}$$

$$\times \int_0^\infty \frac{d\kappa\,\kappa}{\kappa^\nu}\sin^2\left(\frac{\kappa^2 z\sec\vartheta}{2k}\right)$$

$$\times \int_0^{2\pi} d\omega\,\frac{\exp[i\kappa\rho(\sin\omega\sin\beta + \cos\omega\cos\beta\cos\vartheta)]}{[1+(A^2-1)\sin^2\gamma\cos^2(\omega-\alpha)]^{\frac{\nu}{2}}}$$

$$\tag{4.66}$$

Further progress depends on identifying the spectral index ν and defining the vertical distribution of irregularities.

The spectral index was estimated in Section 3.4 using the wavelength and zenith-angle scaling of $\langle\chi^2\rangle$. A value near $\nu=4$ characterizes the signal fluctuations created by ionospheric irregularities and is consistent with *in situ* measurements. We assume that the irregularities are contained in a thin layer at height H and change from distance along the path to height above the surface:

$$\langle\chi(\mathbf{R})\chi(\mathbf{R}+\rho)\rangle = \tfrac{1}{2}(\pi r_e \lambda)^2 \, \Delta H \sec\vartheta \, \langle\delta N^2\rangle\kappa_0 Q_4$$

$$\times \int_0^\infty \frac{d\kappa}{\kappa^3}\left[1 - \cos\left(\frac{\kappa^2 H \sec\vartheta}{k}\right)\right]$$

$$\times \int_0^{2\pi} d\omega \, \frac{\exp[i\kappa\rho(\sin\omega\sin\beta + \cos\omega\cos\beta\cos\vartheta)]}{[1 + (A^2 - 1)\sin^2\gamma\cos^2(\omega - \alpha)]^2}$$

$$(4.67)$$

This can be simplified if we rescale the inter-receiver separation as follows:

$$q = \rho\sqrt{\frac{2\pi\cos\vartheta}{\lambda H}}$$

and change the wavenumber-integration variable to $x = \kappa^2 H \sec\vartheta/k$. The covariance can now be written

$$\langle\chi(\mathbf{R})\chi(\mathbf{R}+\rho)\rangle = \mathcal{B}\int_0^\infty \frac{dx}{x^2}(1 - \cos x)$$

$$\times \int_0^{2\pi} d\omega \, \frac{\exp[iq\sqrt{x}(\sin\omega\sin\beta + \cos\omega\cos\beta\cos\vartheta)]}{[1 + (A^2 - 1)\sin^2\gamma\cos^2(\omega - \alpha)]^2}$$

$$(4.68)$$

where the various parameters in (4.67) have been swept into a single constant:

$$\mathcal{B} = \tfrac{1}{8}\pi\lambda^3 r_e^2 \, \Delta H \sec^2\vartheta \, \langle\delta N^2\rangle\kappa_0 Q_4 \qquad (4.69)$$

We are primarily concerned with the normalized spatial correlation. The mean-square amplitude fluctuation was calculated in (3.150) and corresponds to setting $q = 0$ in (3.68). Dividing one expression by the other gives

$$C_\chi(q) = \frac{(1 + \mathbf{a}^2)^{\frac{3}{2}}}{\pi^2(1 + \mathbf{a}^2/2)}\int_0^\infty \frac{dx}{x^2}(1 - \cos x)$$

$$\times \int_0^{2\pi} d\omega \, \frac{\exp[iq\sqrt{x}(\sin\omega\sin\beta + \cos\omega\cos\beta\cos\vartheta)]}{[1 + \mathbf{a}^2\cos^2(\omega - \alpha)]^2} \qquad (4.70)$$

where

$$\mathbf{a} = \sin\gamma\sqrt{A^2 - 1} \qquad (4.71)$$

One can evaluate this expression numerically for arbitrary values of q, $\mathbf{a}$, α, β, γ and Crane has published such curves [41].

One can usually simplify the calculation of $C_\chi(q)$ by exploiting the large values of axial ratio that characterize the highly elongated ionospheric eddies. Since the angle γ between the magnetic field and the propagation directions is large for most observations, one can assume that the parameter $\mathbf{a}$ is itself quite large. In this case the influential term in the azimuth integration of (4.70) can be rewritten using the following identity:

$$\frac{1}{[1 + \mathbf{a}^2 \cos^2(\omega - \alpha)]^2} = \frac{1}{2\mathbf{a}} \frac{\partial}{\partial \mathbf{a}} \left[\mathbf{a} \left(\frac{1/\mathbf{a}}{(1/\mathbf{a})^2 + \cos^2(\omega - \alpha)} \right) \right]$$

The algebraic combination in large parentheses is well approximated by a delta function when $\mathbf{a}$ is large, as noted in Appendix F:

$$\mathbf{a} \text{ large:} \quad \frac{1}{[1 + \mathbf{a}^2 \cos^2(\omega - \alpha)]^2} \simeq \frac{\pi}{2\mathbf{a}} \delta[\cos(\omega - \alpha)]$$

The delta function selects two values for the azimuth angle: $\omega = \alpha + \pi/2$ and $\omega = \alpha + 3\pi/2$. The covariance is then expressed by a relatively simple integration:

$$\mathcal{A} \gg 1: \quad C_\chi(p) = \frac{2}{\pi} \int_0^\infty \frac{dx}{x^2} (1 - \cos x) \cos(p\sqrt{x}) \qquad (4.72)$$

where the following unique combination of inter-receiver spacing, Fresnel length, zenith angle, receiver orientation and magnetic-field direction defines the parameter:

$$p = 2.51\rho|\cos\alpha\sin\beta - \sin\alpha\cos\beta\cos\vartheta| \sqrt{\frac{\cos\vartheta}{\lambda H}} \qquad (4.73)$$

Numerical values for this function are plotted in Figure 4.20. They are similar to the predictions for optical shadow bands plotted in Figure 4.11. They would approach that result if the axial ratio were $\mathcal{A} = 1$. The difference between these predictions represents the influence of anisotropic irregularities. Notice that the correlation in Figure 4.20 falls to a negative value at approximately $p = 2$, which is four times as large as the isotropic result in Figure 4.11. The later oscillations are similarly enlarged. This comparison emphasizes the importance of using the present description for propagation through the ionosphere, solar wind and interstellar plasma.

4.4.2 Experimental Tests

The predictions of Figure 4.20 can be tested against satellite and radio-astronomy data. Signals from spacecraft in synchronous orbit are influenced only by plasma irregularities in the ionosphere. The 254-MHz signal from a military communication satellite (Tacsat I) was observed at several sites near the equator [42]. Spaced-receiver measurements were made with a separation of 200 m, first on a north–south

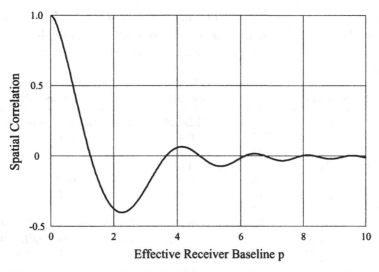

Figure 4.20: The spatial correlation predicted by (4.72) for weak scintillations imposed on radio and microwave signals by ionospheric irregularities whose axial ratios are large.

bearing and then on an east–west axis. The elevation to the satellite was about 70°, so

$$p \approx 2.51\rho \sin(\beta - \alpha)\sqrt{\frac{\cos \vartheta}{\lambda H}} \qquad (4.74)$$

The east–west baseline was rotated relative to the local magnetic field by approximately 70°, giving[7]

$$p \simeq 0.8$$

and the predicted correlation from Figure 4.20 is 0.4. This agrees with the measured values. In the north–south direction, the experimental correlation was 0.85, corresponding to $p \simeq 0.3$. Changing the value of $\beta - \alpha$ by 90° in (4.74) gives precisely this value. We thus find good agreement between theory and experiment for this series of satellite measurements.

Interpretation of spaced-receiver measurements using radioastronomical signals is more complicated because they can also be influenced by the solar wind and interstellar plasma. The earliest radio-astronomy measurements of spatial correlation were made at Cambridge and Jodrell Bank using 3.7- and 6.7-m signals from Cassiopeia [43]. These experiments showed that there was perfect correlation for

---

[7] The ratio

$$p(\text{ew})/p(\text{ns}) = \tan(\beta - \alpha)$$

was inferred from the measurements to be 0.27.

Table 4.1: *Spatial correlation of intensity measured by Little and Maxwell with* $\lambda = 3.7$ *m for various positions of Cassiopeia [40]*

Hours after transit	Elevation (degrees)	Azimuth (degrees)	Effective baseline (km)	Correlation between fluctuation records
0–9	85–23		11.2–6.4	Correlation not significant
10	23	344	6.4	0.4
11	22	352	5.3	0.6
12	21	0	4.4	0.7
13	22	8	4.1	0.9
14	23	16	4.6	0.8
15	27	24	5.6	0.5
16	31	32	6.8	Correlation not significant
17–24	31–85		6.7–11.2	Correlation not significant

a separation of 100 m but none for receivers 200 km apart. Further measurements showed that there was little correlation for spacings of 20 km or more. These findings are consistent with our prediction.

The experiments of Little and Maxwell [40] were the first to investigate this problem systematically. They used the 3.7-m signal from Cassiopeia with a pair of receivers at Cambridge separated by 11 km on a north–south axis. The rising and setting of the source generated different effective baselines. A sample of their data is reproduced in Figure 4.19 and the measured correlations are presented in Table 4.1. The dominant factor in the expression for p is quite large for a scattering height of 300 km:

$$ p = 2.51\rho\sqrt{\frac{\cos\vartheta}{\lambda H}} \simeq 16 $$

The azimuth and elevation angles were quite modest for the source locations that produced significant spatial correlations, so

$$ p \approx 16\beta^2\left(\frac{E^2}{2}\right) $$

The estimates for this parameter are less than unity and direct one's attention to the falling portion of the curve in Figure 4.20. One can claim only general agreement between theory and measurements in this retrospective comparison.

4.5 Problems

Problem 1

Derive an expression similar to (4.10) for the amplitude correlation of signals measured at two receivers placed along the line of sight. Use the paraxial approximation. Why is the longitudinal measurement less valuable than the lateral correlation?

Problem 2

The relationship (4.10) is an integral equation relating the spatial correlation and the turbulence spectrum. Show that it can be inverted as follows:

$$\Phi_n(\kappa)2\pi^2 k^2 R\left(1 - \frac{\sin(\kappa^2 R/k)}{\kappa^2 R/k}\right) = \int_0^\infty d\rho \, \rho J_0(\kappa\rho)\langle\chi(\mathbf{R})\chi(\mathbf{R}+\rho)\rangle$$

One can apparently determine the spectrum if the correlation is measured simultaneously with a large number of receivers. Estimate the number of inter-receiver separations that would be required in order to define the spectrum for different wavenumbers to a given level of accuracy. What error bounds can you place on this inversion if the experimental data is noisy?

Problem 3

Assume that the same constant additive noise source occurs in each receiver of an array used to measure the spatial correlation of a stellar signal that has passed through the atmosphere. What effect does this error have on the profile of C_n^2 that is inferred from the measured data?

Problem 4

Show that the first moment of the amplitude covariance vanishes for spherical waves, just as (4.9) demonstrates for plane waves. How do you reconcile this result with Figure 4.2 for the spatial correlation?

Problem 5

Assume that the spatial spectrum of the arriving wave-front is accurately measured by placing a series of spatial filters in front of an astronomical telescope. Combine (4.55), (4.56) and (4.58) to establish the following relationship between this spectrum and the C_n^2 profile:

$$\int_z^\infty dy \, C_n^2(y) = \frac{1.956}{\pi k^2} \int_0^\infty d\kappa \, \kappa^{\frac{8}{3}} \sin\left(\frac{\kappa^2 z}{k}\right)$$
$$\times \int_0^\infty d\rho \, \rho J_0(\kappa\rho) \int_0^\infty d\mu \cos(\mu\rho) S(\mu)$$

Show that this triple integral can be expressed as a weighted average of the wave-front spectrum:

$$k^2 \int_z^\infty dy\, C_n^2(y) = 0.593 \int_0^\infty d\mu\, S(\mu)\Im\left[\mu^{\frac{5}{3}}M\left(\frac{11}{6},\frac{4}{3},\frac{i\mu^2 z}{k}\right)\right]$$

The kernel defined by the Kummer function weights the contributions of different spatial components in a way that depends on altitude. It was evaluated numerically and its asymptotic behavior for large μ examined in [30].

It is important to remember that this solution does not include aperture-averaging effects. These are important if the wave-front passes through spatial filters that obscure a significant fraction of the downcoming light. As a practical matter, one must use numerical techniques to establish the profile from spatially filtered telescope observations.

References

[1] V. I. Tatarskii, *Wave Propagation in a Turbulent Medium*, translated by R. A. Silverman (Dover, New York, 1967), 142.

[2] See [1], page 113.

[3] D. L. Fried and J. D. Cloud, "Propagation of an Infinite Plane Wave in a Randomly Inhomogeneous Medium," *Journal of the Optical Society of America*, **56**, No. 12, 1667–1676 (December 1966).

[4] V. I. Tatarskii, *The Effects of the Turbulent Atmosphere on Wave Propagation* (translated from the Russian and issued by the National Technical Information Office, U. S. Department of Commerce, Springfield, VA 22161, 1971), 294.

[5] S. M. Rytov, Yu. A. Kravtsov and V. I. Tatarskii, *Principles of Statistical Radiophysics 4, Wave Propagation Through Random Media* (Springer-Verlag, Berlin, 1989), 24.

[6] D. L. Fried, "Propagation of a Spherical Wave in a Turbulent Medium," *Journal of the Optical Society of America*, **57**, No. 2, 175–180 (February 1967).

[7] V. N. Karavainikov, "Fluctuations of Amplitude and Phase in a Spherical Wave," *Akusticheskii Zhurnal (Soviet Physics–Acoustics)*, **3**, No. 2, 175–186 (February 1958).

[8] K. C. Yeh, "Propagation of Spherical Waves Through an Ionosphere Containing Anisotropic Irregularities," *Journal of Research of the NBS–D. Radio Propagation*, **66D**, No. 5, 621–636 (September–October 1962).

[9] A. S. Gurvich, V. I. Tatarskii and L. R. Tsvang, "Scintillation of Light Sources on the Ground," *Trudy Soveshchaniya po Issledovaniya Mertsaniya Zvezd. Izdatel'stvo* (ANSSSR 1958), 33–46. (This paper is not available in English but its results are summarized in [3] on pages 293 and 294.)

[10] V. I. Tatarskii, private communication on 15 March 1994.

[11] M. E. Gracheva, A. S. Gurvich and A. S. Khrupin, "Correlation Functions of the Light Intensity in a Turbulent Atmosphere," *Izvestiya Vysshikh Uchebnykh Zavendenii, Radiofizika (Radiophysics and Quantum Electronics)*, **17**, No. 1, 120–122 (January 1974).

[12] R. W. Lee and A. T. Waterman, "Space Correlations of 35-GHz Transmissions Over a 28-km Path," *Radio Science*, **3**, No. 2, 135–139 (February 1968).

[13] R. J. Hill and S. F. Clifford, "Modified Spectrum of Atmospheric Temperature Fluctuations and Its Application to Optical Propagation," *Journal of the Optical Society of America*, **68**, No. 7, 892–899 (July 1978).

[14] R. G. Frehlich, "Laser Scintillation Measurements of the Temperature Spectrum in the Atmospheric Surface Layer," *Journal of the Atmospheric Sciences*, **49**, No. 16, 1494–1509 (15 August 1992).

[15] R. G. Frehlich, "Estimation of the Parameters of the Atmospheric Turbulence Spectrum Using Measurements of the Spatial Intensity Covariance," *Applied Optics*, **27**, No. 11, 2194–2198 (1 June 1988).

[16] A. Ishimaru, *Wave Propagation and Scattering in Random Media*, vol. 2 (Academic Press, San Diego, CA, 1978), 382–383.

[17] L. C. Andrews and R. L. Phillips, *Laser Beam Propagation through Random Media* (SPIE Optical Engineering Press, Bellingham, Washington, 1998), 214–215.

[18] E. Gaviola, "On Shadow Bands at Total Eclipses of the Sun," *Popular Astronomy*, **56**, 353–360 (August 1948).

[19] J. L. Codona, "The Scintillation Theory of Eclipse Shadow Bands," *Astronomy and Astrophysics*, **164**, No. 2, Part 2, 415–427 (August 1986).

[20] D. L. Fried, "Remote Probing of the Optical Strength of Atmospheric Turbulence and of Wind Velocity," *Proceedings of the IEEE*, **57**, No. 4, 415–420 (April 1969).

[21] R. E. Hufnagel and N. R. Stanley, "Modulation Transfer Function Associated with Image Transmission Through Turbulent Media," *Journal of the Optical Society of America*, **54**, No. 1, 52–61 (January 1964).

[22] R. R. Beland, "Propagation through Atmospheric Optical Turbulence," Chapter 2 in *Atmospheric Propagation of Radiation*, edited by F. R. Smith, vol. 2 of the *Infrared and Electro-Optical Systems Handbook* (SPIE Optical Engineering Press, Bellingham, WA, 1993), 171–174.

[23] P. E. Barnhart, G. Keller and W. E. Mitchell, "Investigation of Upper Air Turbulence by the Method of Analyzing Stellar Scintillation Shadow Patterns," (Report No. TR 59-291 by Ohio State University for the Air Force Cambridge Research Center, Bedford, Massachusetts, 1959).

[24] G. Keller, "Relation Between the Structure of Stellar Shadow Band Patterns and Stellar Scintillation," *Journal of the Optical Society of America*, **45**, No. 10, 845–851 (October 1955).

[25] W. M. Protheroe, "Determination of Shadow Band Structure from Stellar Scintillation Measurements," *Journal of the Optical Society of America*, **45**, No. 10, 851–855 (October 1955).

[26] D. C. Cox, H. W. Arnold and H. H. Hoffman, "Observations of Cloud-Produced Amplitude Scintillation on 19- and 28-GHz Earth–Space Paths," *Radio Science*, **16**, No. 5, 885–907 (September–October 1981).

[27] J. Vernin, "Atmospheric Turbulence Profiles," in *Wave Propagation in Random Media (Scintillation)*, edited by V. I. Tatarskii, A. Ishimaru and V. U. Zavorotny (SPIE and Institute of Physics Publishing, Bristol, 1993), 248–260.

[28] J. Vernin and F. Roddier, "Experimental Determination of Two-Dimensional Spatiotemporal Power Spectra of Stellar Light Scintillation. Evidence for a Multilayer Structure of the Air Turbulence in the Upper Troposphere," *Journal of the Optical Society of America*, **63**, No. 3, 270–273 (March 1973).

[29] A. Rocca, F. Roddier and J. Vernin, "Detection of Atmospheric Turbulent Layers by Spatiotemporal and Spatioangular Correlation Measurements of Stellar-Light Scintillation," *Journal of the Optical Society of America*, **64**, No. 7, 1000–1004 (July 1974).

[30] A. Peskoff, R. S. Margulies and J. M Andres, "Feasibility Study for Remote Sensing of Atmospheric Turbulence Profiles," TRW System Group Report No. 10 636-6002-RO-00, prepared under contract No. NAS 12-2023 for the Electronics Research Center of NASA (June 1970).

[31] A. Peskoff, "Theory of Remote Sensing of Clear-Air Turbulence Profiles," *Journal of the Optical Society of America*, **58**, No. 8, 1032–1040 (August 1968).

[32] G. N. Watson, *A Treatise on the Theory of Bessel Functions* (Cambridge University Press, Cambridge, 1952), 576–617.

[33] J. W. Strohbehn, "Remote Sensing of Clear-Air Turbulence," *Journal of the Optical Society of America*, **60**, No. 7, 948–950 (July 1970).

[34] R. Barakat and G. Newsam, "Remote Sensing of the Refractive Index Structure Parameter via Inversion of Tatarskii's Integral Equation for Both Spherical and Plane Wave Situations," *Radio Science*, **19**, No. 4, 1041–1056 (July–August 1984).

[35] G. R. Ochs, Ting-i Wang, R. S. Lawrence and S. F. Clifford, "Refractive-Turbulence Profiles Measured by One-Dimensional Filtering of Scintillations," *Applied Optics*, **15**, No. 10, 2504–2510 (October 1976).

[36] G. R. Ochs, R. S. Lawrence, T. Wang and P. Zieske, "Stellar Scintillation Measurement of the Vertical Profile of Refractive Index Turbulence in the Atmosphere," *SPIE, Imaging Through the Atmosphere*, **75**, 48–54 (1976).

[37] M. Azouit and J. Vernin, "Remote Investigation of Tropospheric Turbulence by Two-Dimensional Analysis of Stellar Scintillation," *Journal of the Atmospheric Sciences*, **37**, No. 7, 1550–1557 (July 1980).

[38] J. L. Caccia, M. Azouit and J. Vernin, "Wind and C_n^2 Profiling by Single-Star Scintillation Analysis," *Applied Optics*, **26**, No. 7, 1288–1294 (1 April 1987).

[39] M. Spencer, "The Shape of Irregularities in the Upper Atmosphere," *The Proceedings of the Physical Society*, Section B, **68**, 493–503 (August 1955).

[40] C. G. Little and A. Maxwell, "Fluctuations in the Intensity of Radio Waves from Galactic Sources," *Philosophical Magazine* (Taylor and Francis Ltd, http://www.tandf.co.uk/journals), **42**, No. 326, 267–278 (March 1951).

[41] R. K. Crane, "Ionospheric Scintillation," *Proceedings of the IEEE*, **65**, No. 2, 180–199 (February 1977).

[42] M. R. Paulson and R. U. F. Hopkins, "Effects of Equatorial Scintillation Fading on SATCOM Signals," Naval Electronics Center, NELC/TR 1875, San Diego, CA (8 May 1973).

[43] F. G. Smith, G. C. Little and A. C. B. Lovell, "Origin of the Fluctuations in the Intensity of Radio Waves from Galactic Sources," *Nature*, **165**, No. 4194, 422–424 (18 March 1950).

5

The Power Spectrum and Autocorrelation

The amplitudes of signals passing through the atmosphere change with time in a random manner. The amplitude time history for a microwave signal is reproduced in Figure 5.1 from an early experiment. This rapid variability is due primarily to scattering by irregularities that are being carried through the illuminated volume on prevailing winds. To a lesser extent, it is caused by rearrangement of the irregularities by turbulent mixing and the intrinsic process of subdivision that creates the spectrum, as suggested in Figure 2.5 of Volume 1. Variability of signal amplitude is an important consideration for communication systems and target-location systems. It provides a significant tool for atmospheric scientists and astronomers who exploit it to study the properties of remote regions.

We presented several techniques for describing a moving and evolving turbulent atmosphere in Section 6.1 of Volume 1. The vast majority of propagation studies rely on Taylor's hypothesis to describe the temporal changes that occur in a random medium. This approximation involves two assumptions. It is postulated that the entirety of the turbulent medium is frozen during the measurement interval. It is also assumed that one can ignore the variable component of velocity and that the wind velocity is constant at each location. In combination, these two assumptions imply that the entire air mass is transported at constant speed without being deformed as suggested in Figure 6.2 of Volume 1. When these assumptions are valid, the bulk motion of the random medium is equivalent to moving the line of sight parallel to itself through a stationary atmosphere. The amplitude fluctuation measured for a delay time $t + \tau$ should therefore be identical to the amplitude measured at t for a different position defined by $\rho = v\tau$. The temporal covariance should be directly related to the spatial covariance through the wind vector:

$$\langle \chi(\mathbf{R}, t)\chi(\mathbf{R}, t + \tau) \rangle = \langle \chi(\mathbf{R}, t)\chi(\mathbf{R} + \mathbf{v}\tau, t) \rangle \qquad (5.1)$$

This will be the starting point for our descriptions of amplitude variability. More accurate descriptions of evolving random media were also presented in Section 6.1

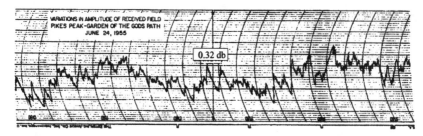

Figure 5.1: A time history of amplitude fluctuations measured at 1046 MHz by Herbstreit and Thompson on a 16-km path [1] © 1995, IRE.

of Volume 1. They can be introduced into the descriptions that we will establish but they present substantially greater analytical challenges.

We concentrate on the variability of amplitude fluctuations measured at a single receiver. In Volume 1 we introduced two techniques for describing signals that change randomly with time. The power spectrum is the most common description – after the amplitude variance. It emerges naturally when the stream of amplitude readings is passed through a spectrum analyzer. The power spectrum is intimately connected to the temporal covariance of the same data stream by the Wiener–Khinchine theorem:

$$W_\chi(\omega) = \int_{-\infty}^{\infty} d\tau\, e^{i\omega\tau} \langle \chi(t)\chi(t+\tau) \rangle \qquad (5.2)$$

The temporal covariance and power spectrum provide equivalent descriptions of the amplitude fluctuations.[1] One can calculate the spectrum for all frequencies if one knows the correlation for all time separations. Conversely, one can calculate the covariance as the inverse Fourier integral if one knows the power spectrum:

$$\langle \chi(t)\chi(t+\tau) \rangle = \frac{1}{2\pi} \int_{-\infty}^{\infty} d\omega\, e^{-i\omega\tau} W_\chi(\omega) \qquad (5.3)$$

The same expressions can be exploited to represent amplitude-difference measurements.

We shall find that the autocorrelation and power spectrum of amplitude fluctuations depend primarily on the inertial range of the turbulence spectrum and large eddies need not be addressed. In this regard, they are similar to the amplitude variance. It also means that aperture averaging and the dissipation region are sometimes

---

[1] Some authors have used the convention

$$W_\chi(\omega) = 4 \int_0^{\infty} d\tau \cos(\omega\tau)\, \langle \chi(t)\chi(t+\tau) \rangle$$

because data was often analyzed in this format in early experiments [2]. The development of fast Fourier transforms means that (5.2) is a natural convention for modern data reduction. It is also more consistent with symmetrical descriptions of Fourier integrals.

important. We can usually ignore anisotropy because eddies in the inertial range have lost the legacy of the large-eddy stage, for which turbulence is generated by atmospheric winds.

5.1 Terrestrial Links

We first consider propagation along horizontal paths near the surface. In this case, one has a second reason to assume that isotropy applies, because eddies near the surface are nearly symmetrical. In addition, they are relatively homogeneous along the path. We are concerned only with the steady wind that blows parallel to the surface because our description is based on Taylor's hypothesis.

5.1.1 Autocorrelation of Terrestrial Signals

Our first description of signal-amplitude variability is based on the *temporal covariance*. It can be deduced from the spatial covariance using (5.1). In blowing the frozen medium across the scattering region, the horizontal wind motion is equivalent to the lateral displacement indicated in Figure 4.1.

$$\rho = \mathbf{i}_y v \tau$$

The wind component along the line of sight does not influence the result because a term $\delta(\kappa_x)$ occurs in the original spatial-covariance expression (4.7). Ignoring aperture averaging, the temporal covariance can be written

$$\langle \chi(\mathbf{R}, t)\chi(\mathbf{R}, t + \tau)\rangle = 4\pi^2 R k^2 \int_0^\infty d\kappa\, \kappa\, \Phi_n(\kappa) J_0(\kappa v \tau) F_\chi(\kappa) \qquad (5.4)$$

where $F_\chi(\kappa)$ is the amplitude-variance weighting function for the reference signal employed.

The *autocorrelation of amplitude fluctuations* is often the preferred quantity because it is normalized to unity for zero time delay:

$$\mu_\chi(\tau) = \frac{\langle \chi(\mathbf{R}, t)\chi(\mathbf{R}, t + \tau)\rangle}{\langle \chi(\mathbf{R}, t)\chi(\mathbf{R}, t)\rangle} \qquad (5.5)$$

We have already evaluated it for plane waves since the autocorrelation is identical to the spatial correlation in that case:

$$\text{Plane wave:} \qquad \mu_\chi(\tau) = C_\chi(\tau v \sqrt{k/R}) \qquad (5.6)$$

Numerical values for the Kolmogorov model were presented in Figure 4.2 and are repeated in Figure 5.2 for reference. One can enlarge that description by including

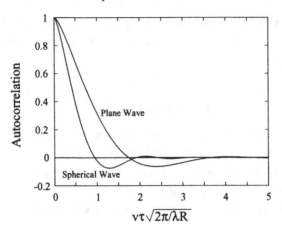

Figure 5.2: Autocorrelation functions for plane and spherical waves, assuming that the random medium is blown past the line of sight by a constant wind. These curves are based on the Kolmogorov model alone.

aperture averaging and inner-scale effects in $C_\chi(\rho)$ using the results developed in Sections 4.1.4 and 4.1.5.

One *cannot* use the spatial correlation for spherical waves to describe the autocorrelation. The reason is that the term $J_0(\kappa v \tau)$ has replaced the path-dependent term $J_0(\kappa u v \tau)$ that now occurs in the double integration of (4.19). Had we used the actual spatial correlation for a spherical wave in this context, we would be describing a convective wind that carries the irregularities *around* the transmitter in a cyclonic pattern. The actual situation is a simple horizontal translation of the eddies and we must use (5.4) with $F_\chi(\kappa)$ replaced by the amplitude-variance weighting function (3.50). The resulting integral was evaluated numerically and is plotted in Figure 5.2. The plane- and spherical-wave autocorrelations are evidently quite similar.

An optical experiment was done to test this prediction [3]. Signals from He–Ne and CO_2 lasers were transmitted simultaneously along the same path. Small detectors were used so that aperture averaging could be ignored. Autocorrelation functions for both signals were computed as functions of delay time from their time histories. Results for the two wavelengths are reproduced in Figure 5.3.

These measurements should be compared with the spherical-wave autocorrelation plotted in Figure 5.2. The path length was 1.2 km and the horizontal wind speed normal to the line of sight was reported to be 3 m.p.h. The wavelengths of the laser signals were 0.632 and 10.6 µm, so the argument of the spatial correlation is proportional to the delay time as follows:

$$v\tau\sqrt{\frac{2\pi}{\lambda R}} = \tau \begin{cases} 0.03 \text{ ms for } CO_2 \\ 0.12 \text{ ms for He–Ne} \end{cases}$$

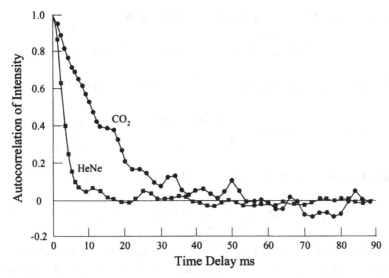

Figure 5.3: Autocorrelation of intensity fluctuations measured with He–Ne- and CO_2-laser signals on a 1.2-km path by Fitzmaurice, Bufton and Minott [3]. Both signals were diverging waves and small receivers were used in order to avoid aperture averaging.

The predicted curve matches the He–Ne-laser measurements remarkably well below in the region of strong correlation $\mu(\tau) > 0.1$ and is not far wrong for data taken beyond this value. A comparison of the CO_2-laser data with its prediction is not as good but gives a generally correct impression of the autcorrelation. This agreement is surprising since one might expect strong scattering on this relatively long path and the spatial correlation is considerably narrowed in that situation, as we will learn in Volume 3.

5.1.2 The Power Spectrum for Terrestrial Links

The power spectrum of amplitude fluctuations is the more familiar description. It can be estimated by taking the Fourier transform of the temporal covariance, as noted in (5.2). One could introduce the analytical expressions for $C_\chi(q)$ established in Chapter 4 and carry out the time-delay integration. That leads to a difficult and unnecessary analysis. Instead, we return to the general expression for temporal covariance based on the wavenumber decomposition and take its Fourier transform. We used this strategy in estimating the single-path phase power spectrum in Section 6.2.3.1 of Volume 1 and it is even more helpful in the present situation.

The first step is to describe the temporal covariance for any model of the turbulent irregularities. Using the general expression (5.4) and noting that the Bessel function is an even function, the power spectrum can be written

$$W_\chi(\omega) = 8\pi^2 Rk^2 \int_0^\infty d\kappa\, \kappa\, \Phi_n(\kappa) F_\chi(\kappa) \int_0^\infty d\tau\, \cos(\tau\omega)\, J_0(\kappa v\tau)$$

The time-delay integration is done with a result from Appendix D:

$$\int_0^\infty d\tau\, \cos(\tau\omega)\, J_0(\kappa v\tau) = \begin{cases} (\kappa^2 v^2 - \omega^2)^{-\frac{1}{2}} & \omega < \kappa v < \infty \\ 0 & 0 < \kappa v < \omega \end{cases} \tag{5.7}$$

and this gives the following description for the power spectrum:

$$W_\chi(\omega) = 8\pi^2 Rk^2 \int_{\omega/v}^\infty d\kappa\, \kappa\, \Phi_n(\kappa) \frac{F_\chi(\kappa)}{\sqrt{\kappa^2 v^2 - \omega^2}} \tag{5.8}$$

The restricted wavenumber integration has a simple physical meaning that we pause to understand. An eddy of size ℓ is carried completely past the line of sight in a time ℓ/v. It can influence only frequencies that are smaller than the reciprocal transit time:

$$\omega < v/\ell$$

This means that the spectrum at frequency ω is determined by wavenumbers greater than ω/v. That is exactly what our description requires.

5.1.2.1 The Plane-wave Power Spectrum

The power spectrum for plane waves was first calculated by Tatarskii [4]. If we substitute the plane-wave amplitude-variance weighting function (3.30) into this expression, we see that

$$W_\chi(\omega) = 4\pi^2 Rk^2 \int_{\omega/v}^\infty d\kappa\, \kappa\, \Phi_n(\kappa) \frac{1}{\sqrt{\kappa^2 v^2 - \omega^2}} \left(1 - \frac{\sin(\kappa^2 R/k)}{\kappa^2 R/k}\right) \tag{5.9}$$

The wavenumber integration in (5.9) depends primarily on small eddies, so one can use the Kolmogorov model to evaluate it:

$$W_\chi(\omega) = 1.303 Rk^2 C_n^2 \int_{\omega/v}^\infty \frac{d\kappa}{\kappa^{\frac{8}{3}}\sqrt{\kappa^2 v^2 - \omega^2}} \left(1 - \frac{\sin(\kappa^2 R/k)}{\kappa^2 R/k}\right) \tag{5.10}$$

Both integrals are convergent at the lower limit, so they can be examined separately. The first is found in Appendix B:

$$\mathcal{J}_1 = \int_{\omega/v}^\infty \frac{d\kappa}{\kappa^{\frac{8}{3}}\sqrt{\kappa^2 v^2 - \omega^2}} = 0.841 \frac{v^{\frac{5}{3}}}{\omega^{\frac{8}{3}}}$$

The second term is more complicated. We make the substitution

$$u = \left(\frac{\kappa v}{\omega}\right)^2$$

and write the trigonometric term as the imaginary part of an exponential:

$$\mathcal{J}_2 = \int_{\omega/v}^{\infty} \frac{d\kappa}{\kappa^{\frac{8}{3}}\sqrt{\kappa^2 v^2 - \omega^2}} \left(\frac{\sin(\kappa^2 R/k)}{\kappa^2 R/k}\right)$$

$$= \frac{kv^{\frac{5}{3}}}{2R\omega^{\frac{8}{3}}} \Im \left\{ \left[\int_1^{\infty} \frac{du}{u^{\frac{17}{6}}\sqrt{u-1}} \exp\left[-u\left(\epsilon - i\frac{\omega^2 R}{kv^2}\right)\right]\right]\right\}_{\epsilon \to 0}$$

Recalling the integral definition for the Kummer function in Appendix G,

$$\int_1^{\infty} e^{-zx} x^{b-a-1} (x-1)^{a-1} \, dx = e^{-z} \Gamma(a) U(a, b, z)$$

we can express the second term as

$$\mathcal{J}_2 = \frac{\sqrt{\pi}kv^{\frac{11}{3}}}{2R\omega^{\frac{14}{3}}} \Im \left[\exp\left(i\frac{\omega^2 R}{kv^2}\right) U\left(\frac{1}{2}, -\frac{4}{3}; -i\frac{\omega^2 R}{kv^2}\right)\right]$$

This expression introduces the *Fresnel frequency* that is defined by the Fresnel length and the wind speed:

$$\omega_F = v\sqrt{\frac{k}{R}} = v\sqrt{\frac{2\pi}{\lambda R}} \tag{5.11}$$

This provides the natural metric for scaling the scintillation frequencies of amplitude fluctuations. It was completely absent from the geometrical-optics expressions for phase spectra established in Volume 1. It is a unique consequence of diffraction. With it we can express the power spectrum as

$$\text{Plane wave:} \quad W_\chi(\omega) - 1.096 \frac{RC_n^2 k^2 v^{\frac{5}{3}}}{\omega^{\frac{8}{3}}} \mathbb{P}\left(\frac{\omega}{\omega_F}\right) \tag{5.12}$$

where a new function is defined by[2]

$$\mathbb{P}(\xi) = 1 - \frac{1.053}{\xi^2} \Im[\exp(i\xi^2) U(\tfrac{1}{2}, -\tfrac{4}{3}; -i\xi^2)] \tag{5.13}$$

We can use this result to establish the power spectrum's behavior for large scintillation frequencies. For large values of ξ one can ignore the Kummer function,

---

[2] One can show that this result is equivalent to the result first derived by Tatarskii [5] by using the connection between the Kummer functions of the first and second kind given in Appendix G.

giving

$$\omega \gg \omega_F \qquad W_\chi(\omega) = 1.096 \frac{RC_n^2 k^2 v^{\frac{5}{3}}}{\omega^{\frac{8}{3}}} \qquad (5.14)$$

Notice that this can also be written in terms of the amplitude variance and Fresnel frequency:

$$\omega \gg \omega_F \qquad W_\chi(\omega) = 3.570(\omega_F)^{\frac{5}{3}} \langle \chi^2 \rangle \frac{1}{\omega^{\frac{8}{3}}} \qquad (5.15)$$

This same frequency dependence will be established in Section 7.3 for the power spectrum of phase fluctuations.

The behavior for phase and amplitude fluctuations is quite different for small scintillation frequencies. The amplitude power spectrum approaches a constant value for very small frequencies but does not depend on the outer-scale region of large eddies. This value can be estimated by taking the limit of (5.13) for small ξ, but that is a delicate process. It is far simpler to return to the defining expression (5.10) and set $\omega = 0$:

$$\lim_{\omega \to 0} [W_\chi(\omega)] = 1.303 \frac{Rk^2 C_n^2}{v} \int_0^\infty \frac{d\kappa}{\kappa^{\frac{11}{3}}} \left(1 - \frac{\sin(\kappa^2 R/k)}{\kappa^2 R/k} \right)$$

With the usual substitution this yields an integral found in Appendix B:

$$\text{Plane wave:} \qquad W_\chi(0) = 0.425 \frac{R^{\frac{7}{3}} k^{\frac{2}{3}} C_n^2}{v} \qquad (5.16)$$

This zero-frequency limit provides an appropriate normalization of the power spectrum. We have computed numerical values for the power spectrum ratio:

$$\frac{W_\chi(\omega)}{W_\chi(0)} = 2.577 \frac{1}{\xi^{\frac{8}{3}}} \left(1 - \frac{1.053}{\xi^2} \Im[\exp(i\xi^2) U(\tfrac{1}{2}, -\tfrac{4}{3}; -i\xi^2)] \right) \qquad (5.17)$$

The results are plotted in Figure 5.4 as a function of the scaled scintillation frequency:

$$\xi = \omega/\omega_F \qquad (5.18)$$

This ratio starts at unity and converts into the traditional 8/3 law when ω exceeds the Fresnel frequency. These results can be used to interpret results of experiments performed with a collimated beam if the receiver is located near the centerline. The same solution provides the basis for describing signals from satellites and stars that pass completely through the atmosphere. Most of the terrestrial-link data has been taken with diverging waves and we turn now to that problem.

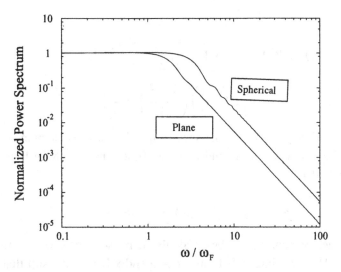

Figure 5.4: Power spectra of amplitude fluctuations for spherical and plane waves propagating near the surface. These curves ignore aperture averaging and are based on the Kolmogorov model.

5.1.2.2 The Spherical-wave Power Spectrum

We follow the same route as that used above to calculate the power spectrum for a plane wave. The temporal covariance for a spherical wave is described by (5.4) with $F_\chi(\kappa)$ replaced by (3.50):

$$\langle \chi(\mathbf{R}, t)\chi(\mathbf{R}, t + \tau)\rangle = 4\pi^2 R k^2 \int_0^1 du \int_0^\infty d\kappa\, \kappa \Phi_n(\kappa) J_0(\kappa v\tau)$$
$$\times \sin^2\left(\frac{\kappa^2 Ru(1 - u)}{2k}\right) \tag{5.19}$$

On taking the Fourier transform and using (5.7) we find that

$$W_\chi(\omega) = 8\pi^2 R k^2 \int_0^1 du \int_{\omega/v}^\infty \frac{d\kappa\, \kappa \Phi_n(\kappa)}{\sqrt{\kappa^2 v^2 - \omega^2}} \sin^2\left(\frac{\kappa^2 Ru(1 - u)}{2k}\right) \tag{5.20}$$

This expression depends primarily on the inertial-range spectrum and the Kolmogorov model gives

$$W_\chi(\omega) = 2.606 R k^2 C_n^2 \int_0^1 du \int_{\omega/v}^\infty \frac{d\kappa}{\kappa^{\frac{8}{3}}\sqrt{\kappa^2 v^2 - \omega^2}} \sin^2\left(\frac{\kappa^2 Ru(1 - u)}{2k}\right) \tag{5.21}$$

Clifford showed how to evaluate this integral in terms of a hypergeometric function and a Kummer function [6]:

$$\text{Spherical wave:} \quad W_\chi(\omega) = 1.096 \frac{R C_n^2 k^2 v^{\frac{5}{3}}}{\omega^{\frac{8}{3}}} \mathbb{S}\left(\frac{\omega}{\omega_F}\right) \tag{5.22}$$

where

$$S(\xi) = \Re\left[1 - {_2F_2}\left(1, -\frac{5}{6}; \frac{3}{2}, -\frac{1}{3}; \frac{i\xi^2}{4}\right)\right.$$

$$\left. - \frac{4}{11}\Gamma\left(-\frac{4}{3}\right)\left(\frac{-i\xi^2}{4}\right)^{\frac{4}{3}} M\left(\frac{1}{2}, \frac{17}{6}; \frac{i\xi^2}{4}\right)\right] \qquad (5.23)$$

The high-frequency portion of the spectrum is the same as the plane-wave result (5.14) since the hypergeometric and Kummer functions vanish for large values of their arguments [6]:

$$\omega \gg \omega_F \qquad W_\chi(\omega) = 1.096\frac{RC_n^2 k^2 v^{\frac{5}{3}}}{\omega^{\frac{8}{3}}} \qquad (5.24)$$

The low-frequency behavior is best calculated by setting $\omega = 0$ in the integral expression (5.21). Two integrals found in Appendix B give a result that is similar to the plane-wave case but with a different numerical coefficient:

$$\text{Spherical wave:} \qquad W_\chi(0) = 0.096\frac{R^{\frac{7}{3}}k^{\frac{2}{3}}C_n^2}{v} \qquad (5.25)$$

We used this value to normalize the power spectrum defined by (5.21). Numerical values for that ratio are plotted in Figure 5.4, where they can be compared with the plane-wave spectrum.

5.1.3 Power-spectrum Measurements

It is relatively easy to make accurate measurements of the power spectrum over the following range of scintillation frequencies:

$$0.01 < \omega/\omega_F < 100$$

by Fourier analyzing the amplitude time history. This was first done using a He–Ne laser signal with path lengths of 50 and 100 m [7]. Small receivers were used in order to avoid aperture averaging of the diverging wave. The agreement between prediction and experiment was satisfactory. It is difficult to extend optical measurements to longer paths without encountering strong-scattering conditions, for which the prediction can be quite different than Figure 5.4 suggests.

One usually encounters weak-scattering conditions at microwave frequencies and a substantial number of power-spectrum measurements has been made at various locations. A summary of those experiments is provided in Table 5.1. All of them were performed with diverging reference signals and their data should be compared with the spherical-wave prediction in Figure 5.4. The earliest power-spectrum measurements were made in conjunction with phase measurements on

Table 5.1: *A summary of power-spectra measurements made with microwave signals on terrestrial paths*

Year	Location	R (km)	Frequency (GHz)	Range (Hz)	Ref.
1970	Hawaii	65	9.6 and 34.5	$0.01 < \omega < 5$	8
1973	Stanford	28	35	$0.1\omega_F < \omega < 80\omega_F$	9, 10
1975	Hawaii	64	9.6 and 22	$0.002 < \omega < 0.4$	11
1975	Hawaii	150	9.6	$0.007 < \omega < 0.4$	12
1979	London	4.1	36 and 110	$0.01 < \omega < 4$	13, 14
1988	Illinois	1.4	142 and 230	$0.001 < \omega < 40$	15
1996	New Mexico	0.6	94	$0.05 < \omega < 25$	16

relatively long paths that were intended to simulate transatmospheric propagation [8][11][12]. Those signals necessarily encountered a wide range of C_n^2 values and wind speed along the paths. Two of these experiments had the advantage of using several wavelengths simultaneously and the results were consistent with our prediction.

The millimeter-wave research program at Stanford used the low- and high-frequency spectral behaviors to estimate C_n^2 and the average wind speed along the path [9]. Comparison of calculated and locally measured wind speeds gave good agreement. This program also examined the amplitude-difference spectrum measured between adjacent receivers. The influence of aperture averaging was investigated by comparing the scintillation experienced at closely spaced antennas. The reduction was more than 10 dB at 35 GHz, indicating the profound influence of aperture averaging for amplitude fluctuations [10].

A comprehensive series of measurements near London extended these measurements in two important ways [13][14]. Reference signals at 36 and 110 GHz were employed, thereby providing data to test the wavelength scaling of the power-spectrum prediction. Surface wind speed and temperature were measured at the location of the receiver. Typical power spectra from this campaign are reproduced in Figure 5.5. The data is entirely consistent with Figure 5.4 and exhibits the predicted scaling with wavelength.

Subsequent millimeter-wave experiments in Illinois extended the frequency range to 230 GHz [15]. A second program in New Mexico used 94-GHz signals [16]. In both experiments extensive micro-meteorological measurements were made with several instrumented towers near the path. In these experiments modern data-processing techniques were used and a wide range of scintillation frequencies was examined. With this enhanced capability the effects of fog, rain, snow and dust on the spectrum were investigated. Their influence was apparent in

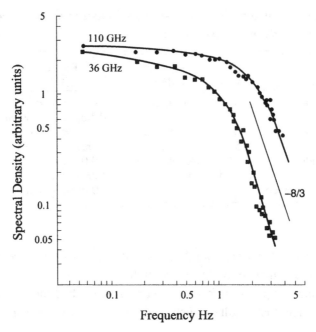

Figure 5.5: Power spectra of amplitude fluctuations measured at 36 and 110 GHz by Ho, Mavrokoukoulakis and Cole on a 4.1-km path [13].

modifications of the power spectrum normally associated with clear-air turbulence conditions.

5.2 Astronomical Observations

We have learned that weak scattering usually describes stellar scintillation observed by optical and infrared telescopes. This means that we should be able to characterize the autocorrelations and power spectra of such signals with the tools we have developed. Because the lower atmosphere is carried along on prevailing winds, the actual configuration of the scattering medium changes with time. This induces amplitude fluctuations like those reproduced in Figure 5.1. The arriving signals are plane waves for astronomical sources and that simplifies matters. On the other hand, the signals must pass through the entire troposphere, so the scintillation is sensitive to the height profiles both of C_n^2 and of the wind speed.

5.2.1 Autocorrelation of Astronomical Signals

We rely on Taylor's hypothesis to describe the changing configuration of tropospheric irregularities encountered by the arriving stellar signals. That leads one to the description of spatial covariance illustrated in Figure 4.10. We assume that the wind blows normal to the plane in which the signal travels:

$$\beta = 0 \quad \text{and} \quad \rho_e = \tau v(z) \tag{5.26}$$

With this simplification the spatial-covariance expression (4.38) yields

$$\langle \chi(t)\chi(t+\tau) \rangle = 4\pi^2 k^2 \sec \vartheta \int_0^\infty dz\, C_n^2(z) \int_0^\infty d\kappa\, \kappa \Omega(\kappa) J_0[\kappa \tau v(z)]$$

$$\times \sin^2\left(\frac{\kappa^2 z \sec \vartheta}{2k}\right) \left(\frac{2J_1(\kappa a_r)}{\kappa a_r}\right)^2 \tag{5.27}$$

In writing this out we have included the familiar aperture-averaging factor because astronomical telescopes use large mirrors to collect light from faint stars. We have also recognized that the horizontal wind speed varies with altitude, since that is the situation established by the meteorological measurements indicated in Section 6.1.1 of Volume 1.

One is reasonably confident that the influential eddies for amplitude fluctuations lie in the inertial range, so we can use

$$\Omega(\kappa) = 0.033\kappa^{-\frac{11}{3}} \tag{5.28}$$

to generate the following expression for the temporal covariance:

$$\langle \chi(t)\chi(t+\tau) \rangle = 1.303 k^2 \sec \vartheta \int_0^\infty dz\, C_n^2(z) \int_0^\infty \frac{d\kappa}{\kappa^{\frac{8}{3}}} J_0[\kappa \tau v(z)]$$

$$\times \sin^2\left(\frac{\kappa^2 z \sec \vartheta}{2k}\right) \left(\frac{2J_1(\kappa a_r)}{\kappa a_r}\right)^2 \tag{5.29}$$

The remaining integrations must be evaluated numerically when the height profiles of C_n^2 and $v(z)$ are specified.

To gain analytical insight into the process we ignore the aperture-averaging term. The remaining wavenumber integration can then be expressed in terms of the thin-layer weighting function defined by the integral (4.43) or its analytical expression (4.45):

$$\langle \chi(t)\chi(t+\tau) \rangle = 0.563 k^{\frac{7}{6}} (\sec \vartheta)^{\frac{11}{6}} \int_0^\infty dz\, z^{\frac{5}{6}} C_n^2(z) \mathbb{T}\left(\tau v(z)\sqrt{k \cos \vartheta / z}\right) \tag{5.30}$$

The function $\mathbb{T}(q)$ is plotted in Figure 4.11. The factor involving the 5/6 power of altitude emphasizes upper-atmosphere conditions. There one often encounters strongly turbulent layers, like those illustrated by the measurements of Figure 2.18 in Volume 1. In that case one can approximate:

$$\langle \chi(t)\chi(t+\tau) \rangle = 0.563 k^{\frac{7}{6}} (\sec \vartheta)^{\frac{11}{6}} \, \Delta H \, H^{\frac{5}{6}} C_n^2(H) \mathbb{T}\left(\tau v(H)\sqrt{k \cos \vartheta /H}\right)$$

$$(5.31)$$

This result can be used in conjunction with Figure 4.11 to suggest how the temporal covariance changes with zenith angle. On the other hand, we have compounded several assumptions and ignored aperture averaging. We therefore cannot be confident that the resulting scaling law accurately reflects the physics.

To learn how the autocovariance depends jointly on zenith angle and aperture size we turn to astronomical observations. One can make such measurements with bright stars and photon-counting techniques. An early experiment done on Mauna Kea in Hawaii (4200 m) used a 61-cm telescope to measure the autocovariance on ten nights [17]. The results varied widely from night to night, representing changing wind patterns and C_n^2 profiles. The autocovariance fell to its $1/e$ value in less than 5 ms at this site. That contrasts with data taken at lower observatories, for which one typically finds 10 ms for the decay time.

More elaborate measurements were recently conducted at the La Palma Observatory in the Canary Islands (2360 m) [18]. A 60-cm telescope was employed to observe Vega. Autocovariance functions measured in this way are reproduced in Figure 5.6 for a variety of telescope openings and two zenith angles on the same night. The data reveals significant changes from early evening to midnight,

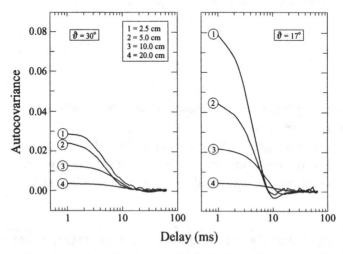

Figure 5.6: Temporal covariance of scintillation observed at the La Palma Observatory by Dravins, Lindegren, Mezey and Young [18]. The measurements at the left were taken early in the evening and those on the right close to midnight. The separate curves correspond to the indicated aperture sizes. © 1997, Astronomical Society of the Pacific: reproduced with permission.

suggesting that there are substantial changes in the profiles of C_n^2 and $v(z)$. A significant reduction in scintillation level with increasing aperture size is also evident.

The report of these measurements also provides a comprehensive survey of temporal variability for astronomical signals. The reader who wishes to pursue this subject should explore this valuable reference [18]. It examines the role of the inner-scale region in modifying the autocovariance for very small delay times and finds evidence for $\ell_0 \simeq 3$ mm, in agreement with measurements on terrestrial links. It also examines the influence of large eddies at the outer-scale boundary and suggests that $L_0 \simeq 5$ km, which is consistent with airborne temperature measurements and microwave phase measurements.

5.2.2 The Power Spectrum of Astronomical Signals

The power spectrum is the alternative description of astronomical scintillation. It can be described by combining the temporal-covariance expression (5.27) with (5.2). The time delay integration is readily done using (5.7):

$$W_\chi(\omega) = 8\pi^2 k^2 \sec \vartheta \int_0^\infty dz\, C_n^2(z) \int_{\omega/v(z)}^\infty \frac{d\kappa\, \kappa\Omega(\kappa)}{\sqrt{\kappa^2 v^2(z) - \omega^2}} \sin^2\left(\frac{\kappa^2 z \sec \vartheta}{2k}\right)$$

$$(5.32)$$

We use the Kolmogorov model (5.28) for the reasons presented earlier, so

$$W_\chi(\omega) = 2.606 k^2 \sec \vartheta \int_0^\infty dz\, C_n^2(z) \int_{\omega/v(z)}^\infty \frac{d\kappa}{\kappa^{\frac{8}{3}}\sqrt{\kappa^2 v^2(z) - \omega^2}}$$

$$\times \sin^2\left(\frac{\kappa^2 z \sec \vartheta}{2k}\right) \qquad (5.33)$$

The wavenumber integration can be done by setting $\kappa = \ell\omega/v(z)$, giving

$$W_\chi(\omega) = 2.606 \frac{k^2 \sec \vartheta}{\omega^{\frac{8}{3}}} \int_0^\infty dz\, C_n^2(z)(v(z))^{\frac{5}{3}} \mathbb{Q}\left(\frac{\omega}{v(z)}\sqrt{\frac{z \sec \vartheta}{2k}}\right) \qquad (5.34)$$

where

$$\mathbb{Q}(\xi) = \int_1^\infty \frac{d\ell}{\ell^{\frac{8}{3}}\sqrt{\ell^2 - 1}} \sin^2(\ell^2 \xi^2)$$

This function can be evaluated with the same techniques as those we used to establish (5.13) and we find

$$\mathbb{Q}(\xi) = 0.421\Re[1 - 1.053\, exp(i\xi^2)\, U(\tfrac{1}{2}, \tfrac{5}{6}; -i\xi^2)] \qquad (5.35)$$

The power-spectrum description is completed when one combines this result with (5.34). One can use numerical values of this function to establish the weighting function in the required integration over all altitudes. We shall not describe that program because the results are similar to the curve for a terrestrial plane wave that is plotted in Figure 5.4.

The small- and large-frequency limits of the astronomical power spectrum can be extracted from the foregoing expressions. For large frequencies one can ignore the Kummer function to establish

$$\omega \text{ large:} \qquad W_\chi(\omega) = 1.096 \frac{k^2 \sec \vartheta}{\omega^{\frac{8}{3}}} \int_0^\infty dz\, C_n^2(z)(v(z))^{\frac{5}{3}} \qquad (5.36)$$

On taking $\omega = 0$ in the integral expression (5.34) we see that the spectrum approaches a constant value:

$$W_\chi(0) = 2.501 k^{\frac{2}{3}} (\sec \vartheta)^{\frac{7}{3}} \int_0^\infty dz\, z^{\frac{4}{3}} \frac{C_n^2(z)}{v(z)} \qquad (5.37)$$

Notice that different combinations of the C_n^2 and v profiles define the coefficients in these scaling laws.

It is instructive to compare these limiting expressions with power spectra measured with millimeter wave-satellite signals. The ionosphere plays no role for signals above 10 GHz and we can use our astronomical expressions to describe their passage through the troposphere. In a recent experiment 20- and 30-GHz signals from the Olympus satellite in geosynchronous orbit were used to measure their amplitude power spectra simultaneously [19]. Experimental results from this program are reproduced in Figure 5.7. The data confirms the constant spectrum level at low

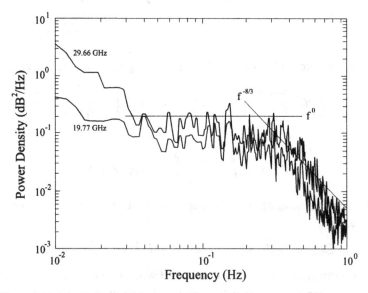

Figure 5.7: Power spectra of scintillation measured with geosynchronous satellite signals at 20 and 30 GHz by Otung, Al-Nuaimi and Evans [19]. © 1998, IEEE.

frequencies predicted by (5.37). The behavior above $\omega = 0.3$ Hz agrees with the $-8/3$ decline indicated by (5.36).

Comparison of the spectra at two frequencies also showed that one can identify rainfall along the satellite-signal path [19], thereby providing a diagnostic tool for communication operators. For large ground stations one must include the aperture-averaging effect, which suppresses the high-frequency portion of the scintillation spectrum [20][21].

5.3 The Ionospheric Influence

Propagation through the ionosphere is influenced by electron-density irregularities if the carrier frequency is less than approximately 10 GHz. Radio-astronomy observations were all made in this range until millimeter-wave signals were recently discovered. The first generation of communication satellites used the assigned pair of frequencies at 4 and 6 GHz. Satellite service to mobile users is provided at 240 and 1500 MHz. Amplitude variations are important for these services. Scintillation experiments have been mounted using signals radiated both by commercial satellites and by research spacecraft specially designed to explore such phenomenon.

The amplitude power spectrum is the common measure of variability with time in this context. The most accurate and plentiful measurements have been made with satellite signals. They are sensitive only to conditions in the ionosphere, not to the solar wind and the interstellar plasma. The lines of sight for navigation satellites sweep through such irregularities as the spacecraft move in their orbits at speeds of 2 km s^{-1} or more. This source motion selects changing configurations of the scattering volume. Most observations have been made with geosynchronous satellites whose lines of sight do not move relative to observing ground stations. In those cases, horizontal winds in the ionosphere of 100–300 m s^{-1} are responsible for moving the irregular structure through the scattering volume.

Predicting the power spectrum is a challenge for several reasons. The first is that we have been able to make only a few *in situ* measurements of the turbulence spectrum of electron-density variations. As a practical mater, we are forced to rely on models with adjustable parameters – like that presented in (3.149). From spaced correlation observations of radio stars we know that electron-density fluctuations are elongated in the direction of the earth's magnetic field. This means that three vector references are important: (a) the line-of-sight vector for the source, (b) the wind and/or satellite velocity vector and (c) the terrestrial magnetic-field vector. We confronted the same problems in our efforts to understand the spatial correlation of amplitude fluctuations imposed by the ionosphere in Section 4.4.1.

With Taylor's hypothesis we can base our current study on the approximate expression (4.72) that was established for large axial ratios. Substituting $\rho = v\tau$ in (4.73), we write the temporal covariance as

$$\mathcal{A} \gg 1: \qquad \langle \chi(t)\chi(t+\tau) \rangle = \langle \chi^2 \rangle \frac{2}{\pi} \int_0^\infty \frac{dx}{x^2}(1 - \cos x)\cos(v\tau m\sqrt{x}) \quad (5.38)$$

where

$$m = 2.51|\cos\alpha \sin\beta - \sin\alpha \cos\beta \cos\vartheta|\sqrt{\frac{\cos\vartheta}{\lambda H}} \qquad (5.39)$$

We can take the Fourier transform of (5.38) to find

$$W_\chi(\omega) = 2\langle \chi^2 \rangle \int_0^\infty \frac{dx}{x^2}(1 - \cos x)\delta(vm\sqrt{x} - \omega)$$

If a function $g(x)$ vanishes for $x = x_0$, the property of the delta function

$$\delta[g(x)] = \frac{1}{|g'(x_0)|}\delta(x - x_0)$$

from Appendix F allows one to express the power spectrum as

$$\mathcal{A} \gg 1: \qquad W_\chi(\omega) = 4\langle \chi^2 \rangle m^2 \frac{1 - \cos(\omega^2/n^2)}{\omega^3} \qquad (5.40)$$

It is important to remember that this ω^{-3} scaling law results from our assumption that the exponent $\nu = 4$ in the model (3.149) that we used to establish (5.38). Had we used a different value for ν, the negative exponent for frequency would be $\nu - 1$. This suggests that we can determine ν by comparing such expressions with scintillation spectra.

A summary of results from satellite experiments in which the power spectrum of amplitude fluctuations was measured is provided in Table 5.2. In an early experiment the scintillation imposed on the 4-GHz signal from ATS-5 was measured [22]. A typical power spectrum from that experiment is reproduced in Figure 5.8. The measurement is accurately fitted by the ω^{-3} prediction for frequencies above 0.1 Hz. This confirms (5.40) in the high-frequency or asymptotic range and indicates that the exponent $\nu \simeq 4$ is close to the correct value for the turbulence spectrum of electron-density variations. This agreement has been confirmed by a number of satellite measurements at quite different frequencies [23].

It is significant that the low-frequency behavior of the spectrum in Figure 5.8 is either flat or declining. That feature is not predicted by (5.40). Crane developed theoretical expressions that do predict both the low- and the high-frequency behavior of satellite power spectra. His results are in agreement with measurements made

Table 5.2: *A summary of satellite experiments in which the scintillation power spectrum was measured*

Year	Frequencies	Satellite	Receiver	Ref.
1976	1.5 and 4 GHz	ATS-5	Brazil	22
1976	150 and 400 MHz	NSSS	Massachusetts	24
1977	150 and 400 MHz	NSSS	Massachusetts	23
	254 MHz	Tacsat	Peru	
	4 GHz	ATS-5	Brazil	
1978	138 MHz	DNA	Alaska	25
1982	40, 140 and 360 MHz	ATS-6	Colorado	26
1992	40 and 140 MHz	ATS-6	Colorado	27
	249 MHz	LES-6	Peru	

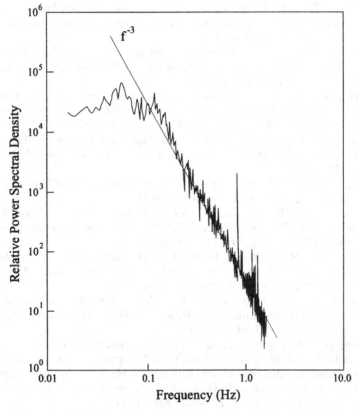

Figure 5.8: A scintillation power spectrum measured by Taur in Brazil using the 4-GHz signal from ATS-5 [22].

at many different frequencies. The reader who wishes to learn more about this complicated problem should begin with Crane's original papers [23][24].

References

[1] J. W. Herbstreit and M. C. Thompson, "Measurements of the Phase of Radio Waves Received over Transmission Paths with Electrical Lengths Varying as a Result of Atmospheric Turbulence," *Proceedings of the IRE*, **43**, No. 10, 1391–1401 (October 1955).

[2] V. I. Tatarskii, *The Effects of the Turbulent Atmosphere on Wave Propagation* (translated from the Russian and issued by the National Technical Information Office, U. S. Department of Commerce, Springfield, VA 22161, 1971), 261.

[3] M. W. Fitzmaurice, J. L. Bufton and P. O. Minott, "Wavelength Dependence of Laser-Beam Scintillation," *Journal of the Optical Society of America*, **59**, No. 1, 7–10 (January 1969).

[4] See [2], page 259.

[5] See [2], page 263.

[6] S. F. Clifford, "Temporal-Frequency Spectra for a Spherical Wave Propagating through Atmospheric Turbulence," *Journal of the Optical Society of America*, **61**, No. 10, 1285–1292 (October 1971).

[7] A. S. Gurvich and N. S. Time, "Frequency Spectrum of Intensity Fluctuations of a Spherical Wave Propagating in the Atmosphere," *Radio Engineering and Electronic Physics*, **15**, No. 4, 683–686 (April 1970).

[8] H. B. Janes, M. C. Thompson, D. Smith and A. W. Kirkpatrick, "Comparison of Simultaneous Line-of-Sight Signals at 9.6 and 34.5 GHz," *IEEE Transactions on Antennas and Propagation*, **AP-18**, No. 4, 447–451 (July 1970).

[9] P. A. Mandics, R. W. Lee and A. T. Waterman, "Spectra of Short-term Fluctuations of Line-of-sight Signals: Electromagnetic and Acoustic," *Radio Science*, **8**, No. 3, 185–201 (March 1973).

[10] R. W. Lee, "Aperture Effects on the Spectrum of Amplitude Scintillation," *Radio Science*, **6**, No. 12, 1059–1060 (December 1971).

[11] M. C. Thompson, L. E. Wood, H. B. Janes and D. Smith, "Phase and Amplitude Scintillations in the 10 to 40 GHz Band," *IEEE Transactions on Antennas and Propagation*, **AP-23**, No. 6, 792–797 (November 1975).

[12] M. C. Thompson, H. B. Janes, L. E. Wood and D. Smith, "Phase and Amplitude Scintillations at 9.6 GHz on an Elevated Path," *IEEE Transactions on Antennas and Propagation*, **AP-23**, No. 6, 850–854 (November 1975).

[13] K. L. Ho, N. D. Mavrokoukoulakis and R. S. Cole, "Propagation Studies on a Line-of-Sight Microwave Link at 36 GHz and 110 GHz," *Microwaves, Optics and Acoustics*, **3**, No. 3, 93–98 (May 1979).

[14] C. G. Helmis, D. N. Asimakopoulos, C. A. Caroubalos, R. S. Cole, F. C. Medeiros Filho and D. A. R. Jayasuriya, "A Quantitative Comparison of the Refractive Index Structure Parameter Determined from Refractivity Measurements and Amplitude Scintillation Measurements at 36 GHz," *IEEE Transactions on Geoscience and Remote Sensing*, **GE-21**, No. 2, 221–224 (April 1983).

[15] R. J. Hill, S. F. Clifford, R. J. Lataitis and A. D. Sarma, "Scintillation of Millimeter-Wave Intensity and Phase Caused by Turbulence and Precipitation," AGARD Conference Proceedings No. 454, Electromagnetic Wave Propagation Panel Specialists' Meeting held in Copenhagen, Denmark (9–13 October 1989).

[16] W. D. Otto, R. J. Hill, A. D. Sarma, J. J. Wilson, E. L. Andreas, J. R. Gosz and D. I. Moore, "Results of the Millimeter-wave Instrument Operated at Sevilleta, New Mexico," NOAA Technical Memorandum ERL ETL-262, Boulder, Colorado (February 1996).

[17] J. C. Dainty, B. M. Levine, B. J. Brames and K. A. O'Donnell, "Measurements of the Wavelength Dependence and Other Properties of Stellar Scintillation at Mauna Kea, Hawaii," *Applied Optics*, **21**, No. 7, 1196–1200 (1 April 1982).

[18] D. Dravins, L. Lindegren, E. Mezey and A. T. Young, "Atmospheric Intensity Scintillation of Stars. I. Statistical Distributions and Temporal Properties," *Publications of the Astronomical Society of the Pacific*, **109**, No. 732, 173–207 (February 1997).

[19] I. E. Otung, M. O. Al-Nuaimi and B. G. Evans, "Extracting Scintillations from Satellite Beacon Propagation Data," *IEEE Transactions on Antennas and Propagation*, **AP-46**, No. 10, 1580–1581 (October 1998).

[20] E. Vilar, J. Haddon, P. Lo and T. J. Moulsley, "Measurement and Modelling of Amplitude and Phase Scintillations in an Earth–Space Path," *Journal of the Institution of Electronic and Radio Engineers*, **55**, No.3, 87–96 (March 1985).

[21] D. C. Cox, H. W. Arnold and H. H. Hoffman, "Observations of Cloud-Produced Amplitude Scintillation on 19- and 28-GHz Earth–Space Paths," *Radio Science*, **16**, No. 5, 885–907 (September–October 1981).

[22] R. R. Taur, "Simultaneous 1.5- and 4-GHz Ionospheric Scintillation Measurement," *Radio Science*, **11**, No.12, 1029–1036 (December 1976).

[23] R. K. Crane, "Ionospheric Scintillation," *Proceedings of the IEEE*, **65**, No. 2, 180–199 (February 1977).

[24] R. K. Crane, "Spectra of Ionospheric Scintillation," *Journal of Geophysical Research*, **81**, No. 13, 2041–2050 (1 May 1976).

[25] E. J. Fremouw, R. L. Leadabrand, R. C. Livingston, M. D. Cousins, C. L. Rino, B. C. Fair and R. A. Long, "Early Results from the DNA Wideband Satellite Experiment – Complex-signal Scintillation," *Radio Science*, **13**, No. 1, 167–187 (January–February 1978).

[26] K. C. Yeh and C. H. Liu, "Radio Wave Scintillations in the Ionosphere," *Proceedings of the IEEE*, **70**, No. 4, 324–360 (April 1982).

[27] A. Bhattacharyya, K. C. Yeh and S. J. Franke, "Deducing Turbulence Parameters from Transionospheric Scintillation Measurements," *Space Science Reviews*, **61**, 335–386 (1992).

6

Frequency Correlations

The wavelength scaling of amplitude variance discussed in Section 3.2.4 was found to give good agreement between theory and experiment. We can take this comparison one step further and compare the detailed variations of amplitude at two wavelengths. The amplitude histories for radio-astronomical signals at 81.5 and 118.5 MHz received from a radio source in Cassiopeia [1] are reproduced in Figure 6.1. Investigations of this correlated signal behavior were originally driven by practical considerations. Radio astronomers working with very faint signals wanted to know how wide a frequency band they could use to improve the signal-to-noise ratio. Engineers designing terrestrial and satellite communication relay links wondered whether turbulence scattering would limit the information bandwidth that could be transmitted through the atmosphere.

Both questions are related to the *medium bandwidth* which measures the frequency separation over which received signals behave in the same way. Severe medium-bandwidth limitations had been observed on tropospheric and ionospheric scatter-propagation circuits. That experience raised the possibility that only a narrow slice of frequency spectrum would be coherent and thus useful for line-of-sight transmissions. Experimental data showed instead that a very large bandwidth should be available [2][3][4][5][6][7]. For weak scattering the signal strength is correlated over a wide range of frequencies because the strong coherent field provides a powerful stabilizing reference – as it cannot for scatter propagation.

Attention soon shifted to using comparisons of multifrequency signals as a scientific tool for probing the atmosphere. This is attractive because one can do so without disturbing the conditions one is trying to measure. Such results are relatively easy to interpret because the two signals can be made to travel along the same path. One can change the frequency separation rapidly at the transmitter and receiver for microwave links. That flexibility is seldom possible in spaced-receiver experiments. The frequency correlation is sensitive to inner-scale effects at optical

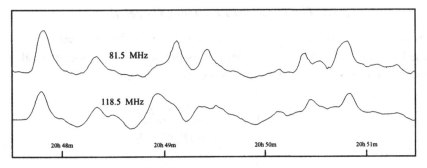

Figure 6.1: Amplitude records measured simultaneously at 81.5 and 118.5 MHz by Burrows and Little using the source in Cassiopeia [1].

wavelengths and thus provides a way to investigate the dissipation range of the turbulence spectrum.

Incoherent sources can also be used for these studies. Comparisons of starlight collected at different wavelengths by a single telescope provide a method by which to establish the profile of turbulent activity with crude accuracy. Satellite signals at different frequencies are regularly used to establish the spectrum of electron-density fluctuations in the ionosphere. More recently, very short microwave pulses have been received from distant quasars. The lack of correlation among their constituent frequency components causes significant pulse distortion. The frequency correlation of the wideband signals received from radio stars provides a unique tool for studying the interstellar plasma.

Our results are applicable only to some propagation situations because we are working here with weak scattering. These predictions provide a good description of frequency correlations for terrestrial microwave links. They also describe short-path optical links that satisfy the condition for weak scattering. Two-color astronomical comparisons can be analyzed with these techniques. So too can radio-astronomy and satellite observations unless the frequencies are too low or the source is too close to the horizon.

6.1 Terrestrial Links

Careful measurements of the amplitude–frequency correlation have been made with horizontal transmissions. These experiments exploit both microwave and optical signals. Valuable scientific descriptions of the lower atmosphere have been generated in this way. The same measurements provide a firm basis for predicting the performance of wideband radar and communication systems whose signals pass through the troposphere.

6.1.1 The Predicted Frequency Correlation

The *frequency correlation* of amplitude fluctuations is the simplest description of these bandwidth limitations. It is defined as the amplitude covariance taken at different wavenumbers divided by the respective rms values:

$$\mathcal{R}(k_1, k_2) = \frac{\langle \chi(k_1)\chi(k_2) \rangle}{\sqrt{\langle \chi^2(k_1) \rangle \langle \chi^2(k_2) \rangle}} \tag{6.1}$$

We have introduced a special symbol to differentiate the frequency correlation from the spatial and temporal correlations defined previously. Some authors refer to this as the *wavelength correlation*, but we shall always call it the frequency correlation in order to avoid confusion. Some microwave experiments measured the *amplitude coherence*, which contains more information and is better behaved statistically.

It is important to identify several approximations that are consistently made in order to simplify the calculation required to predict these quantities. Refractive-index irregularities are relatively homogeneous near the surface and one can usually ignore variations of the spectrum along a horizontal path. If the transmitter and receiver are close to the surface, one can also approximate the real atmosphere by an isotropic random medium. All examinations of frequency correlation thus far have exploited the paraxial approximation which describes small-angle forward scattering. This is a valid assumption for optical and infrared signals because their wavelengths are much less than the size of the smallest eddy. The paraxial approximation is also consistently used in the literature to describe microwave links. We will use it here since these signals are influenced most strongly by the large eddies which do satisfy the requirements for small-angle scattering of microwaves.

It is a straightforward matter to calculate the correlation between signal fluctuations at different wavelengths if the scattering is weak. The first-order expression for logarithmic amplitude fluctuations (2.26) shows that the electromagnetic wavenumber occurs (a) in the unperturbed field, (b) in Green's function, and (c) as the factor k^2. That description for χ is enlarged here to remind us of this three-fold dependence:

$$\chi(\mathbf{R}, k) = -2k^2 \int d^3r \, A(\mathbf{R}, \mathbf{r}, k) \, \delta n(\mathbf{r}, t) \tag{6.2}$$

where

$$A(\mathbf{R}, \mathbf{r}, k) = \Re \left(\frac{G(\mathbf{R}, \mathbf{r}, k) E_0(\mathbf{r}, k)}{E_0(\mathbf{R}, k)} \right) \tag{6.3}$$

For a homogeneous random medium the covariance of two signals measured at different wavenumbers becomes

$$\langle \chi(k_1)\chi(k_2) \rangle = 4k_1^2 k_2^2 \int d^3 r_1 \int d^3 r_2\, A(\mathbf{R}, \mathbf{r}_1, k_1) A(\mathbf{R}, \mathbf{r}_2, k_2)$$

$$\times \int d^3 \kappa\, \Phi_n(\kappa, k_1, k_2) \exp[i\kappa \cdot (\mathbf{r}_1 - \mathbf{r}_2)] \tag{6.4}$$

This expression depends on the co-spectrum of refractive-index variations taken at different frequencies $\Phi_n(\kappa, k_1, k_2)$. It allows us to describe the effects of fluctuations both in temperature and in concentration of water vapor that influence the combination differently depending on the wavelength. With this description one can also recognize the effect of dispersion, which is important in comparing optical and infrared scintillations [8].

In this chapter we shall concentrate on optical and microwave experiments, for which dispersion is not an important consideration. That choice allows us to simplify the amplitude covariance as follows:

$$\langle \chi(k_1)\chi(k_2) \rangle = 4k_1^2 k_2^2 \int d^3 \kappa\, \Phi_n(\kappa) \int d^3 r_1 \int d^3 r_2 \exp[i\kappa \cdot (\mathbf{r}_1 - \mathbf{r}_2)]$$

$$\times A(\mathbf{R}, \mathbf{r}_1, k_1) A(\mathbf{R}, \mathbf{r}_2, k_2)$$

$$= 4 \int d^3 \kappa\, \Phi_n(\kappa) D(\kappa, k_1) D(-\kappa, k_2) \tag{6.5}$$

where

$$D(\kappa, k) = k^2 \int d^3 r\, A(\mathbf{R}, \mathbf{r}, k) \exp(i\kappa \cdot \mathbf{r}) \tag{6.6}$$

This volume integration was evaluated for a plane wave incident on the random medium using the paraxial approximation in (3.18):

$$D(\kappa, k) = \frac{k}{2} \int_0^R dx \sin\left(\frac{x\kappa_r^2}{2k}\right) \exp[i\kappa_z(R - x)] \tag{6.7}$$

In the corresponding expression for a spherical wave (3.48) we simply replace x by $x(1 - x/R)$ in the trigonometric term. On introducing this expression into (6.5) we find that

$$\langle \chi(k_1)\chi(k_2) \rangle = k_1 k_2 \int d^3 \kappa\, \Phi_n(\kappa) \int_0^R dx_1 \int_0^R dx_2 \exp[i\kappa_x(x_1 - x_2)]$$

$$\times \sin\left(\frac{x_1\kappa_r^2}{2k_1}\right) \sin\left(\frac{x_2\kappa_r^2}{2k_2}\right)$$

where x_1 and x_2 are different points on the common horizontal path. The path integrations are collapsed if one makes the usual assumptions about the transmission

distance and eddy size. The following result emerges when the irregularities are
also isotropic:

$$\langle \chi(k_1)\chi(k_2) \rangle = 4\pi^2 k_1 k_2 \int_0^\infty d\kappa \, \kappa \, \Phi_n(\kappa) \left(\frac{2J_1(\kappa a_r)}{\kappa a_r} \right)^2$$

$$\times \int_0^R dx \, \sin\left(\frac{x\kappa^2}{2k_1} \right) \sin\left(\frac{x\kappa^2}{2k_2} \right)$$

Here we have introduced the aperture-averaging weighting function that was derived
in Section 3.2.3. A standard trigonometric relation allows this expression to be
written as the difference of two single-frequency variances:

$$\langle \chi(k_1)\chi(k_2) \rangle = 2\pi^2 k_1 k_2 \int_0^\infty d\kappa \, \kappa \, \Phi_n(\kappa) \left(\frac{2J_1(\kappa a_r)}{\kappa a_r} \right)^2 \int_0^R dx$$

$$\times \left\{ \left\{ 1 - \cos\left[\frac{x\kappa^2}{k} \left(\frac{r+1}{2} \right) \right] \right\} \right.$$

$$\left. - \left\{ 1 - \cos\left[\frac{x\kappa^2}{k} \left(\frac{r-1}{2} \right) \right] \right\} \right\} \qquad (6.8)$$

The following frequency and wavelength ratio will play a central role in our discussions:

$$r = \frac{k_2}{k_1} = \frac{f_2}{f_1} = \frac{\lambda_1}{\lambda_2} \qquad (6.9)$$

The integrations along the path are readily performed for the plane-wave case:

$$\langle \chi(k_1)\chi(k_2) \rangle = 2\pi^2 R k_1 k_2 \int_0^\infty d\kappa \, \kappa \, \Phi_n(\kappa) \left(\frac{2J_1(\kappa a_r)}{\kappa a_r} \right)^2$$

$$\times \left\{ \frac{\sin\left[\frac{R\kappa^2}{k_2} \left(\frac{r-1}{2} \right) \right]}{\frac{R\kappa^2}{k_2} \left(\frac{r-1}{2} \right)} - \frac{\sin\left[\frac{R\kappa^2}{k_2} \left(\frac{r+1}{2} \right) \right]}{\frac{R\kappa^2}{k_2} \left(\frac{r+1}{2} \right)} \right\} \qquad (6.10)$$

The weighting function in braces is plotted in Figure 6.2 as a function of $\kappa \sqrt{R/k}$
for various frequency ratios. It reduces to the weighting function for amplitude
variance when the two frequencies are identical. Aperture averaging and the inner
scale length limit the contributions of large wavenumbers in that case. The situation changes dramatically when the frequencies are different. The influence of
large wavenumbers is suppressed in this case by mutual interference between the
two waves. The separated frequencies provide a *spectral window*, which eliminates
wavenumbers above a specified value. One can move this window by changing

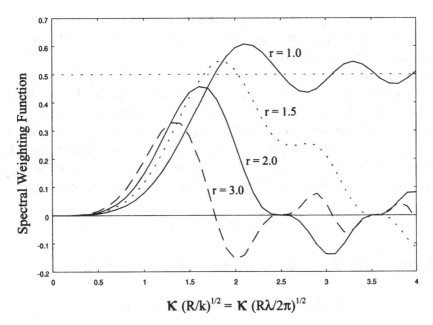

Figure 6.2: Spectral weighting functions for the amplitude covariance measured between different frequencies plotted as a function of wavenumber times the Fresnel length. These curves correspond to the weighting function defined in (6.10). They correpsond to a plane wave incident on the random medium.

the frequency ratio. This selective elimination makes it possible to investigate the small-eddy portion of the spectrum with multifrequency experiments.

Further progress depends on specifying the spectrum of refractive-index irregularities. In the troposphere we use the general model introduced in Section 2.2.6.2 of Volume 1. The dissipation region is characterized by the function $\mathcal{F}(\kappa \ell_0)$ which depends on the inner scale length and multiplies the basic Kolmogorov model for the inertial range. We approximate the outer-scale region by a sharp cutoff at κ_0. Different portions of this spectrum are important depending on the relative values of four important quantities: (1) the Fresnel length, (2) the outer scale length, (3) the inner scale length and (4) the aperture radius of the receiver. The relationships of these quantities to one another depend on the wavelength of the radiation. For that reason we shall treat optical and microwave propagation separately.

6.1.2 Microwave-link Experiments

Simultaneous transmissions at two microwave frequencies along the same path provide an effective way to investigate turbulent irregularities in the lower atmosphere. Microwave signals have several advantages. Their amplitude fluctuations are usually small and thus accurately described by the Rytov approximation. Phase and amplitude can both be measured in these bands without great effort. The frequency separation can be changed rapidly by tuning the transmitter and receiver. These experiments are usually performed on horizontal paths, for which conditions along the path are relatively constant.

The frequencies used in these experiments have ranged from 10 to 110 GHz. The path lengths varied from 4 to 64 km and the corresponding Fresnel lengths lie between 15 and 45 m. These values are usually greater than the antenna size so one can ignore aperture averaging. They are also considerably larger than the inner scale length and the dissipation region of the spectrum need not be included. On the other hand, these Fresnel lengths are comparable to the outer scale length. These experiments are therefore characterized by the following relationships:

$$\ell_0 \ll a_r \ll \sqrt{\lambda R} \simeq L_0 \tag{6.11}$$

which simplifies the description of such experiments considerably.

6.1.2.1 Frequency-correlation Measurements

Most microwave measurements are made with transmitters that are best characterized as omnidirectional. The free-space electric field can therefore be approximated by a spherical wave and the covariance is expressed by (6.8) with the path location x replaced by $x(1 - x/R)$ in the trigonometric terms. The aperture-averaging factor there is dropped in view of the inequality (6.11). These conditions also mean that the spectrum can be replaced by the basic Kolmogorov model with an outer-scale cutoff:

$$\langle \chi(k_1)\chi(k_2) \rangle = 2\pi^2 k_1 k_2 \int_0^R dx \int_{\kappa_0}^\infty d\kappa\, \kappa \frac{0.033 C_n^2}{\kappa^{\frac{11}{3}}} Q(\kappa, x) \tag{6.12}$$

where

$$Q(\kappa, x) = \left\{ 1 - \cos\left[\frac{\kappa^2 x(1 - x/R)}{k_2} \left(\frac{r+1}{2} \right) \right] \right\}$$
$$- \left\{ 1 - \cos\left[\frac{\kappa^2 x (1 - x/R)}{k_2} \left(\frac{r-1}{2} \right) \right] \right\} \tag{6.13}$$

The function $Q(\kappa, x)$ behaves like κ^4 for small wavenumbers so the lower limit κ_0 is not influential. With the substitutions

$$u = \kappa^2 \frac{x(1 - x/R)}{k} \left(\frac{r \pm 1}{2} \right)$$

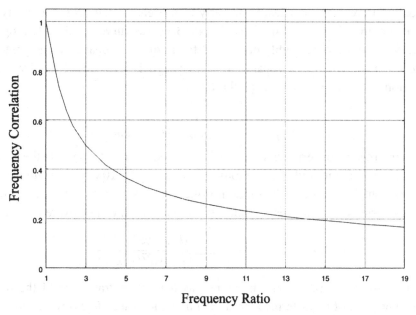

Figure 6.3: Frequency correlation of amplitude fluctuations predicted with the Kolmogorov model by ignoring inner-scale effects and aperture averaging.

the wavelength-dependent terms are separated out from the integrations. The transformed wavenumber integral can be found in Appendix B:

$$\langle \chi(k_1)\chi(k_2) \rangle = 2\pi^2(0.033C_n^2)k_1k_2^{\frac{1}{6}}\left[\left(\frac{r+1}{2}\right)^{\frac{5}{6}} - \left(\frac{r-1}{2}\right)^{\frac{5}{6}}\right]$$

$$\times 1.729 \int_0^R dx \, x^{\frac{5}{6}}(1 - x/R)^{\frac{5}{6}}$$

This expression is divided by the rms values of the amplitude fluctuations measured at the separate frequencies to give the following result:

$$\mathcal{R}(k_1, k_2) = \frac{1}{r^{\frac{5}{12}}}\left[\left(\frac{r+1}{2}\right)^{\frac{5}{6}} - \left(\frac{r-1}{2}\right)^{\frac{5}{6}}\right] \tag{6.14}$$

which depends on the frequency ratio alone. The same expression[1] describes plane-wave microwave signals traveling on horizontal paths [9]. This basic relationship is plotted in Figure 6.3 and suggests that there are surprisingly large correlations

---

[1] For plane waves the equation above (6.14) contains

$$\int_0^R dx \, x^{\frac{5}{6}} \qquad \text{rather than} \qquad \int_0^R dx \, x^{\frac{5}{6}}(1 - x/R)^{\frac{5}{6}}$$

but these terms drop out during normalization.

for widely different frequencies. In an early measurement campaign the utility of terrestrial communications in the band 27–40 GHz was surveyed [10]. The signal strength was found to be highly correlated for frequency separations as great as 6 GHz when the frequency separation of two signals was varied rapidly. That observation is consistent with our prediction.

6.1.2.2 Amplitude-coherence Measurements

Detailed microwave experiments were then mounted in order to measure the coherence of phase and amplitude fluctuations in the range 10–110 GHz. The *amplitude coherence* contains substantially more information than does the simple two-color frequency correlation and is better behaved statistically. It can be expressed as

$$\text{Coh}(k_1, k_2; \omega) = \frac{W(k_1, k_2; \omega)}{\sqrt{W(k_1, k_1; \omega)W(k_2, k_2; \omega)}} \tag{6.15}$$

where the cross-spectral density is defined as the cosine transform of the time-displaced product of logarithmic amplitude fluctuations at different wavenumbers:

$$W_\chi(k_1, k_2; \omega) = 2 \int_0^\infty d\tau \cos(\omega\tau) \langle \chi(k_1, t)\chi(k_2, t + \tau)\rangle \tag{6.16}$$

This definition assumes that the power spectrum is real, but that is the case in our applications.[2] If the irregularities can be characterized by a frozen medium moving across the transmission path, the techniques developed in Chapter 5 yield the following expression for a plane wave:

$$\langle \chi(k_1, t)\chi(k_2, t + \tau)\rangle = 2\pi^2 k_1 k_2 \int_0^\infty d\kappa \, \kappa \, \Phi_n(\kappa) J_0(\kappa U \tau)$$

$$\times \int_0^R dx \sin\left(\frac{\kappa^2 x}{2k_1}\right) \sin\left(\frac{\kappa^2 x}{2k_2}\right) \tag{6.17}$$

Using the integral

$$\int_0^\infty d\tau \cos(\omega\tau) J_0(\kappa U \tau) = \begin{cases} \left(\kappa^2 U^2 - \omega^2\right)^{-\frac{1}{2}} & 0 < \omega < \kappa U \\ 0 & 0 < \kappa U < \omega \end{cases}$$

from Appendix D one can calculate the cosine transform of (6.17) and thus establish a general expression for the cross-spectral density. The Kolmogorov spectrum is

---

[2] The authors of [11] and [12] used the square of this ratio to define the coherence. That poses no problem since data from those experiments was expressed as the square root of their coherence function.

appropriate in view of the inequality (6.11):

$$W_\chi(k_1, k_2; \omega) = 0.132\pi^2 C_n^2 k_1 k_2 \int_{\omega/U}^{\infty} \frac{d\kappa}{\kappa^{\frac{8}{3}} \sqrt{\kappa^2 U^2 - \omega^2}}$$

$$\times \int_0^R dx \sin\left(\frac{\kappa^2 x}{2k_1}\right) \sin\left(\frac{\kappa^2 x}{2k_2}\right) \qquad (6.18)$$

Ishimaru showed how to express these integrals for plane-wave signals in terms of Kummer functions [13], but it is just as easy to compute them numerically. For spherical waves one replaces x by $x(1 - x/R)$. The Stanford group calculated the amplitude coherence from (6.18) for spherical waves and two wavelength ratios [14]. Their results are reproduced in Figure 6.4 and show that scintillation frequencies are suppressed above a distinctive level that depends on the wavelength ratio. This behavior is caused by interference between the two waves. They measured amplitude coherence on a 28-km path using simultaneous transmissions at 11.63 and 34.89 GHz. To enhance path commonality the links were crossed and their average separation was less than 1 m. This is small relative to the Fresnel length and one can safely ascribe any loss of signal coherence to the wavelength separation.[3] The amplitude-coherence data agreed with the theoretical prediction provided by Figure 6.4.

A second experiment was conducted using simultaneous transmissions at 9.6 and 34.52 GHz on a 64-km path in Hawaii [13]. The transmitters were separated by 2.4 m and the receivers by 3.0 m. The phase and amplitude coherences were measured as functions of the scintillation frequency and typical values are reproduced in Figure 6.5. Ishimaru showed that these measurements agree with the predictions of weak scattering [13]. That was confirmed by comparing signals at 9.55, 19.1, 22.2, 25.4 and 33.3 GHz on the same path [12]. Similar measurements were made in the UK using 36 and 110 GHz on a 4-km path [15] and they too agreed with the prediction. There is thus good agreement between microwave measurements and weak-scattering theory based on the Kolmogorov model.

6.1.3 Laser Experiments

One can also make accurate measurements for the frequency correlation of variations in optical intensity. Irradiance fluctuations are linearly related to the logarithmic amplitude variation χ when the scattering is weak, so their frequency

---

[3] Hill and Lataitis considered the case of separated sources and receivers. Their results allow one to describe the results of crossed-path-propagation experiments exactly [8].

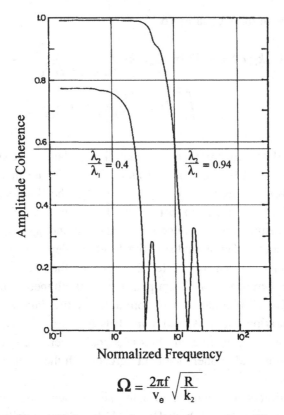

$$\Omega = \frac{2\pi f}{v_e}\sqrt{\frac{R}{k_2}}$$

Figure 6.4: Predictions of the amplitude coherence for a spherical wave using a Kolmogorov spectrum. These results were computed by Mandics, Harp, Lee and Waterman and are expressed as a function of the normalized scintillation frequency for two wavelength ratios [14].

correlations should be the same. In these experiments one uses laser sources and dichroic mirrors to force the waves to travel along the same path, as illustrated in Figure 6.6. The transmitted wave is collimated by lenses and is approximated by a plane wave close to the beam axis where the measurements are usually made. The values of C_n^2 are constant for horizontal paths and one can use the expression (6.10) to describe such measurements. The path should be less than a few hundred meters for the weak-scattering assumption to be valid. This means that the Fresnel length is only a few centimeters and one can safely ignore the outer-scale limit. On the other hand, this value is comparable to the inner scale length and the dissipation range should play an important role. It is this feature which allows chromatic scintillometers to measure the inner scale length. Aperture averaging can often be ignored and these experiments are characterized by the following inequality:

$$\ell_0 \simeq \sqrt{\lambda R} \ll L_0 \tag{6.19}$$

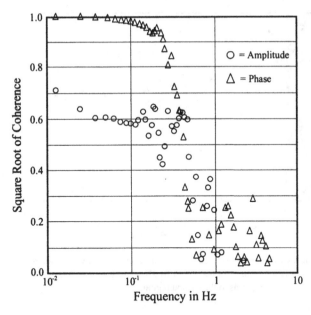

Figure 6.5: Measurements of the phase and amplitude coherences as functions of the scintillation frequency for a 64-km path using simultaneous transmissions at 9.6 and 34.52 GHz made by Janes, Thompson, Smith and Kirkpatrick [11]. © 1970, IEEE.

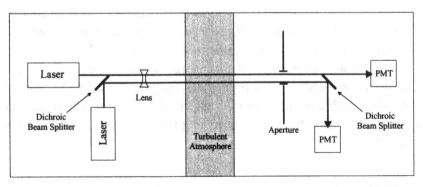

Figure 6.6: The experimental arrangement for measuring the correlation of two laser signals with different wavelengths traveling along the same path.

6.1.3.1 Ignoring Inner-scale Effects

The interpretation of early optical measurements of frequency correlation ignored inner-scale effects. Experimental data was compared with the simplified expression (6.14) that is valid only when $a_r = \ell_0 = 0$. In the first two-color experiment He–Ne and argon lasers were used, but the results were not published in the open literature and attracted little attention [16]. Careful experiments were done later using He–Ne and He–Cd lasers [17]. Path lengths between 650 and 1750 m were used, with conditions ranging from weak to strong scattering. Experimental data from

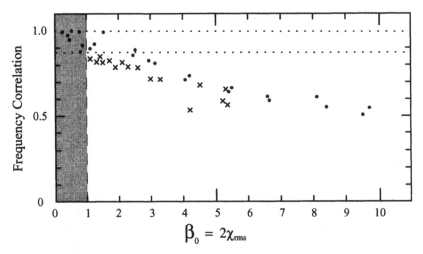

Figure 6.7: Measurements of the frequency correlation of optical intensity fluctuations plotted versus the rms logarithmic amplitude variation. This data was taken by Gurvich, Kan and Pokasov using He–Ne and He–Cd lasers [17]. Weak scattering corresponds to the shaded region.

this study is reproduced in Figure 6.7. The theoretical result $\mathcal{R} = 0.77$ represents reasonable agreement with the data obtained under weak-scattering conditions.

No such agreement was found when a similar experiment was done later using He–Ne and Nd:YAG lasers on a 150-m path [18]. Values ranging from 0.82 to 0.92 were measured under weak-scattering conditions, in contrast to the prediction $\mathcal{R} = 0.70$ for this wavelength ratio. A much larger frequency separation was later achieved with He–Ne and CO_2 lasers [19][20] and produced correlations between 0.20 and 0.45, which are well above the predicted value. It was thus recognized that (6.14) is not a good basis for interpreting two-color optical experiments on horizontal paths.

6.1.3.2 The Inner-scale Influence

This brings one back to the basic relationship (6.19) between the parameters that define the propagation. We noted before that the dissipation range of the spectrum ought to play an important role in the short-path experiments which are often used to ensure that conditions of weak scattering pertain. The question is that of whether its inclusion can close the gap between theory and experiment. The dissipation region is described by the function $\mathcal{F}(\kappa \ell_0)$ introduced in (2.63) of Volume 1 and it must be included in the integration of (6.10). That complicates the calculation considerably and hence one must resort to numerical techniques. Another group used a Gaussian model to calculate $\mathcal{R}$ for various values of the inner scale length [18]. The impact of aperture averaging for various diaphragm openings was also investigated [21][22].

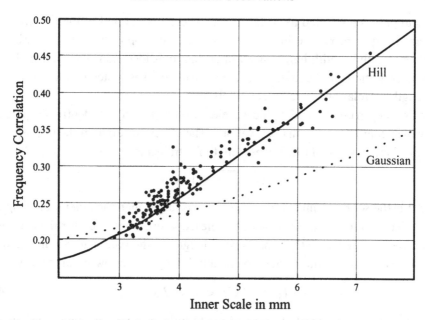

Figure 6.8: A comparison of inner-scale measurements and bichromatic correlation made by Thiermann [20]. The wavelength correlation was measured by comparing the amplitudes of He–Ne and CO_2-laser signals traveling along a common path. The inner scale length was measured with ultrasonic anemometer/thermometers. The solid curves represent two common models used to describe the dissipation region.

These refined predictions agreed with their measurements if the inner scale length was 3–4 mm.

Subsequent experiments went a step further [19][20]. The authors measured the inner scale length with meteorological instruments and used those values to interpret simultaneous measurements of the two-color correlation. They calculated the correlation as a function of the inner scale length using both the Hill bump model and the Gaussian model. The results are reproduced in Figure 6.8 and favor the Hill bump model portrayed in Figure 2.7 of Volume 1.

The early experiment using 0.442- and 0.633-μm lasers was repeated in Boulder [23]. Various path lengths were used in order to investigate both weak- and strong-scattering regimes. The data was compared with the correlation prediction of the Hill bump model and gave good agreement for an inner scale length of 8 mm. These chromatic experiments now provide a sensitive way to measure the inner scale length of atmospheric turbulence in a path-averaged manner and are favored because they do not disturb the medium [24].

6.2 Astronomical Observations

The chromatic flickering of stars at various wavelengths can be observed by placing color filters at the focal plane of a large telescope. The arriving signals are incoherent and accurately represented as plane waves. These optical and infrared waves are influenced only by refractive-index irregularities in the troposphere and the variation of turbulent activity with height is an important consideration. We shall find that turbulent conditions in the lower atmosphere can be investigated by measuring the wavelength correlation of stellar scintillation.

The first step in understanding the wavelength correlation is to identify the parameters which characterize such observations. The effective height of the troposphere H is less than 10 km and the Fresnel length for transmission is

$$\sqrt{\lambda H \sec \vartheta} \approx 10 \text{ cm}$$

where ϑ is the source zenith angle. This value is larger than the inner scale length[4] but smaller than the outer scale length. Telescope openings used for these observations are usually larger than the Fresnel length so the parameters which define these observations are thus related to one another by the following inequality:

$$\ell_0 \ll \sqrt{\lambda H} < a_r \ll L_0 \tag{6.20}$$

This indicates that the inertial range of the turbulence spectrum should provide a reasonable model for calculating chromatic stellar scintillation.

6.2.1 Wavelength Correlation of Stellar Scintillation

Astronomical observations are made at night when the atmosphere is usually stable. If the source is not too close to the horizon, intensity fluctuations can be described by the first-order Rytov solution. This description is complicated by two factors that were not important for propagation along terrestrial links.

The first problem is caused by the differential ray bending that occurs in the upper atmosphere. The paths followed by waves of different wavelengths are displaced from one another when they reach the turbulent region because of dispersion, as illustrated in the exaggerated diagram of Figure 6.9. The path separation ρ depends on four factors: (1) the effective height of the scattering region H, (2) the zenith angle of the star ϑ, (3) the refractive index $n(\lambda)$ and (4) the scale height h_0 of the stable atmosphere above the irregularities:

$$\rho(\lambda_1, \lambda_2, \vartheta) = D_0(\vartheta)[n(\lambda_1) - n(\lambda_2)]\left[1 - \exp\left(-\frac{H}{h_0}\right)\right] \tag{6.21}$$

The function $D_0(\vartheta)$ is a standard astronomical module [25]. This spreading means that we must calculate the joint correlation for different wavelengths at different positions in order to estimate the wavelength correlation.

---

[4] Although there is considerable uncertainty regarding ℓ_0 in the upper atmosphere as noted in Section 2.3.4 of Volume 1.

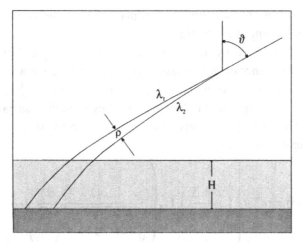

Figure 6.9: Spreading of waves with different wavelengths that is caused by dispersion in the atmosphere above the turbulent region.

The second complication arises because stellar signals travel through the entire atmosphere before reaching ground-based telescopes and turbulent activity varies significantly with altitude. The usual procedure is to represent the spectrum as the product of a locally homogeneous wavenumber spectrum $\Omega(\kappa)$ that does not contain the structure constant and a function that describes the variation of C_n^2 with location. The review of tropospheric measurements in Section 2.3.3 of Volume 1 showed that C_n^2 changes rapidly with altitude but does not vary appreciably in the horizontal plane. This suggests that we can replace the spectrum of inhomogeneous refractive-index fluctuations by

$$\Phi_n\left(\kappa, \frac{\mathbf{r}_1 + \mathbf{r}_2}{2}\right) = \Omega(\kappa) C_n^2\left(\frac{z_1 + z_2}{2}\right) \tag{6.22}$$

where the structure function now depends only on the average height.

To describe the frequency correlation of signals that have traveled along the displaced paths shown in Figure 6.9 we use the techniques that were introduced in Section 4.2.1:

$$\langle \chi(k_1)\chi(k_2)\rangle = 2\pi^2 k_1 k_2 \sec\vartheta \int_0^\infty d\kappa \, \kappa\Omega(\kappa) J_0(\kappa\rho)\left(\frac{2J_1(\kappa a_r)}{\kappa a_r}\right)^2$$

$$\times \int_0^\infty dz \, C_n^2(z)\left\{\cos\left[\frac{z\kappa^2 \sec\vartheta}{k}\left(\frac{r-1}{2}\right)\right]\right.$$

$$\left. - \cos\left[\frac{z\kappa^2 \sec\vartheta}{k}\left(\frac{r+1}{2}\right)\right]\right\} \tag{6.23}$$

When the wavelengths are the same this reduces to the expression (4.38) for the spatial correlation of amplitude fluctuations.

This result was exploited by Tatarskii and Zhukova to express the wavelength correlation as the difference of two spatial correlations at different wavelengths [26]. They used several approximations in doing so. They modeled $C_n^2(z)$ as a uniform slab of thickness H, as suggested in Figure 6.9. They used the inertial range of the spectrum to evaluate the spatial correlations and neglected aperture averaging to find the following expression:

$$
\mathcal{R}(\lambda_1, \lambda_2, \vartheta) = r^{-\frac{5}{12}} \left[\left(\frac{r+1}{2} \right)^{\frac{5}{6}} C_\chi \left(\frac{\rho(\lambda_1, \lambda_2, \vartheta)}{\sqrt{2\pi H \lambda_2 \sec \vartheta \left(\dfrac{r+1}{2} \right)}} \right) \right.
$$
$$
\left. - \left(\frac{r-1}{2} \right)^{\frac{5}{6}} C_\chi \left(\frac{\rho(\lambda_1, \lambda_2, \vartheta)}{\sqrt{2\pi H \lambda_2 \sec \vartheta \left(\dfrac{r-1}{2} \right)}} \right) \right] \qquad (6.24)
$$

In this result $C_\chi(D)$ is the monochromatic spatial correlation function for a Kolmogorov spectrum derived in (4.43). This result can be written in terms of the difference in wavelength:

$$
\epsilon = \frac{\lambda_1 - \lambda_2}{2\lambda_2} = \frac{\Delta\lambda}{2\lambda_2} \qquad (6.25)
$$

The separation $\Delta\lambda$ is usually small compared with the average value in the visible range, so

$$
\mathcal{R}(\Delta\lambda, \vartheta) = \frac{1}{(1 - \epsilon^2)^{\frac{5}{12}}} \left[C_\chi \left(\frac{\rho(\lambda_1, \lambda_2, \vartheta)}{\sqrt{2\pi H \lambda \sec \vartheta}} \right) \right.
$$
$$
\left. - \epsilon^{\frac{5}{6}} C_\chi \left(\frac{\rho(\lambda_1, \lambda_2, \vartheta)}{\epsilon^{\frac{1}{2}} \sqrt{2\pi H \lambda \sec \vartheta}} \right) \right] \qquad (6.26)
$$

This expression was evaluated for several values of the zenith angle and the results are reproduced in Figure 6.10. They are in reasonably good agreement with chromatic observations made at the Pulkovo Observatory in Russia [27].

6.2.2 Inversion of Stellar Correlations

Successful descriptions of stellar scintillation have an important corollary. It should be possible to determine the vertical profile of C_n^2 from ground level by measuring the wavelength correlation of intensity fluctuations. To understand how this is

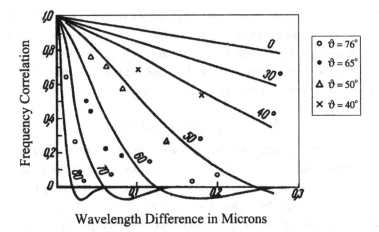

Figure 6.10: Correlation of stellar scintillations expressed as a function of the difference in wavelength for several values of the zenith angle calculated by Tatarskii and Zhukova [26]. The experimental points were measured by Zhukova at the Pulkovo Observatory [27].

possible we return to the general expression (6.23) and reverse the order of integration. Recalling that the separation ρ depends on known quantities through (6.21), the frequency correlation can be written as

$$\mathcal{R}(\Delta\lambda, \vartheta) = \int_0^\infty dz\, C_n^2(z) K(z, \Delta\lambda, \vartheta) \tag{6.27}$$

The kernel is expressed in terms of the plane-wave spatial correlation defined in (4.43):

$$K(z, \Delta\lambda, \vartheta) = \frac{0.545 z^{\frac{5}{6}}}{r^{\frac{5}{12}} \mathcal{N}} \left[\left(\frac{r+1}{2}\right)^{\frac{5}{6}} C_x \left(\frac{\rho(\lambda_1, \lambda_2, \vartheta)}{\sqrt{2\pi z \lambda_2 \sec \vartheta \left(\frac{r+1}{2}\right)}} \right) \right.$$

$$\left. - \left(\frac{r-1}{2}\right)^{\frac{5}{6}} C_x \left(\frac{\rho(\lambda_1, \lambda_2, \vartheta)}{\sqrt{2\pi z \lambda_2 \sec \vartheta \left(\frac{r-1}{2}\right)}} \right) \right] \tag{6.28}$$

where

$$\mathcal{N} = \int_0^\infty dz\, z^{\frac{5}{6}} C_n^2(z) \tag{6.29}$$

This relationship is an integral equation connecting the unknown profile of C_n^2 and the frequency correlation. It can be inverted to give the profile if the correlation is measured for many different wavelength separations or for many different zenith angles. The usual procedure is to choose two wavelengths by inserting optical filters and let the setting of a star generate different zenith angles. This was done using starlight at 0.4 and 0.8 μm from Sirius and the profile was determined from a few hundred meters to 20 km with modest accuracy [28].

6.3 Ionospheric Influences

Now let us return to the galactic radio signals with which we introduced the concept of frequency correlation. The signals from radio stars arrive at the earth as plane waves. Phase and amplitude fluctuations are imposed on them as they pass through the interstellar plasma and through the ionosphere. We concentrate here on ionospheric influences and use the analytical description developed in Section 3.4.1.

6.3.1 The Predicted Correlation

We begin with the expression (3.142) for logarithmic amplitude variations induced by random changes in the electron density:

$$\chi(R, k) = 4\pi^2 r_e \int d^3 r \, A(R, r, k) \, \delta N(r, t) \tag{6.30}$$

In doing so we have made explicit the role played by the electromagnetic wavenumber in that expression. The paraxial approximation can be used to describe the scattering of radio waves in the ionosphere. If we represent the spatial correlation of δN by the Fourier integral of its spectrum Ψ_N we can describe the correlation of amplitude fluctuations measured at different frequencies by the following expression:

$$\langle \chi(k_1)\chi(k_2)\rangle = \pi^2 r_e^2 \lambda_1 \lambda_2 \int d^3\kappa \int_0^\infty ds_1 \int_0^\infty ds_2 \, \Psi_N\left[\kappa, \tfrac{1}{2}(s_1+s_2)\right]$$

$$\times \exp[i\kappa_s(s_1 - s_s)] \sin\left(\frac{s_1\kappa_r^2}{2k_1}\right) \sin\left(\frac{s_2\kappa_r^2}{2k_2}\right) \tag{6.31}$$

The electron-density fluctuations are neither homogeneous nor isotropic, which means that we must deal with the full spectral form indicated here. When we change to sum and difference coordinates it becomes apparent that the wavenumber component along the line of sight must vanish:

$$\langle \chi(k_1)\chi(k_2)\rangle = 2\pi^3 r_e^2 \lambda_1 \lambda_2 \int d^3\kappa \, \delta(\kappa_s)$$

$$\times \int_0^\infty ds \, \Psi_N(\kappa, s) \sin\left(\frac{s\kappa^2}{2k_1}\right) \sin\left(\frac{s\kappa^2}{2k_2}\right) \tag{6.32}$$

The next task is to estimate the four basic parameters which define the propagation conditions. The Fresnel length for ionospheric transmission falls in the range

$$500 \text{ m} < \sqrt{\lambda H} < 2000 \text{ m}$$

for scattering irregularities at 300 km and the wavelengths used in early satellite and radio-astronomy experiments. These values are larger than the sizes of most radio telescopes and one can ignore aperture averaging. The inner scale length is a few meters, which is considerably less than the sizes of the antennas which gather these signals. We examined various estimates for the outer scale length of ionospheric irregularities in Section 2.4.3.2 of Volume 1 and concluded that L_0 is best understood as the transition from horizontal traveling waves to the field-aligned irregularities. The effective outer scale length is probably a kilometer or more in the ionosphere, which is comparable to the Fresnel length. The following relationships therefore characterize these transmission measurements:

$$\ell_0 \ll a_r < \sqrt{\lambda H} \approx L_0 \tag{6.33}$$

This means that we can ignore the upper wavenumber limit for the spectrum. On the other hand, the outer-scale wavenumber κ_0 may play an important role if the spectrum is sufficiently steep.

We will use the power-law spectrum of electron-density fluctuations introduced by (2.135) in Volume 1 to describe the irregularities. That model is combined with the propagation expression (6.31) using the geometry illustrated in Figure 4.16 of Volume 1:

$$\langle \chi(k_1)\chi(k_2) \rangle = 2\pi^3 r_e^2 \lambda_1 \lambda_2 \int_{\kappa_0}^{\infty} d\kappa \, \kappa^2 \int_0^{\pi} d\psi \, \sin\psi \int_0^{2\pi} d\omega \, \delta(\kappa \cos\psi)$$

$$\times \int_0^{\infty} ds \, \sin\left(\frac{s\kappa^2 \sin^2\psi}{2k_1}\right) \sin\left(\frac{s\kappa^2 \sin^2\psi}{2k_2}\right)$$

$$\times \frac{\langle \delta N^2(s) \rangle Q_\nu \kappa_0^{\nu-3}}{\kappa^\nu (\sin^2\Theta + A^2 \cos^2\Theta)^{\nu/2}} \tag{6.34}$$

The angle Θ between the magnetic field and the wavenumber vector κ was established in (3.148).

$$\cos\Theta = \cos\psi \cos\gamma + \sin\psi \sin\gamma \cos\omega$$

The polar integration in (6.34) is collapsed by the delta function and the other integrations can be separated:

$$\langle \chi(k_1)\chi(k_2) \rangle = 8\pi^3 r_e^2 Q_\nu \kappa_0^{\nu-3} \int_0^{\frac{\pi}{2}} d\omega \left[1 + (A^2 - 1) \sin^2\gamma \cos^2\omega \right]^{-\nu/2}$$

$$\times \lambda_1 \lambda_2 \int_0^{\infty} \frac{d\kappa \, \kappa}{\kappa^\nu} \int_0^{\infty} ds \, \langle \delta N^2(s) \rangle \sin\left(\frac{s\kappa^2}{2k_1}\right) \sin\left(\frac{s\kappa^2}{2k_2}\right)$$

$$\tag{6.35}$$

The anisotropic parameters A and γ appear only in the multiplicative integral and therefore do not influence the outcome. The frequency correlation can now be written as

$$\langle \chi(k_1)\chi(k_2) \rangle = C_0 \lambda_1 \lambda_2 \int_{\kappa_0}^{\infty} \frac{d\kappa}{\kappa^{\nu-1}} \int_0^{\infty} ds \, \langle \delta N^2(s) \rangle \sin\left(\frac{s\kappa^2}{2k_1}\right) \sin\left(\frac{s\kappa^2}{2k_2}\right)$$

where C_0 summarizes all terms that do not depend on frequency. We convert to altitude with $s = z \sec \vartheta$ and use a thin-phase-screen model to represent the scattering region:

$$\langle \chi(k_1)\chi(k_2) \rangle = C_1 \lambda_1 \lambda_2 \int_{\kappa_0}^{\infty} \frac{d\kappa}{\kappa^{\nu-1}} \left\{ \left\{ 1 - \cos\left[\frac{\kappa^2 H \sec \vartheta}{k_2}\left(\frac{r+1}{2}\right)\right] \right\} \right.$$
$$\left. - \left\{ 1 - \cos\left[\frac{\kappa^2 H \sec \vartheta}{k_2}\left(\frac{r-1}{2}\right)\right] \right\} \right\}$$

The ionospheric spectral index has been estimated thus far in two ways. From the *in situ* probe measurements reviewed in Section 2.4.2.3 of Volume 1 and the frequency scaling of amplitude fluctuations we judge that this index lies in the range $3 < \nu < 4$. The integrals above converge at small wavenumbers for these values and one can drop the lower limit. A change of variable then brings both integrals into the same form. The normalized frequency correlation becomes

$$\mathcal{R}(k_1, k_2) = \frac{1}{r^{\frac{\nu-2}{4}}} \left[\left(\frac{r+1}{2}\right)^{\frac{\nu}{2}-1} - \left(\frac{r-1}{2}\right)^{\frac{\nu}{2}-1} \right] \tag{6.36}$$

This predicted wavelength correlation is plotted in Figure 6.11 for $\nu = 3$ and $\nu = 4$. Both curves fall slowly with increasing frequency separation. This means that the correlation should be significant for the pairs of frequencies ordinarily used in radio-astronomy and satellite experiments. Since the curves are substantially different, we see that frequency-correlation measurements provide a third way to measure the spectral index of ionospheric irregularities.

The previous description does not recognize bending by ionization above the level of scattering. The overlying ambient ionization bends the two waves differently because the ionosphere is dispersive. In that refinement, one must consider the correlation of waves with both frequency separation and path separation. Source size can also influence these measurements and detailed comparisons should include that refinement. Combining all these factors poses a difficult computational challenge and will not be pursued here.

6.3.2 Radio-astronomy Measurements

A number of pioneering experiments used signals from galactic sources to measure the frequency correlation and they are summarized in Table 6.1. Burrows

Table 6.1: *Frequency correlation measurements using galactic radio sources*

Location	f_1 (MHz)	f_2 (MHz)	Ratio	Observed correlation	Ref.
Cambridge	81.5	118.5	1.45	0.61 ± 0.13	1
Cambridge	45	81.5	1.81	Considerable	29
Boston	63	113	1.79	0.43 (0.2–0.8)	30
	113	228	2.02	0.68 (0.5–0.9)	
Cambridge	26	79	3.04	0.16 (−0.11 to 0.42)	31
Alaska	68	223	3.25	0.30 (0.15–0.50)	32
Sydney	60	200	1.23	Negligible	33

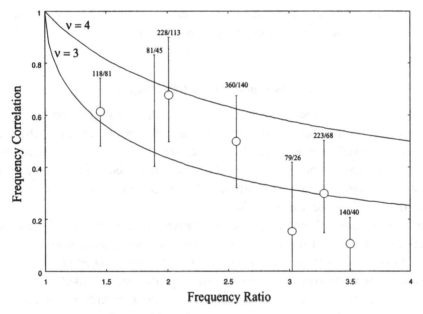

Figure 6.11: Predicted frequency correlations for two values of the spectral index. Experimental data points for satellite and galactic sources are overplotted with experimental error bars.

and Little were the first to do so, using 81.5- and 118.5-MHz signals from Cassiopeia [1]. Signals at these frequencies were received with a 30-foot parabolic reflector using cross-polarized feeds. A typical segment of data from this experiment is reproduced in Figure 6.1 and the correlation was measured over many hours to be $\mathcal{R} = 0.61 \pm 0.13$. That result is consistent with Smith's comparison of 45- and 81.5-MHz signals, for which he found considerable correlation [29].

This technique was later extended to other galactic sources and different frequencies. Aarons, Allen and Elkins compared 113- and 228-MHz signals [30] to find correlations between 0.5 and 0.9, with a mean value of 0.7. Succeeding measurements in this program showed that the correlation for 113- and 228-MHz signals was approximately 0.45, whereas that between 40- and 228-MHz signals was about 0.35. Chivers found negligible correlation between 26- and 79-MHz signals measured at adjacent receivers [31]. Lansinger and Fremouw measured the correlation between 68- and 223-MHz signals from Cassiopeia at an auroral site and found values ranging from 0.15 to 0.50, depending on the zenith angle of the source [32]. These observations are generally consistent with the correlations predicted for weak-scattering conditions that were plotted in Figure 6.11. An exception was noted by Bolton and Stanley, who found no significant correlation between signals in the range 60–200 MHz. On the other hand, their observations were made from Australia and used sources very close to the horizon, for which strong scattering was probably an important factor [33].

6.3.3 Observations of Satellite Signals

A different approach to measuring frequency correlation employs coherent signals transmitted by orbiting spacecraft. These signals are not influenced by the solar plasma and interstellar conditions. Rather, they represent the unique consequence of propagation through ionospheric irregularities. Simultaneous observations of 54- and 150-MHz signals from a low-altitude satellite [34] gave correlations between 0.2 and 0.3, in fair agreement with Figure 6.11.

Satellites in geosynchronous orbit provide excellent platforms for these experiments because they do not move relative to receiving ground stations. In the ATS-6 Radio Beacon experiment coherent signals were radiated at 40, 140 and 360 MHz. Amplitude fluctuations for all three signals were recorded in Colorado. Frequency correlations were later computed from these records for two pairs of signals, 40 versus 140 MHz and 140 versus 360 MHz [35]. Results from these comparisons are reproduced in Figure 6.12 and indicate that the correlation declines as the scintillation index S_4 increases. That makes sense because the frequency correlation should be smaller for strong amplitude fluctuations than it is for weak scattering. At the other extreme, these estimates become sensitive to receiver noise when the scintillation level is very low. We therefore emphasize points in the unshaded boxes of Figure 6.12. Correlations for the first pair of signals, at 140 and 360 MHz, agree fairly well with the prediction. Correlations of the 40- and 140-MHz signals do not agree so well and probably indicate the influence of strong scattering.

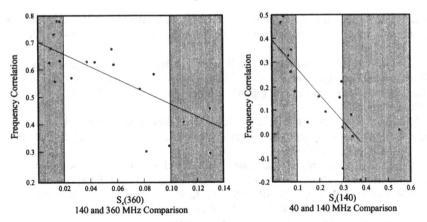

Figure 6.12: Frequency correlation of amplitude fluctuations measured with coherent signals received from the ATS-6 geosyncrhonous satellite at 40, 140 and 360 MHz by Umeki, Liu and Yeh [35]. The results are expressed as a function of the scintillation index S_4 measured at the higher frequency.

6.4 Problem

Problem 1

For a horizontal path and the Kolmogorov spectrum show that the amplitude coherence at zero scintillation frequency is given by

$$\text{Coh}(k_1, k_2; 0) = \frac{1}{r^{\frac{2}{3}}}\left[\left(\frac{r+1}{2}\right)^{\frac{4}{3}} - \left(\frac{r-1}{2}\right)^{\frac{4}{3}}\right]$$

Compute appropriate values for the microwave data reproduced in Figure 6.5 and compare the results with the measured zero-frequency intercepts.

References

[1] K. Burrows and C. G. Little, "Simultaneous Observations of Radio Star Scintillations on Two Widely Spaced Frequencies," *Jodrell Bank Annals*, **1**, No. 2, 29–35 (December 1952).

[2] M. F. Bakhareva, "The Correlation Between Waves of Different Frequencies Travelling Through a Layer of Statistically Inhomogeneous Medium," *Radiotekhnika i Electronika (Radio Engineering and Electronic Physics)*, **4**, No. 1, 141–155 (1959).

[3] R. B. Muchmore and A. D. Wheelon, "Frequency Correlation of Line-of-Sight Signal Scintillations," *IEEE Transactions on Antennas and Propagation*, **AP-11**, No. 1, 46–51 (January 1963).

[4] K. G. Budden, "The Theory of the Correlation of Amplitude Fluctuations of Radio Signals at Two Frequencies, Simultaneously Scattered by the Ionosphere," *Journal of Atmospheric and Terrestrial Physics*, **27**, No. 8, 883–897 (August 1965).

[5] E. N. Bramley, "Correlation of Signal Fluctuations at Two Frequencies in Propagation Through an Irregular Medium," *Proceedings of the IEE*, **115**, No. 10, 1439–1442 (October 1968).

[6] D. L. Fried, "Spectral and Angular Covariance of Scintillation for Propagation in a Randomly Inhomogeneous Medium," *Applied Optics*, **10**, No. 4, 721–731 (April 1971).

[7] I. M. Fuks, "Correlation of the Fluctuations of Frequency-Spaced Signals in a Randomly Inhomogeneous Medium," *Izvestiya Vysshikh Uchebnykh Zavedenii Radiofizika (Radiophysics and Quantum Electronics)*, **17**, No. 11, 1272–1276 (November 1974).

[8] R. J. Hill and R. J. Lataitis, "Effect of Refractive Dispersion on the Bichromatic Correlation of Irradiances for Atmospheric Scintillation," *Applied Optics*, **28**, No. 19, 4121–4125 (1 October 1989).

[9] R. W. Lee and J. C. Harp, "Weak Scattering in Random Media, with Applications to Remote Sensing," *Proceedings of the IEEE*, **57**, No. 4, 375–406 (April 1969).

[10] J. F. Roche, H. Lake, D. T. Worthington, C. K. H. Tsao and J. T. de Bettencourt, "Radio Propagation at 27–40 GHz," *IEEE Transactions on Antennas and Propagation*, **AP-18**, No. 4, 452–462 (July 1970).

[11] H. B. Janes, M. C. Thompson, D. Smith and A. W. Kirkpatrick, "Comparison of Simultaneous Line-of-Sight Signals at 9.6 and 34.5 GHz," *IEEE Transactions on Antennas and Propagation*, **AP-18**, No. 4, 447–451 (July 1970).

[12] M. C. Thompson, L. E. Wood, H. B. Janes and D. Smith. "Phase and Amplitude Scintillations in the 10 to 40 GHz Band," *IEEE Transactions on Antennas and Propagation*, **AP-23**, No. 6, 792–797 (November 1975).

[13] A. Ishimaru, "Temporal Frequency Spectra of Multifrequency Waves in Turbulent Atmosphere," *IEEE Transactions on Antennas and Propagation*, **AP-20**, No. 1, 10–19 (January 1972).

[14] P. A. Mandics, J. C. Harp, R. W. Lee and A. T. Waterman, Jr, "Multifrequency Coherences of Short-Term Fluctuations of Line-of-sight Signals – Electromagnetic and Acoustic," *Radio Science*, **9**, No. 8–9, 723–731 (August–September 1974).

[15] N. D. Mavrokoukoulakis, K. L. Ho and R. S. Cole, "Temporal Spectra of Atmospheric Amplitude Scintillations at 110 GHz and 36 GHz," *IEEE Transactions on Antennas and Propagation*, **AP-26**, No. 6, 875–877 (November 1978).

[16] M. W. Fitzmaurice, "Experimental Investigations of Optical Propagation in Atmospheric Turbulence," NASA Technical Report R-370, see especially pages 99–124, NASA Goddard Space Flight Center Report, Greenbelt, MD 20771 (August 1971).

[17] A. S. Gurvich, V. Kan and V. V. Pokasov, "Two-frequency Fluctuations of Light Intensity in a Turbulent Medium," *Optica Acta* (Taylor and Francis Ltd, http://www.tandf.co.uk/journals), **26**, No. 5, 555–562 (May 1979).

[18] N. Ben-Yosef, E. Goldner and A. Weitz, "Two-color Correlation of Scintillations," *Applied Optics*, **25**, No. 19, 3486–3489 (1 October 1986).

[19] E. Azoulay, V. Thiermann, A. Jetter, A. Kohnle and Z. Azar, "Optical Measurement of the Inner Scale of Turbulence," *Journal of Physics D: Applied Physics*, **21**, No. 10, S41–S44 (14 October 1988).

[20] V. Thiermann and E. Azoulay, "Modeling of Structure Constant and Inner Scale of Refractive Index Fluctuations – an Experimental Investigation," *Proceedings of the SPIE 1115, Propagation Engineering*, 124–135 (March 1989). (This work was presented by Thierman for his Ph.D. dissertation at Hamburg in 1990.)

[21] Z. Azar, H. M. Loebenstein, G. Appelbaum, E. Azoulay, U. Halavee, M. Tamir and M. Tur, "Aperture Averaging of the Two-wavelength Intensity Covariance Function in Atmospheric Turbulence," *Applied Optics*, **24**, No. 15, 2401–2407 (1 August 1985).

[22] O. Shoham, E. Goldner, A. Weitz and N. Ben-Yosef, "Aperture Averaging Effects on the Two-color Correlation of Scintillations," *Applied Optics*, **27**, No. 11, 2157–2160 (1 June 1988).

[23] J. H. Churnside, R. J. Lataitis and J. J. Wilson, "Two-color Correlation of Atmospheric Scintillation," *Applied Optics*, **31**, No. 21, 4285–4290 (20 July 1992).

[24] R. J. Hill, "Comparison of Scintillation Methods for Measuring the Inner Scale of Turbulence," *Applied Optics*, **27**, No. 11, 2187–2193 (1 June 1988).

[25] J. M. Pernter and F. M. Exner, *Meteorologische Optik* (W. Braumüller, Vienna and Leipzig, 1910), 173.

[26] V. I. Tatarskii and L. N. Zhukova, "The Chromatic Flickering of Stars," *Doklady Akademii Nauk SSSR (Soviet Physics – Doklady)*, **124**, No. 3, 112–116 (April 1960).

[27] L. N. Zhukova, "Chromatic Scintillations," *Astronomicheskii Zhurnal (Soviet Astronomy)*, **3**, No. 1, 533–534 (January–February 1959).

[28] J. L. Caccia, J. Vernin and M. Azouit, "Structure Function C_n^2 Profiling by Two-color Stellar Scintillation with Atmospheric Dispersion," *Applied Optics*, **27**, No. 11, 2229–2235 (1 June 1988).

[29] F. G. Smith, "Origin of the Fluctuations in the Intensity of Radio Waves from Galactic Sources," *Nature*, **165**, No. 4194, 422–423 (18 March 1950).

[30] J. Aarons, R. S. Allen and T. J. Elkins, "Frequency Dependence of Radio Star Scintillations," *Journal of Geophysical Research*, **72**, No. 11, 2891–2902 (1 June 1967).

[31] H. J. A. Chivers, "The Simultaneous Observation of Radio Star Scintillations on Different Radio-Frequencies," *Journal of Atmospheric and Terrestrial Physics*, **17**, No. 3, 181–187 (1960).

[32] J. M. Lansinger and E. J. Fremouw, "The Scale Size of Scintillation-producing Irregularities in the Auroral Ionosphere," *Journal of Atmospheric and Terrestrial Physics*, **29**, No. 10, 1229–1242 (October 1967).

[33] J. G. Bolton and G. J. Stanley, "Observations on the Variable Source of Cosmic Radio Frequency Radiation in the Constellation of Cygnus," *Australian Journal of Scientific Research, A*, **1**, 58–69 (March 1948).

[34] J. L. Jespersen and G. Kamas, "Satellite Scintillation Observations at Boulder, Colorado," *Journal of Atmospheric and Terrestrial Physics*, **26**, No. 4, 457–473 (April 1964).

[35] R. Umeki, C. H. Liu and K. C. Yeh, "Multifrequency Studies of Ionospheric Scintillations," *Radio Science*, **12**, No. 2, 311–317 (March–April 1977).

7

Phase Fluctuations

This volume has concentrated on amplitude fluctuations so far. We need now to understand how the Rytov approximation modifies the descriptions of phase fluctuations that were generated in Volume 1. We found there that geometrical optics provides a surprisingly accurate model for phase variance, the phase structure function and angle-of-arrival variations. We will find that those descriptions are modified very little by diffraction effects. We have delayed addressing this topic for that reason.

The phase reference of the received signal is established by the unperturbed field. Variations about this reference are large for optical links. In the case of microwave transmission they are usually small and comparable to the amplitude variations. It is necessary to compute the phase only to second order. We base our examination on Yura's version of the second-order Rytov solution (2.64). The phase fluctuation is identified with the imaginary term in the exponent of that expression

$$\varphi = b + d - ab \qquad (7.1)$$

where a and b are the real and imaginary parts of the scattering integral. The term d is second order and is described by the imaginary part of the double-scattering expression (2.63). The first step is to estimate the average phase fluctuation:

$$\langle \varphi \rangle = \langle b \rangle + \langle d \rangle - \langle ab \rangle$$

We note that $\langle b \rangle$ vanishes because it is represented by a volume integral of $\delta\varepsilon$ whose ensemble average is zero. This means that $\langle \varphi \rangle$ depends solely on the second-order terms. The mean value of d is small for most cases of practical interest. In fact, when the signal is a plane or spherical wave, we will show in Chapter 8 that $\langle d \rangle$ actually vanishes for terrestrial and astronomical paths. Therefore to second order

$$\langle \varphi \rangle = -\langle ab \rangle = -\langle \varphi_1 \chi_1 \rangle \qquad (7.2)$$

and the average phase shift of the disturbed field is negative.

This focuses attention on the *correlation of phase and amplitude*. That quantity could not be calculated with geometrical optics because it requires an accurate description of the logarithmic amplitude fluctuation. This combination influences the distribution of phase fluctuations and the bounds placed on their description by the Rytov approximation. One of the important tasks in the present chapter is to estimate this quantity. Expressions for it are developed in Section 7.4 and numerical estimates are made for typical propagation situations. In the Fresnel regime, in which optical signals operate, we shall find that it is proportional to $\langle \chi^2 \rangle$ and therefore modest. For microwave signals the situation can be quite different.

The *phase fluctuation* about the mean level is the quantity that is invariably measured. It is described to first order by the basic relationship (2.28). The phase variance is the primary measurement of such experiments and it can be estimated to first order from (7.1) as

$$\langle \varphi^2 \rangle = \langle b^2 \rangle = k^4 \left\langle \left| \int d^3r \, \delta\varepsilon(\mathbf{r}, t) \, \Im \left(G(\mathbf{R}, \mathbf{r}) \frac{E_0(\mathbf{r})}{E_0(\mathbf{R})} \right) \right|^2 \right\rangle \qquad (7.3)$$

We show how to calculate this quantity in Section 7.1. The task is considerably more difficult than it was with geometrical optics because the line integrals of that approximation[1] are replaced by the volume integral indicated here. Large eddies dominate the outcome in both approximations and the results are surprisingly similar. We shall find that one can circumvent these difficult calculations by using invariance relations in conjunction with the geometrical results generated in Volume 1.

Similar invariance relations can be used to establish the diffraction corrections for the phase structure function. These techniques and their results are presented in Section 7.2. The power spectra of phase fluctuations are examined in Section 7.3. The distribution of signal phase is developed in Section 7.5 and found to be Gaussian in all situations of practical interest. In these examinations we find that diffraction provides only minor refinements to the results of geometrical optics.

7.1 The Phase Variance

Geometrical optics provides a robust description for the variance of phase fluctuations. The discussion in Section 4.1.3 of Volume 1 showed that it is proportional to the first moment of the turbulence spectrum when the irregularities are reasonably isotropic:

$$\text{Geometrical optics:} \qquad \langle \varphi_0^2 \rangle = 4\pi^2 k^2 R \int_0^\infty d\kappa \, \kappa \, \Phi_n(\kappa) \qquad (7.4)$$

This description is strongly influenced by the large-eddy portion of the spectrum.

---

[1] See equation (3.21) in Volume 1.

We now ask how diffraction influences this important parameter. The Rytov approximation describes the phase variance to first order by (7.3). In practice, it is far more convenient to work with the imaginary part of the complex weighting function that was defined in (2.39). With this approach we can express the variance as

$$\text{Rytov:} \qquad \langle \varphi^2 \rangle = 4 \int d^3\kappa \, \Phi_n(\kappa) E(\kappa) E(-\kappa) \qquad (7.5)$$

where the effects of diffraction are contained in the phase weighting function $E(\kappa)$. To estimate its influence we examine the different types of wave separately.

7.1.1 The Phase Variance for Plane Waves

The phase weighting function for plane waves was first calculated with the paraxial approximation in (3.12). For a path near the surface that expression can be recast in terms of coordinates whose origin coincides with the transmitter as illustrated in Figure 3.3:

$$\text{Plane wave:} \qquad E(\kappa) = \frac{k}{2} \int_0^R dx \exp(ix\kappa_x) \cos\left(\kappa_r^2 \frac{(R-x)}{2k} \right) \qquad (7.6)$$

The phase variance is constructed from (7.5) using the shift $x' = R - x$ to give

$$\langle \varphi^2 \rangle = k^2 \int d^3\kappa \, \Phi_n(\kappa) \int_0^R dx_1' \int_0^R dx_2' \exp[i\kappa_x(x_2' - x_1')] \cos\left(\frac{x_1'\kappa_r^2}{2k} \right) \cos\left(\frac{x_2'\kappa_r^2}{2k} \right)$$

The arguments that led to (3.27) allow one to collapse the horizontal integrations into a single integral along the path:

$$\langle \varphi^2 \rangle = 2\pi k^2 \int d^3\kappa \, \Phi_n(\kappa)\delta(\kappa_x) \int_0^R dx \cos^2\left(\frac{x\kappa_r^2}{2k} \right) \qquad (7.7)$$

Here we have dropped the prime on the horizontal distance variable.

If we assume that the irregularities are isotropic, the wavenumber integration is conveniently expressed in spherical coordinates and the angular integrations can be performed to give

$$\langle \varphi^2 \rangle = 4\pi^2 k^2 \int_0^\infty d\kappa \, \kappa \Phi_n(\kappa) \int_0^R dx \cos^2\left(\frac{x\kappa^2}{2k} \right) \qquad (7.8)$$

This is the analog of the amplitude description for plane waves given in (3.28). The

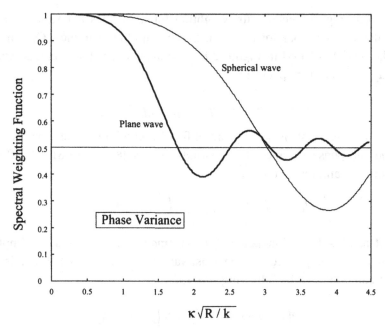

Figure 7.1: Phase-variance weighting functions for plane and spherical waves plotted as functions of the wavenumber times the Fresnel length.

horizontal integration can be done [1][2] to give

$$\langle \varphi^2 \rangle = 4\pi^2 R k^2 \int_0^\infty d\kappa \, \kappa \, \Phi_n(\kappa) F_\varphi(\kappa) \tag{7.9}$$

where

$$\text{Plane wave:} \quad F_\varphi(\kappa) = \frac{1}{2}\left(1 + \frac{\sin(R\kappa^2/k)}{R\kappa^2/k}\right) \tag{7.10}$$

The influence of diffraction is represented by this *spectral weighting function for phase variance*. It depends on the product of the wavenumber and the Fresnel length. This function is plotted in Figure 7.1 together with a similar result for spherical waves that we will soon derive. Small wavenumbers correspond to Fresnel scattering and invariably describe optical propagation. The phase weightings for plane and spherical waves both start at unity but soon diverge. Recall that this was also the case for their amplitude weighting functions plotted in Figure 3.4. Fraunhofer scattering corresponds to large wavenumbers relative to the Fresnel length and there the weighting functions for the two types of wave coalesce.

One can proceed to evaluate the phase-variance expression for various models of the turbulence spectrum. In doing so, one must pay special attention to the large-eddy region because its weighting function does not vanish for small wavenumbers – as it does for the logarithmic amplitude variance. One must use one of the models plotted in Figure 2.8 of Volume 1 that try to characterize the large-eddy region of the spectrum.

7.1.1.1 The Sum Rule

The computational task is greatly simplified by combining the expressions for amplitude and phase variance. If we simply add (3.28) and (7.8) the diffraction terms drop out[2] and one is left with

$$\langle \varphi^2 \rangle + \langle \chi^2 \rangle = 4\pi^2 R k^2 \int_0^\infty d\kappa \, \kappa \, \Phi_n(\kappa)$$

The right-hand side is independent of diffraction effects and is simply the geometrical-optics expression for the phase variance given by (7.4). With this recognition we can establish the following *sum rule*:

$$\text{Plane wave:} \qquad \langle \varphi^2 \rangle + \langle \chi^2 \rangle = \langle \varphi_0^2 \rangle \qquad\qquad (7.11)$$

This is the first of several useful invariance relations connecting phase and amplitude measurements. It does not depend on the assumption of isotropy, since one could have obtained the same answer by adding (3.27) and (7.7). Moreover, we shall learn in Chapter 8 that it does not depend on the paraxial approximation and is also valid for wide-angle scattering.

We will exploit the relationship (7.11) in a powerful way. We can calculate the phase variance from the amplitude-variance estimates made in Chapter 3 and the geometrical-optics results from Chapter 4 in Volume 1 since

$$\langle \varphi^2 \rangle = \langle \varphi_0^2 \rangle - \langle \chi^2 \rangle$$

The geometrical-optics phase variance was estimated in (4.12) of Volume 1 for the von Karman model as follows:

$$\langle \varphi_0^2 \rangle = 0.782 R k^2 C_n^2 \kappa_0^{-\frac{5}{3}} \qquad\qquad (7.12)$$

although other outer-scale models give substantially the same result. By contrast, the amplitude variance depends decisively on the nature of the scattering process and is determined by the wavelength of the radiation.

---

[2] It is no more complicated than

$$\sin^2\left(\frac{\chi \kappa^2}{2k}\right) + \cos^2\left(\frac{\chi \kappa^2}{2k}\right) = 1$$

7.1.1.2 Fresnel Scattering

In the Fresnel region we found in (3.35) that

$$\langle \chi^2 \rangle = 0.307 C_n^2 R^{\frac{11}{6}} k^{\frac{7}{6}}$$

so we can express the phase variance as follows:

Fresnel: $\qquad \langle \varphi^2 \rangle = \langle \varphi_0^2 \rangle \left[1 - 0.393 \left(\frac{R \kappa_0^2}{k} \right)^{\frac{5}{6}} \right]$ $\qquad$ (7.13)

This result is little different than the geometrical-optics estimate because the scattering parameter is small in this regime.

7.1.1.3 Fraunhofer Scattering

For microwave links the Fresnel length is large and can be comparable to the outer scale length of tropospheric irregularities. That situation is sometimes realized on millimeter-wave links near the surface [3]. Fraunhofer scattering characterizes the propagation in those situations and the influential eddies lie in the range

$$\kappa \sqrt{R/k} > 1$$

This shifts attention to the wavenumber range in which the weighting function is oscillating around its asymptotic value 0.5. To examine this region, we replot the plane wave weighting function in Figure 7.2 for larger values of $\kappa \sqrt{R/k}$ than those used in Figure 7.1. For comparison, we have added an expanded version of the amplitude-variance weighting function that was first presented in Figure 3.4.

If the outer scale length is smaller than the Fresnel length, it means that the turbulence spectrum has significant energy in the wavenumber range for which both the phase weighting and the amplitude weighting are approximately one half. In that case we can estimate the phase variance by graphical means:

Fraunhofer: $\qquad \langle \varphi^2 \rangle \simeq 4\pi^2 R k^2 \int_0^\infty d\kappa \, \kappa \, \Phi_n(\kappa) \frac{1}{2} \to \frac{1}{2} \langle \varphi_0^2 \rangle$

Since the amplitude weighting is also one half in this region we can write

Fraunhofer: $\qquad \langle \varphi^2 \rangle = \frac{1}{2} \langle \varphi_0^2 \rangle = \langle \chi^2 \rangle$ $\qquad$ (7.14)

Notice that this outcome confirms the sum rule (7.11).

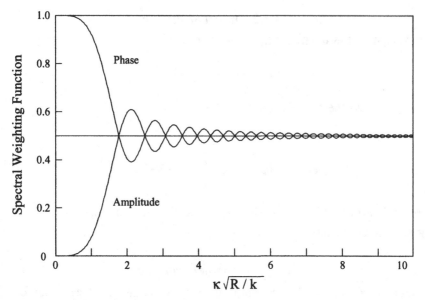

Figure 7.2: Phase- and amplitude-variance weighting functions for plane-wave propagation near the surface. The scale here is considerably expanded relative to that used in Figure 3.4 and Figure 7.1.

7.1.1.4 Intermediate Cases

One should ask what happens between the Fresnel and Fraunhofer scattering regimes where we cannot rely on either (7.13) or (7.14) to estimate the phase variance. This is the region in which the outer-scale scattering parameter

$$\zeta_0 = \frac{\kappa_0^2 R}{k} = \frac{2\pi \lambda R}{L_0^2}$$

is neither large nor small. Fortunately, we laid the foundation for describing this situation in (3.41) where we introduced the outer-scale factor

$$\mathbb{O}(\zeta_0) = \mathbb{O}\left(\frac{2\pi \lambda R}{L_0^2}\right)$$

which multiplies the plane-wave amplitude variance (3.35) in the general case. This function was plotted in Figure 3.6 for two common models that represent the large-eddy region of the spectrum. We can write the amplitude variance in terms of the phase-variance reference as

$$\langle \chi^2 \rangle = 0.393 \eta^{\frac{5}{6}} \mathbb{O}(\zeta_0) \langle \varphi_0^2 \rangle \tag{7.15}$$

With the sum rule (7.11) we can express the phase variance in a way that includes both diffraction and outer-scale effects:

$$\langle \varphi^2 \rangle = \langle \varphi_0^2 \rangle [1 - 0.393 \eta^{\frac{5}{6}} \mathbb{O}(\zeta_0)] \tag{7.16}$$

When ζ_0 is small, $\mathbb{O}(\zeta_0)$ is essentially unity and the Fresnel result is recovered. At the opposite extreme, one can use the asymptotic expression for $\mathbb{O}(\zeta_0)$ given above (3.43) to confirm the Fraunhofer result.

7.1.2 The Phase Variance for Spherical Waves

We can establish similar expressions for spherical waves propagating along terrestrial links. All that we require is the phase weighting function $E(\kappa)$. This can be inferred from $\Lambda(\kappa)$ given by (3.47):

$$\text{Spherical wave:} \quad E(\kappa) = \frac{k}{2} \int_0^R dx \exp(i x \kappa_x) \cos\left(\kappa_r^2 \frac{x(R-x)}{2kR}\right) \quad (7.17)$$

The steps followed to reach expressions for $\langle \chi^2 \rangle$ in Section 3.2.2 then yield [4][5]

$$\langle \varphi^2 \rangle = 4\pi^2 R k^2 \int_0^\infty d\kappa \, \kappa \Phi_n(\kappa) \int_0^1 du \cos^2\left(\frac{\kappa^2 R u (R-u)}{2k}\right) \quad (7.18)$$

when the irregularities are isotropic. If we write this as

$$\langle \varphi^2 \rangle = 4\pi^2 R k^2 \int_0^\infty d\kappa \, \kappa \Phi_n(\kappa) F_\varphi(\kappa) \quad (7.19)$$

the *phase-variance weighting function for spherical waves* can be expressed in terms of Fresnel integrals [6]:

$$\text{Spherical wave:} \quad F_\varphi(\kappa) = \frac{1}{2}\left\{ 1 + \sqrt{\frac{2\pi k}{R\kappa^2}} \left[\cos\left(\frac{R\kappa^2}{4k}\right) C\left(\sqrt{\frac{R\kappa^2}{2\pi k}}\right) \right.\right.$$

$$\left.\left. + \sin\left(\frac{R\kappa^2}{4k}\right) S\left(\sqrt{\frac{R\kappa^2}{2\pi k}}\right) \right] \right\} \quad (7.20)$$

This function is also plotted in Figure 7.1 as a function of $\kappa \sqrt{R/k}$. One notes there that it behaves differently than the corresponding plane-wave weighting function. The two curves begin and end at the same values. In between they represent quite different responses to eddies of different sizes.

On adding (3.51) and (7.20) one observes that the sum rule is also valid for this type of signal:

$$\text{Spherical wave:} \quad \langle \varphi^2 \rangle + \langle \chi^2 \rangle = \langle \varphi_0^2 \rangle \quad (7.21)$$

where the reference value is again given by (7.12). One can exploit this relation to describe the phase variance for optical links:

$$\text{Fresnel:} \qquad \langle \varphi^2 \rangle = \langle \varphi_0^2 \rangle \left[1 - 0.165 \left(\frac{R\kappa_0^2}{k} \right)^{\frac{5}{6}} \right] \qquad (7.22)$$

In the Fraunhofer regime one has the same result as that found for plane waves:

$$\text{Fraunhofer:} \qquad \langle \varphi^2 \rangle = \langle \chi^2 \rangle = \tfrac{1}{2} \langle \varphi_0^2 \rangle$$

7.1.3 The Phase Variance for Beam Waves

The foundation for describing beam waves was set forth in Section 3.2.5. Using that basis, a general expression for the amplitude variance was established in (3.92). Ishimaru showed that the phase variance is given by a similar result that differs from that expression by a simple change of sign [7]:

$$\langle \varphi^2(\rho, R) \rangle = 2\pi^2 k^2 \int_0^R dx \int_0^\infty d\kappa \, \kappa \, \Phi_n(\kappa) \left\{ J_0(2i\kappa\rho\gamma_2) \exp\left(-\frac{\kappa^2 \gamma_2 (R - x)}{k} \right) \right.$$
$$\left. + \Re \left[\exp\left(-i\frac{\kappa^2 \gamma (R - x)}{k} \right) \right] \right\} \qquad (7.23)$$

In using this description we must remember that γ and γ_2 depend on the distance from the transmitter through the definitions (3.89).

It is again helpful to separate a term that contains the entire influence of the outer-scale region of the spectrum. To this end we add (3.92) and (7.23):

$$\langle \varphi^2(\rho, R) \rangle + \langle \chi^2(\rho, R) \rangle = 4\pi^2 k^2 \int_0^R dx \int_0^\infty d\kappa \, \kappa \, \Phi_n(\kappa)$$
$$\times J_0(2i\kappa\rho\gamma_2) \exp\left(-\frac{\kappa^2 \gamma_2(x)(R - x)}{k} \right)$$

If we rearrange the integrand in the following way:

$$J_0(b\kappa) \exp(-a\kappa^2) = 1 - \left[1 - \exp(-a\kappa^2) \right] - \left[1 - J_0(b\kappa) \right] \exp(-a\kappa^2)$$

we can write the sum of the phase and amplitude variances as

$$\langle \varphi^2(\rho, R) \rangle + \langle \chi^2(\rho, R) \rangle = 4\pi^2 R k^2 \int_0^\infty d\kappa \, \kappa \, \Phi_n(\kappa) - T_2 - T_3$$

where

$$T_2 = 4\pi^2 k^2 \int_0^R dx \int_0^\infty d\kappa \, \kappa \, \Phi_n(\kappa) \left[1 - \exp\left(-\frac{\kappa^2 \gamma_2(x)(R - x)}{k} \right) \right]$$

and

$$T_3 = 4\pi^2 k^2 \int_0^R dx \int_0^\infty d\kappa\, \kappa\, \Phi_n(\kappa)[1 - J_0(2i\kappa\rho\gamma_2(x))] \exp\left(-\frac{\kappa^2\gamma_2(x)(R-x)}{k}\right)$$

The integrands of the second and third terms are well-behaved for small wavenumbers and we can use the Kolmogorov model to evaluate them. The resulting integrals were evaluated in the steps that led to the amplitude-variance expression (3.94). The phase variance for beam waves can therefore be estimated from the following combination:

$$\langle \varphi^2(\rho, R)\rangle = 4\pi^2 R k^2 \int_0^\infty d\kappa\, \kappa\, \Phi_n(\kappa) - \langle \chi^2(\rho, R)\rangle$$

$$- 1.632 C_n^2 R^{\frac{11}{6}} k^{\frac{7}{6}} \left(\frac{\lambda R}{\pi w_0^2(R)}\right)^{\frac{5}{6}} M\left(-\frac{5}{6}, 1; \frac{2\rho^2}{w_0^2(R)}\right) \qquad (7.24)$$

The first term is the familiar phase variance for a plane wave estimated with geometrical optics, which carries the entire burden of representing the influence of large eddies. The amplitude variance $\langle \chi^2 \rangle$ is defined analytically by (3.94) and graphically by Figures 3.15 and 3.16. This result should be tested experimentally with the variety of beam-wave shapes that one can now generate using laser sources. Notice that the phase variance for a focused beam wave diverges at the focus, in the same way as that in which the first-order amplitude-variance expression did.

7.2 The Phase Structure Function

Our discussions of phase fluctuations in Volume 1 attached considerable importance to the *phase structure function*. It is defined by the mean-square phase difference

$$\mathcal{D}_\varphi(\rho) = \langle |\varphi(\mathbf{R}) - \varphi(\mathbf{R} + \rho)|^2\rangle \qquad (7.25)$$

that is constructed from phase measurements made at separated receivers. The receivers are usually deployed normal to the line-of-sight vector in terrestrial experiments because that is the most sensitive arrangement. Astronomical observations must cope with other geometries because the source moves relative to fixed ground installations. We will concentrate here on propagation paths near the surface, for which the atmospheric irregularities can be approximated by a homogeneous, isotropic random medium.

Phase-difference measurements are preferred for several reasons. Phase trends are usually common to the two signals and disappear on taking their difference. This means that one can often deal with a stationary data stream. In addition, large-scale inhomogeneities and intermittent structures can enclose both paths and their influence is suppressed when one takes the phase difference. This means that large eddies in the turbulence spectrum are a good deal less important than they are in phase-variance measurements.

The phase structure function is the natural performance measure for many instruments. It describes the measurement accuracy of interferometers that are widely used for precise source location and imaging. We learned in Section 5.2 of Volume 1 that the angular resolution of an interferometer boresighted on the source is described by

$$\langle \delta\theta^2 \rangle = \frac{1}{k^2 \rho^2} \langle |\varphi(R + \rho) - \varphi(R)|^2 \rangle \qquad (7.26)$$

and the following important relationship emerges:

$$\langle \delta\theta^2 \rangle = \frac{1}{k^2 \rho^2} \mathcal{D}_\varphi(\rho) \qquad (7.27)$$

The description of the phase structure function developed in Volume 1 was based on geometrical optics and gave the following description for a plane wave traveling close to the surface[3]

$$\text{Plane wave:} \qquad \mathcal{D}(\rho) = 8\pi^2 R k^2 \int_0^\infty d\kappa \, \kappa \, \Phi_n(\kappa)[1 - J_0(\kappa\rho)] \qquad (7.28)$$

In the wider context of diffraction theory this combination is called the *wave structure function*. In this usage, we drop the subscript φ to differentiate it from the diffraction-corrected version of the phase structure function. Similar expressions for spherical and beam waves were developed in (5.8) and (5.31) of Volume 1. The wave structure function is surprisingly different for the three primary types of wave even when the random medium is the same.

In those discussions, we also learned several important things about the variation of $\mathcal{D}(\rho)$ with spacing. It was initially assumed that the inertial range of the turbulence spectrum is uniquely important since the wavenumber integration in (7.28) converges at small wavenumbers for the Kolmogorov model. If one ignores the inner- and outer-scale wavenumber boundaries of that model

$$\mathcal{D}(\rho) = \text{constant} \times \rho^{\frac{5}{3}} \qquad (7.29)$$

and this expression is widely assumed to be an adequate description. On the other hand, we should notice that the integration in (7.28) barely converges for the Kolmogorov model. This precarious situation means that large eddies do influence the structure function throughout the range of inter-receiver spacings, as is

---

[3] See Section 5.1.2 of Volume 1.

evident in Figure 5.8 of Volume 1. By contrast, we found that the smallest eddies are not important unless the receivers are very close to one another.

In developing the phase structure function with diffraction theory, we will consistently separate out a term that represents the wave structure function. By doing so, we can isolate the entire influence of the outer-scale region for each type of wave. That allows us to evaluate the remaining terms with the Kolmogorov model. It is simplest to illustrate this program with specific types of signal.

7.2.1 The Plane-wave Structure Function

The structure function could be calculated if one knew the spatial correlation of phase fluctuations since (7.25) can be written as

$$D_\varphi(\rho) = 2[\langle \varphi^2(\mathbf{R}) \rangle - \langle \varphi(\mathbf{R})\varphi(\mathbf{R} + \rho) \rangle]$$

We developed an expression for the spatial covariance of amplitude fluctuations in terms of the weighting function $D(\kappa)$ in (4.6). In a completely analogous way we can write the spatial covariance for phase in terms of $E(\kappa)$:

$$\langle \varphi(\mathbf{R})\varphi(\mathbf{R} + \rho) \rangle = 4 \int d^3\kappa \, \Phi_n(\kappa) \exp(i\kappa \cdot \rho) \, E(\kappa)E(-\kappa) \qquad (7.30)$$

The phase variance term needed above is just this expression with $\rho = 0$ and the phase structure function becomes

$$D_\varphi(\rho) = 8 \int d^3\kappa \, \Phi_n(\kappa)[1 - \exp(i\kappa \cdot \rho)]E(\kappa)E(-\kappa) \qquad (7.31)$$

To complete the description we need the weighting function $E(\kappa)$. We developed the expression (7.6) for it with the paraxial approximation. Following the steps that led to (7.7), we can write

$$D_\varphi(\rho) = 4\pi k^2 \int d^3\kappa \, \Phi_n(\kappa)[1 - \exp(i\kappa \cdot \rho)]\delta(\kappa_x) \int_0^R dx \cos^2\left(\frac{x\kappa_r^2}{2k}\right)$$

For paths close to the surface one can assume that isotropy applies. The structure function then depends on the magnitude of the separation vector but not on its orientation:

Plane wave: $\quad D_\varphi(\rho) = 8\pi^2 k^2 \int_0^R dx \int_0^\infty d\kappa \, \kappa \Phi_n(\kappa)[1 - J_0(\kappa\rho)] \cos^2\left(\frac{x\kappa^2}{2k}\right)$

$$(7.32)$$

This can be integrated along the path to give the following description [8]:

Plane wave: $\quad D_\varphi(\rho) = 8\pi^2 k^2 R \int_0^\infty d\kappa \, \kappa \Phi_n(\kappa)[1 - J_0(\kappa\rho)]F_\varphi(\kappa) \quad (7.33)$

Here $F_\varphi(\kappa)$ is the phase-variance weighting function (7.10). We need to calculate numerical values from this expression in order to learn how the structure function varies with separation when diffraction effects are included.

There are several ways to proceed. The simplest approach is to rearrange the phase weighting function as follows:

$$F_\varphi(\kappa) = \frac{1}{2}\left(1 + \frac{\sin(R\kappa^2/k)}{R\kappa^2/k}\right) = 1 - \frac{1}{2}\left(1 - \frac{\sin(R\kappa^2/k)}{R\kappa^2/k}\right) = 1 - F_\chi(\kappa)$$

(7.34)

With this separation the phase structure function becomes

$$\mathcal{D}_\varphi(\rho) = 8\pi^2 R k^2 \left(\int_0^\infty d\kappa\, \kappa\, \Phi_n(\kappa)[\,1 - J_0(\kappa\rho)] \right.$$
$$\left. - \int_0^\infty d\kappa\, \kappa\, \Phi_n(\kappa)F_\chi(\kappa) + \int_0^\infty d\kappa\, \kappa\, \Phi_n(\kappa)J_0(\kappa\rho)F_\chi(\kappa) \right)$$

where $F_\chi(\kappa)$ is the plane-wave amplitude weighting function defined by (3.30). This partitioning allows one to identify the first integral in large parentheses with the wave structure function. The second integral is just twice the basic expression for the amplitude variance established in (3.29). The third integral is twice the spatial correlation for amplitude fluctuations derived in (4.10). We can therefore write the result in terms of plane-wave quantities that have been thoroughly discussed in earlier chapters:

$$\mathcal{D}_\varphi(\rho) = \mathcal{D}(\rho) - 2\langle\chi^2\rangle + 2\langle\chi^2\rangle C_\chi(\rho) \qquad (7.35)$$

This form suggests that we should normalize the result by dividing $\mathcal{D}_\varphi(\rho)$ by $2\langle\chi^2\rangle$ and that is commonly done in presenting numerical values [9]:

$$\text{Plane wave:} \qquad \frac{\mathcal{D}_\varphi(\rho)}{2\langle\chi^2\rangle} = C_\chi(\rho) - 1 + \frac{\mathcal{D}(\rho)}{2\langle\chi^2\rangle} \qquad (7.36)$$

With this foundation we can make further simplifications if we are prepared to ignore several corrections that have been introduced in previous chapters. In its simplest version, the wave structure function is approximated by the similarity expression presented in (5.22) of Volume 1, which we repeat:

$$\mathcal{D}(\rho) = 8\pi^2 R k^2 \int_0^\infty d\kappa\, \kappa\, \Phi_n(\kappa)[1 - J_0(\kappa\rho)]$$
$$= 2.914 R k^2 C_n^2 \rho^{\frac{5}{3}} \qquad \text{for} \qquad \rho \ll L_0$$

The basic expression for the amplitude variance (3.35) ignores aperture averaging and inner-scale effects, but is commonly used in this context to rewrite the last term

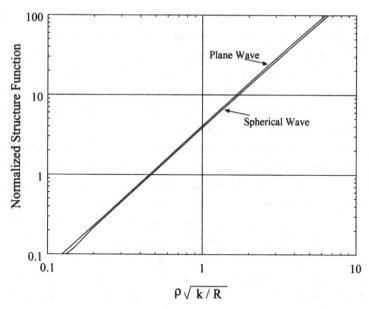

Figure 7.3: Phase structure functions for plane and spherical waves plotted versus the inter-receiver separation divided by the Fresnel length.

as follows:

$$\frac{\mathcal{D}(\rho)}{2\langle \chi^2 \rangle} = 4.746 \left(\rho \sqrt{k/R} \right)^{\frac{5}{3}} \qquad \text{for} \qquad \rho \ll L_0$$

The spatial correlation was evaluated with the Kolmogorov model in (4.11) and is plotted in Figure 4.2. On combining these results the normalized phase structure function becomes

$$\frac{\mathcal{D}_\varphi(\rho)}{2\langle \chi^2 \rangle} = 4.746 q^{\frac{5}{3}} - 1 + C_\chi(q) \qquad \text{for} \qquad \rho \ll L_0 \qquad (7.37)$$

where q is the ratio of the inter-receiver spacing and the Fresnel length [10]. Computed values of this function are plotted on logarithmic coordinates in Figure 7.3. The resulting curve is dominated by the term

$$4.746 q^{\frac{5}{3}}$$

over most of the range. A level doubling occurs when the spacing is comparable to the Fresnel length. The phase structure function is approximately

$$\mathcal{D}_\varphi(\rho) = 1.457 R k^2 C_n^2 \rho^{\frac{5}{3}} \qquad \text{for} \qquad \ell_0 < \rho < \sqrt{\lambda R} \qquad (7.38)$$

but is twice this value above the break:

$$\mathcal{D}_\varphi(\rho) = 2.914 R k^2 C_n^2 \rho^{\frac{5}{3}} \qquad \text{for} \qquad \sqrt{\lambda R} < \rho < L_0 \qquad (7.39)$$

The second form is valid until outer-scale effects take hold. They have not been included in the wave structure function used for this calculation, as is evident in the simple 5/3 approximation employed. The predicted level doubling is difficult to identify in the logarithmic plots of Figure 7.3 but becomes apparent in the numerical computations for $q < 2$. This transition is very gradual and the full result (7.38) is not realized unless $q < 0.1$. In that region one should also include inner-scale effects, which we have not done.

The substantial advantage of the general description (7.36) is that the corrections required for Figure 7.3 are relatively easy to introduce. The adjustment needed for each correction is supplied by the modules that have been developed in this volume and the first. For example, the important influence of large eddies can be included by using the exact expression for the wave structure function defined in Volume 1 by (5.23) and Figure 5.8. In doing so, one should also adjust the amplitude-variance term in the denominator by applying the outer-scale factor plotted in Figure 3.6. Aperture averaging influences the amplitude variance and one should modify the denominator with the receiver gain factor (3.66) that is plotted in Figure 3.9. The Fresnel length for optical links operating in the weak-scattering regime is usually only a few centimeters and inner-scale effects are often important. In that case, one should also multiply the amplitude variance by the inner-scale factor plotted in Figure 3.5. In making these corrections, one needs to recognize their concurrent influences on the spatial correlation, which were developed in Sections 4.1.4 and 4.1.5. The virtue of our description (7.36) lies in the simplicity with which these corrections can be implemented.

A second approach leads one to the same result. It is fundamentally an expansion of the sum rule introduced in (7.11). By following the steps that led to (7.32), we can express the amplitude structure function in the following form:

$$\text{Plane wave:} \qquad \mathcal{D}_\chi(\rho) = 8\pi^2 k^2 \int_0^R dx \int_0^\infty d\kappa\, \kappa \Phi_n(\kappa)$$

$$\times [1 - J_0(\kappa\rho)] \sin^2\left(\frac{x\kappa^2}{2k}\right) \qquad (7.40)$$

On adding this expression to (7.32) one finds that the sum of the phase and amplitude structure functions is just the wave structure function:

$$\mathcal{D}_\varphi(\rho) + \mathcal{D}_\chi(\rho) = \mathcal{D}(\rho) = 8\pi^2 Rk^2 \int_0^\infty d\kappa\, \kappa \Phi_n(\kappa)[1 - J_0(\kappa\rho)] \qquad (7.41)$$

This is the second of the important invariance relations. One can construct the phase structure function from $\mathcal{D}_\chi(\rho)$ and that route has been taken by some [9].

It is significant that this result is valid even if the irregularities are not isotropic and homogeneous. We shall learn later that it is also true for spherical waves. It is valid for beam waves if one uses the appropriate expression for the wave structure function.

7.2.2 The Spherical-wave Structure Function

One can establish the phase structure functions for spherical waves using the same approach as that which led to the plane-wave result:

$$\mathcal{D}_\varphi(\rho) = 8\pi^2 R k^2 \int_0^1 du \int_0^\infty d\kappa\, \kappa \Phi_n(\kappa)[1 - J_0(\kappa u\rho)] \cos^2\left(\frac{\kappa^2 R u(R-u)}{2k}\right)$$

(7.42)

The corresponding amplitude structure function is

$$\mathcal{D}_\chi(\rho) = 8\pi^2 R k^2 \int_0^1 du \int_0^\infty d\kappa\, \kappa \Phi_n(\kappa)[1 - J_0(\kappa u\rho)] \sin^2\left(\frac{\kappa^2 R u(R-u)}{2k}\right)$$

(7.43)

From these expressions it is clear that [11]

$$\mathcal{D}_\varphi(\rho) + \mathcal{D}_\chi(\rho) = \mathcal{D}(\rho) = 8\pi^2 R k^2 \int_0^1 du \int_0^\infty d\kappa\, \kappa \Phi_n(\kappa)[1 - J_0(\kappa u\rho)]$$

(7.44)

To obtain numerical values of $\mathcal{D}_\varphi(\rho)$ we express $\mathcal{D}_\chi(\rho)$ in terms of the amplitude variance and spatial correlation. Dividing by $2\langle\chi^2\rangle$ yields

Spherical wave: $\qquad \dfrac{\mathcal{D}_\varphi(\rho)}{2\langle\chi^2\rangle} = C_\chi(q) - 1 + \dfrac{\mathcal{D}(\rho)}{2\langle\chi^2\rangle}$ (7.45)

The spatial correlation is identified in (4.19) and plotted in Figure 4.2. The wave structure function for spherical waves was investigated in Section 5.1.1 of Volume 1 and is plotted there in Figure 5.4 as a function of $\kappa_0\rho$. One can use its similarity version given by (5.12) in Volume 1 and the basic amplitude-variance expression for spherical waves (3.53) to write

Spherical wave: $\qquad \dfrac{\mathcal{D}_\varphi(\rho)}{2\langle\chi^2\rangle} = 4.407 q^{\frac{5}{3}} + C_\chi(q) - 1 \qquad$ for $\qquad \rho \ll L_0$

(7.46)

where q has the usual meaning. This simplified expression is plotted in Figure 7.3, where it can be compared with the plane-wave result. The two results are parallel lines on logarithmic coordinates. They are separated in this range of q primarily by the different coefficients of $q^{\frac{5}{3}}$ in the defining relations (7.37) and (7.46). Corrections for inner- and outer-scale effects and aperture averaging can be added in the manner described above.

7.2.3 The Beam-wave Structure Function

Despite its analytical complexity, it is important to consider the phase structure function for beam waves in view of their many applications. Using the description presented in Section 3.2.5, Ishimaru first presented general expressions for the phase and amplitude structure functions of a Gaussian beam wave [12][13]:

$$\left.\begin{array}{c} \mathcal{D}_\chi(\rho_1, \rho_2) \\ \mathcal{D}_\varphi(\rho_1, \rho_2) \end{array}\right\} = 4\pi^2 k^2 \int_0^R dx \int_0^\infty d\kappa\, \kappa \Phi_n(\kappa, x) \mathcal{J}(x, \kappa) \qquad (7.47)$$

where

$$\mathcal{J}(x, \kappa) = \Re\left[A(\kappa, x)\exp\left(-\frac{\kappa^2 \gamma_2 (R - x)}{k}\right) \right.$$
$$\left. \mp [1 - J_0(\kappa Q)]\exp\left(-\frac{i\kappa^2 \gamma (R - x)}{k}\right) \right] \qquad (7.48)$$

and

$$A(\kappa, x) = \tfrac{1}{2} J_0(2i\kappa\rho_1\gamma_2) + \tfrac{1}{2} J_0(2i\kappa\rho_2\gamma_2) - J_0(P\kappa) \qquad (7.49)$$

The parameters P and Q depend on the locations of the receivers and the beam-wave properties $\gamma = \gamma_1 - i\gamma_2$ through the definitions (4.31) and (4.32).

The approach suggested in Section 7.2.1 is the best way to organize the resulting calculations. If we combine the two structure-function expressions contained in (7.47), we find that

$$\mathcal{D}_\chi(\rho_1, \rho_2) + \mathcal{D}_\varphi(\rho_1, \rho_2) = \mathcal{D}(\rho_1, \rho_2) \qquad (7.50)$$

where the *wave structure function for a beam wave* is defined by

$$\mathcal{D}(\rho_1, \rho_2) = 8\pi^2 k^2 \int_0^R dx \int_0^\infty d\kappa\, \kappa \Phi_n(\kappa, x) \Re\left[A(\kappa, x)\exp\left(-\frac{\kappa^2 \gamma_2 (R - x)}{k}\right) \right]$$

$$(7.51)$$

If one knew this function, one could calculate the normalized phase structure function from the following relationship:

Beam wave: $$\frac{\mathcal{D}_\varphi(\rho_1, \rho_2)}{2\langle \chi^2 \rangle} = C_\chi(\rho_1, \rho_2) - 1 + \frac{1}{2\langle \chi^2 \rangle}\mathcal{D}(\rho_1, \rho_2) \qquad (7.52)$$

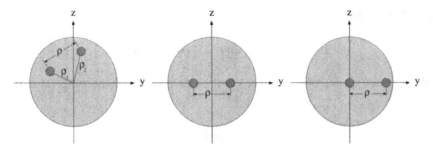

Figure 7.4: Three arrangements of receivers employed to measure the phase structure function of a beam wave in a plane normal to its centerline.

Recall that we established an integral expression for the spatial correlation of amplitude fluctuations in (4.33). It can be reduced to numerical values when the locations of the receivers and beam-wave properties are specified. The amplitude variance is defined analytically by (3.94) and presented graphically in Figures 3.15 and 3.16.

What remains is to estimate the wave structure function for different locations of receivers. A general arrangement is illustrated in the first panel of Figure 7.4. The wavenumber integration in (7.51) can be done analytically for the Kolmogorov spectrum

$$\mathcal{D}(\rho_1, \rho_2) = 2.606 C_n^2 k^2 \int_0^R dx \int_0^\infty \frac{d\kappa}{\kappa^{\frac{8}{3}}} \exp\left(-\frac{\kappa^2 \gamma_2 (R - x)}{k}\right) \Re[A(\kappa, x)]$$

by rearranging the function $A(\kappa, x)$ as follows:

$$A(\kappa, x) = [1 - J_0(P\kappa)] - \tfrac{1}{2}[1 - J_0(2i\kappa\rho_1\gamma_2)] - \tfrac{1}{2}[1 - J_0(2i\kappa\rho_2\gamma_2)]$$

The required integration is now the same for each of the three terms and can be expressed in terms of Kummer functions using the result for $\mathcal{J}$ that was derived in the steps leading to (3.94):

$$\mathcal{D}_\varphi(\rho) = 8.702 k^{\frac{7}{6}} C_n^2 (\gamma_2)^{\frac{5}{6}} \int_0^R dx\, (R - x)^{\frac{5}{6}}$$

$$\times \left\{ \Re\left[M\left(-\frac{5}{6}, 1; -\frac{kP^2}{4\gamma_2(R - x)}\right)\right] - \frac{1}{2} M\left(-\frac{5}{6}, 1; -\frac{\gamma_2 k\rho_1^2}{R - x}\right)\right.$$

$$\left. - \frac{1}{2} M\left(-\frac{5}{6}, 1; -\frac{\gamma_2 k\rho_2^2}{R - x}\right)\right\} \tag{7.53}$$

It is both appropriate and expedient to simplify the alignment of the receivers. Typical deployments used in measurement programs are illustrated in the second and third panel of Figure 7.4. In the second panel the receivers lie on a diameter at equal distances from the centerline:

$$\rho_1 = \tfrac{1}{2}\rho \mathbf{i}_y \quad \text{and} \quad \rho_2 = -\tfrac{1}{2}\rho \mathbf{i} y \tag{7.54}$$

The third panel shows one receiver on the centerline and the other at a radial distance ρ:

$$\rho_1 = \rho \mathbf{i}_{yz} \quad \text{and} \quad \rho_2 = 0 \tag{7.55}$$

Miller, Ricklin, Andrews and Phillips investigated these two cases for collimated and divergent beam waves [14][15]. They obtained analytical and graphical descriptions for the wave structure function. Lukin specialized the result (7.53) for the arrangements of receivers defined by (7.55) and provided graphical descriptions of $\mathcal{D}(\rho)$ for five different beam-wave configurations [16]. An earlier contribution examined the influence of the dissipation region [17]. These calculations give one hope that a more complete description will be possible when analytical and numerical techniques are combined. There is evidently a good deal of research still required in this area to support the measurements that are now under way.

7.3 The Phase Power Spectrum

The power spectrum is the most common description of phase fluctuations. It emerges naturally when a stream of phase measurements is passed through a spectrum analyzer. We described the phase power spectrum for a variety of phase measurements in Chapter 6 of Volume 1 using geometrical optics. For single-path phase measurements we found that the low-frequency portion of the power spectrum is influenced primarily by large eddies. Conversely, the high-frequency behavior is dominated by small eddies in the turbulence hierarchy. We have learned in this volume that diffraction effects become important for eddies that are comparable in extent to the Fresnel length. It is interesting to see whether they modify the high-frequency behavior of the phase spectrum in a measurable way.

In describing the temporal variability of amplitude in Chapter 5, we noted that the power spectrum of a random variable is given by the Fourier transform of its temporal covariance. For single-path phase measurements we can therefore write

$$W_\varphi(\omega) = \int_{-\infty}^{\infty} d\tau \, \langle \varphi(t)\varphi(t+\tau) \rangle \exp(i\omega\tau) \tag{7.56}$$

The temporal covariance and power spectrum provide equivalent descriptions of the phase fluctuations. One can calculate the spectrum for all frequencies if one knows the correlation for all temoporal separations. Conversely, one can calculate the covariance as the inverse Fourier integral if one knows the phase spectrum:

$$\langle \varphi(t)\varphi(t+\tau)\rangle = \frac{1}{2\pi}\int_{-\infty}^{\infty} d\omega\, W_\varphi(\omega)\exp(-i\omega\tau) \tag{7.57}$$

If one is describing phase-difference measurements, one replaces φ by $\Delta\varphi$ in these expressions.

7.3.1 The Plane-wave Power Spectrum

Let us consider single-path phase measurements made with a collimated beam near the surface. This type of signal can be approximated by a plane wave near the beam axis where most measurements are made. For constant values of C_n^2 and wind speed v along the path, we found the following expression using geometrical optics:

$$\text{Geometrical optics:} \quad W_\varphi(\omega) = 8\pi^2 Rk^2 \int_{\omega/v}^{\infty} \frac{d\kappa\, \kappa\, \Phi_n(\kappa)}{(\kappa^2 v^2 - \omega^2)^{\frac{1}{2}}} \tag{7.58}$$

It is necessary to model the large eddies as well as the inertial range in order to generate an expression that is valid for all frequencies. We used the von Karman model[4] in Section 6.2.3.1 of Volume 1 to do so:

$$\text{von Karman model:} \quad W_\varphi(\omega) = \frac{2.192 Rk^2 C_n^2 v^{\frac{5}{3}}}{\left(\omega^2 + \kappa_0^2 v^2\right)^{\frac{4}{3}}} \tag{7.59}$$

The challenge is to enlarge this description to include diffraction effects. It is surprisingly easy to do so. From the representation (7.56) of the power spectrum it is apparent that the temporal covariance is the basic building block. Taylor's hypothesis was introduced in Chapter 5 to relate time delays to lateral displacements by assuming that one has a frozen random medium moving at constant speed. This is usually a good approximation for single-path measurements. It implies that we can make the replacement $\rho = v\tau$ in the spatial covariance, which we infer from the phase-structure-function expressions developed in Section 7.2:

$$\langle \varphi(t)\varphi(t+\tau)\rangle = 4\pi^2 Rk^2 \int_0^{\infty} d\kappa\, \kappa\, \Phi_n(\kappa) J_0(\kappa\tau v) F_\varphi(\kappa) \tag{7.60}$$

where $F_\varphi(\kappa)$ is the phase-variance weighting function defined by (7.10). The integration over τ required in (7.56) operates on the Bessel function and can be

---

[4] The wavenumber spectrum for the von Karman model is

$$\Phi_n(\kappa) = 0.033 C_n^2 \left(\kappa^2 + \kappa_0^2\right)^{-\frac{11}{6}} \quad \text{for} \quad 0 < \kappa < \kappa_s$$

performed analytically using the result (5.7):

$$W_\varphi(\omega) = 8\pi^2 Rk^2 \int_{\omega/v}^\infty \frac{d\kappa\, \kappa\, \Phi_n(\kappa)}{\left(\kappa^2 v^2 - \omega^2\right)^{\frac{1}{2}}} F_\varphi(\kappa) \qquad (7.61)$$

We next use the relationship between the phase and amplitude weighting functions (7.34) to write the power spectrum as a combination of terms:

$$W_\varphi(\omega) = 8\pi^2 Rk^2 \int_{\omega/v}^\infty \frac{d\kappa\, \kappa\, \Phi_n(\kappa)}{\left(\kappa^2 v^2 - \omega^2\right)^{\frac{1}{2}}} - 8\pi^2 Rk^2 \int_{\omega/v}^\infty \frac{d\kappa\, \kappa\, \Phi_n(\kappa)}{\left(\kappa^2 v^2 - \omega^2\right)^{\frac{1}{2}}} F_\chi(\kappa)$$

$$(7.62)$$

The first term is just the geometrical-optics expression for the power spectrum of a plane wave. It must be evaluated with a turbulence model that recognizes the small-wavenumber portion of $\Phi_n(\kappa)$. One obtains (7.59) when one uses the von Karman model to do so. The second term is the general expression for the power spectrum of logarithmic amplitude fluctuations established in (5.8). It can be evaluated with the Kolmogorov model to give (5.12) with (5.13). We can avoid further calculations by combining those results:[5]

$$W_\varphi(\omega) = 2.192 Rk^2 C_n^2 v^{\frac{5}{3}} \left\{ \frac{1}{\left(\omega^2 + \kappa_0^2 v^2\right)^{\frac{4}{3}}} - \frac{1}{2\omega^{\frac{8}{3}}} \right.$$

$$\left. + \frac{0.527}{\omega^{\frac{8}{3}}} \Im\left[\frac{1}{\xi} \exp(i\xi)\, U\left(\frac{1}{2}, -\frac{4}{3}; -i\xi\right)\right] \right\} \qquad (7.63)$$

The dimensionless variable

$$\xi = \frac{\omega^2 R}{v^2 k} = \left(\frac{\omega}{\omega_F}\right)^2 \qquad (7.64)$$

scales the scintillation frequency with the Fresnel frequency:

$$\omega_F = v\sqrt{\frac{2\pi}{\lambda R}}$$

that was introduced in (5.11). One can show that this result is equivalent to that first derived by Tatarskii [18] if one expresses $U(a, b; z)$ in terms of the more familiar Kummer function $M(a, b; z)$ using their relationship given in Appendix G. In general, one must use numerical techniques to generate specific values for the power spectrum since the Kummer function is not tabulated for the required indices.

---

[5] The spectrum expressions obtained by Tatarskii [18] and Clifford [19] ignored the outer-scale wavenumber. They also used a different defintion of the power spectrum, which produced an additional factor of two.

On the other hand, one can extract several important properties of the power spectrum before turning to numerical methods. The influence of diffraction is represented by the second and third terms in (7.63). Their combination behaves differently depending on whether the scintillation frequency is greater than or less than ω_F. They cancel out below this dividing line and the power spectrum is then described by the geometrical-optics result:

$$\omega \ll \omega_F \qquad W_\varphi(\omega) = \frac{2.192 R k^2 C_n^2 v^{\frac{5}{3}}}{\left(\omega^2 + \kappa_0^2 v^2\right)^{\frac{4}{3}}} \qquad (7.65)$$

The Kummer function in (7.62) can be omitted when the frequency is well above ω_F and one finds that

$$\omega \gg \omega_F \qquad W_\varphi(\omega) = \frac{1.097 R k^2 C_n^2 v^{\frac{5}{3}}}{\omega^{\frac{8}{3}}} \qquad (7.66)$$

Diffraction thus alters the high-frequency behavior of the spectrum by halving the geometrical-optics result.

Clifford used similar techniques to describe the power spectrum of spherical waves in terms of hypergeometric functions [19]. He found that the limiting forms (7.65) and (7.66) also characterize diverging waves.

7.3.2 Numerical Results

Lukin and Pokasov generated numerical values for the single-path phase power spectrum [20]. They included both the large and small eddy portion of the turbulent spectrum. The small-wavenumber and inertial ranges were described by the von Karman model. Near the inner-scale wavenumber the dissipation range was modeled by a Gaussian function. The results were expressed as integrals of hypergeometric functions modulated by the beam-wave weighting functions. They specialized this result to describe plane, spherical and collimated beam waves by making appropriate choices for key parameters in the weighting functions. The resulting expressions cannot be reduced to familiar functions and must be evaluated numerically. The resulting values were presented in terms of the dimensionless ratio

$$\frac{[\omega/(\kappa_0 v)] W_\varphi(\omega)}{\langle \varphi^2 \rangle} \qquad (7.67)$$

This ratio is plotted in Figure 7.5 for the three types of wave versus the scaled frequency

$$\omega/(\kappa_0 v)$$

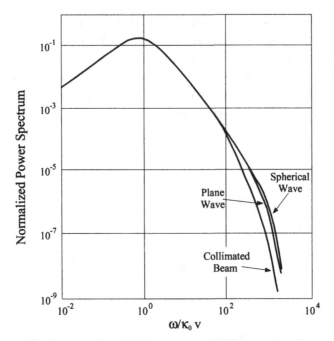

Figure 7.5: Normalized single-path phase power spectra for three types of wave calculated by Lukin and Pokasov [20]. The outer-scale wavenumber is assumed to be $\kappa_0^2 R/k = 10^{-4}$ and the ratio $\kappa_0/\kappa_m = 10^{-3}$ in each case.

and the parameter choices

$$\kappa_0^2 R/k = 10^{-4} \qquad \text{and} \qquad \kappa_0/\kappa_m = 10^{-3}$$

which are reasonable estimates for optical paths near the surface.

Several interesting features emerge from the three power spectra plotted in Figure 7.5. Notice first that they are virtually the same for frequencies below 100 Hz. The initial linear rise with frequency is due to the factor $\omega/(\kappa_0 v)$ in the normalized spectrum that multiplies the flat von Karman spectrum. The fall following the maximum is due to the $\omega^{-\frac{8}{3}}$ behavior and is common to the three types of wave. The curves begin to separate beyond 100 Hz. This is primarily due to their different responses to the very small eddies as one enters the dissipation region. Notice that the spectrum for a collimated beam wave is considerably narrower than the plane- and spherical-wave results.

7.4 Correlation of Phase and Amplitude

The discussions in this volume have so far addressed amplitude or phase measurements as if they were entirely separate problems. In fact, they represent complementary electromagnetic responses to the random medium. There is considerable merit in asking how these two random variables are correlated. For microwave signals one can measure the single-path phase and amplitude of a signal simultaneously without great effort. One can approximate the *cross correlation* by forming the sample average from their individual time series:

$$\langle \varphi \chi \rangle = \frac{1}{T} \int_0^T dt \, \varphi(t)\chi(t) \tag{7.68}$$

The correlation of phase and amplitude is a third measure of the electric field's behavior. It is an important quantity in its own right and should play a more prominent role in propagation research than it has thus far.

Let us pause to note the ways in which the cross correlation influences our work. We found in (7.2) that the average phase fluctuation has a non-vanishing second-order component that is equal to the negative of the cross correlation. We will learn that the phase distribution also depends on the cross correlation. More fundamentally, validity conditions for the Rytov approximation will be established from inequalities that depend on the cross correlation.

It is interesting that the cross correlation can be expressed in terms of the logarithmic amplitude variance both for plane and for spherical waves as

$$\langle \varphi \chi \rangle = \text{constant} \times \langle \chi^2 \rangle \tag{7.69}$$

Notice that one cannot make this connection for the phase variance since it is defined primarily by the outer-scale region of the turbulence spectrum. We shall find that the cross correlation is influenced by the outer scale length more strongly than is the amplitude variance. The inverse of this sensitivity is that simultaneous measurements of amplitude fluctuations and the cross correlation can be used to explore the outer-scale region with a deftness and clarity that is not possible using either quantity alone. It is important to understand a good deal more about the cross correlation before planning such experiments. We can use the Rytov approximation to estimate this quantity. We concentrate on terrestrial links, addressing plane and spherical waves separately.[6]

7.4.1 Plane-wave Cross Correlation

A basic description for the correlation of amplitude and phase was established in (2.40). For links near the surface, we shift to the spectrum of refractive-index irregularities and assume that they are isotropic:

$$\langle \varphi \chi \rangle = 4 \int d^3\kappa \, \Phi_n(\kappa)D(\kappa)E(-\kappa) \tag{7.70}$$

---

[6] The power spectrum of the cross correlation for beam waves was considered briefly by Ishimaru [21].

We introduce the paraxial-approximation expressions (3.17) for $D(\kappa)$ and (7.6) for $E(\kappa)$:

$$\langle \varphi\chi \rangle = k^2 \int_0^\infty d\kappa\, \kappa^2 \Phi_n(\kappa) \int_0^\pi d\psi\, \sin\psi \int_0^{2\pi} d\omega \int_0^R dx_1 \int_0^R dx_2$$

$$\times \exp[i\kappa(x_1 - x_2) \cos\psi] \sin\left(\frac{x_1 \kappa^2 \sin^2\theta}{2k}\right) \cos\left(\frac{x_2 \kappa^2 \sin^2\theta}{2k}\right)$$

Our usual treatment of the horizontal integrations yields the following description [22][23]:

$$\langle \varphi\chi \rangle = 2\pi k^2 \int_0^\infty d\kappa\, \kappa^2 \Phi_n(\kappa) \int_0^\pi d\psi\, \sin\psi \int_0^{2\pi} d\omega\, \delta(\kappa \cos\psi)$$

$$\times \int_0^R dx\, \sin\left(\frac{x\kappa^2}{2k}\right) \cos\left(\frac{x\kappa^2}{2k}\right)$$

$$= 2\pi^2 Rk^2 \int_0^\infty d\kappa\, \kappa \Phi_n(\kappa) \int_0^R dx\, \sin\left(\frac{x\kappa^2}{k}\right)$$

and carrying out the horizontal integration gives

Plane wave: $$\langle \varphi\chi \rangle = 2\pi^2 Rk^2 \int_0^\infty d\kappa\, \kappa \Phi_n(\kappa) F_{\varphi\chi}(\kappa) \tag{7.71}$$

where

$$F_{\varphi\chi}(\kappa) = \frac{2k}{R\kappa^2} \sin^2\left(\frac{R\kappa^2}{2k}\right) \tag{7.72}$$

is the spectral weighting function for the cross correlation that is plotted in Figure 7.6. One sees that it rises rapidly from the origin, reaches a maximum value of 0.36 and then begins an oscillatory decline. The initial rise is more rapid than that for the amplitude weighting function. This is also made evident by comparing their small-argument expansions.

$$F_\chi(\kappa) = 1 - \frac{k}{R\kappa^2} \sin\left(\frac{R\kappa^2}{k}\right) \simeq \kappa^4\left(\frac{R^2}{6k^2}\right)$$

$$F_{\varphi\chi}(\kappa) = \frac{2k}{R\kappa^2} \sin^2\left(\frac{R\kappa^2}{2k}\right) \simeq \kappa^2\left(\frac{R}{2k}\right) \tag{7.73}$$

This contrast suggests that the cross correlation is more sensitive to large eddies. By the same token $\langle \varphi\chi \rangle$ should be less sensitive to the small eddies near the dissipation region.

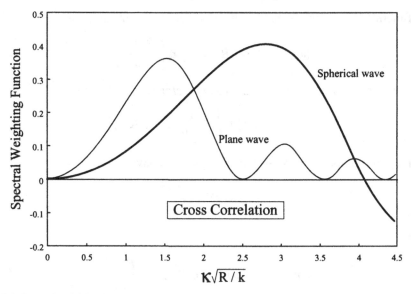

Figure 7.6: Spectral weighting functions that define the correlation of phase and amplitude fluctuations for plane and spherical waves.

We first use the Kolmogorov model with a sharp cutoff at the outer-scale wavenumber to evaluate the cross correlation:

$$\langle \varphi\chi \rangle = 2\pi^2 R k^2 0.033 C_n^2 \int_{\kappa_0}^{\infty} \frac{d\kappa}{\kappa^{\frac{8}{3}}} \left[\frac{2k}{R\kappa^2} \sin^2\left(\frac{R\kappa^2}{2k} \right) \right] \qquad (7.74)$$

For optical and infrared propagation we can ignore the outer-scale cutoff and the substitution $\zeta = R\kappa^2/(2k)$ yields an integral found in Appendix B. The following expression is then appropriate for the Fresnel regime [24]:

$$\text{Fresnel:} \qquad \langle \varphi\chi \rangle_{\text{pl}} = 1.146 R^{\frac{11}{6}} k^{\frac{7}{6}} C_n^2 \qquad (7.75)$$

This combination of independent variables is identical to that found in the amplitude-variance expression (3.35), so one can write

$$\text{Fresnel:} \qquad \langle \varphi\chi \rangle_{\text{pl}} = 3.732 \langle \chi^2 \rangle_{\text{pl}} \qquad (7.76)$$

This relationship will play an important role in our studies.

We noted in (7.73) that the cross correlation is more sensitive to the large eddies than is the amplitude variance. This suggests that we may have been too hasty in setting $\kappa_0 = 0$. The outer scale length can be comparable to the Fresnel length for millimeter-wave links [3], which means that we should consider the influence of the lower limit. We can write the cross correlation as a product of the previous result and a function that depends only on the outer-scale scattering parameter:

$$\langle \varphi \chi \rangle = 1.146 R^{\frac{11}{6}} k^{\frac{7}{6}} C_n^2 \mathbb{K} \left(\frac{R \lambda \kappa_0^2}{2\pi} \right) \tag{7.77}$$

where the *outer-scale factor for the cross correlation* is defined by

$$\mathbb{K}(x) = 0.284 \int_x^\infty \frac{d\zeta}{\zeta^{\frac{17}{6}}} (1 - \cos \zeta) \tag{7.78}$$

For large arguments one can use the asymptotic behavior:

$$\lim_{x \to \infty} [\mathbb{K}(x)] = \frac{0.155}{x^{\frac{11}{6}}}$$

so that

$$\text{Fraunhofer:} \qquad \langle \varphi \chi \rangle = 0.58 \langle \chi^2 \rangle \left(\frac{R \lambda \kappa_0^2}{2\pi} \right)^{-\frac{11}{6}} \tag{7.79}$$

The numerator is always less than unity when the Rytov approximation is valid. The scattering parameter becomes quite large in the Fraunhofer regime, which means that

$$\text{Fraunhofer:} \qquad \langle \varphi \chi \rangle \to 0$$

in contrast to its substantial value in the Fresnel regime.

7.4.2 Spherical-wave Cross Correlation

We next use the spherical-wave descriptions (3.48) for $D(\kappa)$ and (7.17) for $E(\kappa)$ to calculate the cross correlation for a diverging wave. Skipping the steps and explanations that led to the plane-wave result, we find [5]

$$\langle \varphi \chi \rangle = 4\pi^2 R k^2 \int_0^\infty d\kappa \, \kappa \, \Phi_n(\kappa) \left[\frac{1}{2} \int_0^1 du \, \sin \left(\frac{\kappa^2 R u (1-u)}{k} \right) \right] \tag{7.80}$$

The spectral weighting function for a spherical wave can be expressed in terms of the Fresnel integrals identified in (3.52):

$$F_{\varphi\chi}(\kappa) = \frac{1}{2} \sqrt{\frac{2\pi k}{R \kappa^2}} \left[\cos \left(\frac{\kappa^2 R}{4k} \right) S \left(\sqrt{\frac{\kappa^2 R}{2\pi k}} \right) - \sin \left(\frac{\kappa^2 R}{4k} \right) C \left(\sqrt{\frac{\kappa^2 R}{2\pi k}} \right) \right]$$

$$\tag{7.81}$$

This is also plotted in Figure 7.6. One sees there that it behaves much differently than the plane-wave weighting function.

To evaluate $\langle \varphi \chi \rangle$ in the Fresnel regime we introduce the Kolmogorov model into (7.80) and ignore the outer-scale cutoff. It is helpful to reverse the order of integration in doing so:

$$\langle \varphi \chi \rangle = 0.033 \times 2\pi^2 R k^2 C_n^2 \int_0^1 du \int_0^\infty \frac{d\kappa}{\kappa^{\frac{8}{3}}} \sin\left(\frac{\kappa^2 R u (1 - u)}{k}\right)$$

$$= 0.033 \times \pi^2 R^{\frac{11}{6}} k^{\frac{7}{6}} C_n^2 \int_0^1 du \, [u(1 - u)]^{\frac{5}{6}} \int_0^\infty \frac{dw}{w^{\frac{11}{6}}} \sin w$$

Both integrals are found in Appendix B, so we have

$$\text{Fresnel:} \qquad \langle \varphi \chi \rangle_{\text{sph}} = 0.463 R^{\frac{11}{6}} k^{\frac{7}{6}} C_n^2 \tag{7.82}$$

Recalling the basic description (3.53) for the amplitude variance of spherical waves, this can be written as follows:

$$\text{Fresnel:} \qquad \langle \varphi \chi \rangle_{\text{sph}} = 3.734 \langle \chi^2 \rangle_{\text{sph}} \tag{7.83}$$

This is remarkably similar to the plane-wave version (7.76). In this connection, we note that $\langle \varphi \chi \rangle$ and $\langle \chi^2 \rangle$ for spherical waves are only 40% of the plane-wave values, yet their mutual relationship is virtually the same. One can also show that the cross correlation for spherical waves vanishes in the Fraunhofer regime.

7.4.3 Cross-correlation Measurements

The cross correlation seems to have been measured only a few times. In a comprehensive experiment the amplitude and phase along a 64-km path between two Hawaiian islands were recorded [25]. These two components were measured simultaneously using five frequencies in the range 10–40 GHz. In the report of this experiment it was noted that "No correlation was found between the standard deviations of amplitude and phase fluctuations." That observation does not address the correlation of instantaneous values required by (7.68). It is unfortunate that a simple extension of the data-processing program was not undertaken when the raw data was available.

A true cross-correlation measurement was made using a coherent 379-MHz signal radiated by the DNA Wideband Satellite [26]. The complex signal was measured at a ground station in Ancon, Peru. Phase and amplitude histories were extracted from this signal and used to compute the following value:

$$\langle \varphi \chi \rangle = -0.61 \varphi_{\text{rms}} \chi_{\text{rms}}$$

This result is not explained by the predictions.

7.5 The Phase Distribution

To complete our description of phase fluctuations we need to establish their probability density function (PDF). In examining this question with geometrical optics we found that φ is distributed as a Gaussian random variable (GRV):[7]

$$\text{Geometrical optics:} \quad P(\varphi)\,d\varphi = \frac{d\varphi}{\sigma\sqrt{2\pi}} \exp\left(-\frac{\varphi^2}{2\sigma^2}\right) \tag{7.84}$$

where

$$\sigma^2 = \langle \varphi_0^2 \rangle \tag{7.85}$$

This prediction was confirmed by two experiments, one using a laser signal on a terrestrial path [27] and the other using three UHF signals from a satellite [26]. We need now to study how diffraction considerations influence the phase distribution.

7.5.1 The First-order Phase Distribution

If one retains only the first term in the phase expression (7.1) and uses the explicit description (2.24), one finds

$$\text{First order:} \quad \varphi_1 = -k^2 \int d^3r \, \delta\varepsilon(\mathbf{r}) \, \Im\left(G(\mathbf{R}, \mathbf{r})\frac{E_0(\mathbf{r})}{E_0(\mathbf{R})}\right) \tag{7.86}$$

The dielectric variations $\delta\varepsilon$ can be measured at microwave and optical frequencies in the troposphere, as described in Section 2.3.2 of Volume 1. We found there that $\delta\varepsilon$ is often distributed as a GRV. In those cases one can be sure that φ_1 is also Gaussian because it is a linear combination of GRVs represented by this volume integral. On the other hand, we also found that temperature measurements sometimes behave like a log-normal random variable. In those cases, one can use the argument presented in Section 10.1 which suggests that the volume integral of a non-Gaussian random variable rapidly approaches a GRV. One would expect the phase distribution to be described by (7.84) in both situations.

7.5.2 The Second-order Phase Distribution

We can extend this examination to the second-order terms in the phase expression (7.1). Even though a and b are GRVs, we cannot be confident that their product is. Moreover, the second-order term d is described by two volume integrals of the ensemble-averaged product $\langle \delta\varepsilon(\mathbf{r}_1)\,\delta\varepsilon(\mathbf{r}_2)\rangle$ which was found not to be Gaussian even if the $\delta\varepsilon$ are.[8]

---

[7] See Section 8.1 of Volume 1.
[8] In this connection, it is helpful to review Problem 3 in Chapter 2 of Volume 1.

Fortunately, there is a mathematical technique for dealing with random variables that are not Gaussian. It is called the *method of cumulant analysis* and its basic results are summarized in Appendix M. The first step is to express the probability density function for phase in terms of its characteristic function:

$$P(\varphi) = \frac{1}{2\pi} \int_{-\infty}^{\infty} dq \exp(-iq\varphi) \langle \exp(iq\varphi) \rangle \tag{7.87}$$

Cumulant analysis expresses the ensemble-averaged exponential function as a power series in the transform variable q. The coefficients in this series are called cumulants and they are defined by the phase moments. To second order[9]

$$\langle \exp(iq\varphi) \rangle = \exp\left(iq K_1 - \tfrac{1}{2}q^2 K_2\right) \tag{7.88}$$

where the first two cumulants are given by

$$\begin{aligned} K_1 &= \langle \varphi \rangle \\ K_2 &= \langle \varphi^2 \rangle - \langle \varphi \rangle^2 \end{aligned} \tag{7.89}$$

The first cumulant is the average value and is proportional to the cross correlation according to (7.2). When we substitute the second-order phase expression from (7.1) we see that K_2 is just the first-order phase variance:

$$\begin{aligned} K_2 &= \langle (b + d - ab)^2 \rangle - \langle b + d - ab \rangle^2 \\ &= \langle b^2 \rangle + O(\delta\varepsilon^4) \\ &= \sigma^2 = \langle \varphi_0^2 \rangle \end{aligned} \tag{7.90}$$

With these results the characteristic function becomes

$$\langle \exp(iq\varphi) \rangle = \exp\left(-iq\langle\varphi\chi\rangle - \tfrac{1}{2}q^2\sigma^2\right) \tag{7.91}$$

and the phase distribution can be calculated from (7.87) as

$$P(\varphi)\,d\varphi = \frac{d\varphi}{\sigma\sqrt{2\pi}} \exp\left(-\frac{1}{2\sigma^2}(\varphi + \langle\varphi\chi\rangle)^2\right) \tag{7.92}$$

The phase distribution is thus Gaussian to second order. This agrees with the basic result (7.84) but with a negative phase bias. We need to estimate this offset for typical propagation conditions.

---

[9] One could take the calculation to fourth order by keeping the third term in the cumulant expansion which is proportional to q^3. That is done in Section 10.2 to find the PDF of logarithmic intensity fluctuations and produces a measurable change in the distribution. This procedure involves considerable effort and seems not to be justified here because the measurements of phase distribution are so limited.

7.5.2.1 The Fresnel Regime

Even on a short path the rms phase measured for optical signals is more than ten radians [27]. By contrast, the phase offset for plane and spherical waves is given by

$$\text{Fresnel:} \qquad \langle \varphi \rangle = - \langle \varphi_1 \chi_1 \rangle = -3.7 \langle \chi_1^2 \rangle \qquad (7.93)$$

Since the amplitude variance must be less than unity, this phase offset is less than two radians. That is a meaningful shift in the context of the measured excursions that are summarized in Figure 8.2 of Volume 1. On the other hand, changes in the absolute phase shift caused by atmospheric adjustments are probably a good deal larger during a typical measurement period. Most detrending programs used to process phase data would erase this offset along with the phase drift.

Microwave measurements made on elevated paths are often conducted in the Fresnel regime because the outer scale length is quite large in the free atmosphere. The measured phase fluctuations are typically small. For example, the rms phase at 1046 MHz is only a few degrees on a 16-km path [28]. The corresponding logarithmic amplitude values ranged from 0.01 to 0.1 so the phase shift predicted by (7.93) is again comparable to the fluctuating component. However, ambient changes of refractivity generate phase trends which probably dwarf this effect.[10]

7.6 Problems

Problem 1

Convince yourself that the sum rule (7.11) is valid also for propagation through random media that are not homogeneous. This means that one can apply that relationship to astronomical signals passing through the atmosphere.

Problem 2

Use the expressions for the phase structure function established in Section 7.2 to estimate the mean-square angle of arrival for terrestrial links. Express the results as a diffraction correction to the geometrical-optics descriptions for plane and spherical waves presented in Section 7.2 of Volume 1. Notice that $\langle \delta \theta^2 \rangle$ measurements fall midway between the phase variance and the amplitude variance in terms of the portion of the turbulence spectrum that is important. This suggests that diffraction caused by small eddies may be influential. Evaluate your resulting expressions with the Kolmogorov model to see whether this is true. Be sure to include aperture averaging.

---

[10] See the raw phase data reproduced in Figures 4.1 and 4.5 of Volume 1.

Problem 3

Consider the measurements of phase difference made by a microwave interferometer with a receiver baseline ρ. For an arriving plane wave show that the power spectrum of $\Delta\varphi$ is related to the single-path power spectrum by [29]

$$W_{\Delta\varphi}(\omega) = 4\sin^2\left(\frac{\omega\rho}{2v}\right)W_\varphi(\omega)$$

Notice that this is the same relationship as that established with geometrical optics in (6.88) of Volume 1. For a spherical wave show that the power spectrum is also related to the single-path spectrum by [19]

$$W_{\Delta\varphi}(\omega) = 2\left[1 - \frac{v}{\rho\omega}\sin\left(\frac{\rho\omega}{v}\right)\right]W_\varphi(\omega)$$

which is the same connection as that made by geometrical optics in (6.75) of Volume 1.

Problem 4

Using geometrical optics, develop a general expression for the single-path phase power spectrum of a spherical wave. If the large eddies are properly described by the von Karman model of the turbulence spectrum, show that

$$W_\varphi(\omega) = 2.192Rk^2C_n^2v^{\frac{5}{3}}\int_0^1 \frac{du\,u^{\frac{5}{3}}}{\left(\omega^2 + u^2\kappa_0^2v^2\right)^{\frac{4}{3}}}$$

Prove that the integral over all frequencies is equal to that over the plane-wave spectrum given by (7.59). Why does this spherical-wave result diverge for very small frequencies, in contrast to that expression, which is finite?

References

[1] V. I. Tatarskii, *Wave Propagation in a Turbulent Medium*, translated by R. A. Silverman (Dover, New York, 1967), 136.

[2] V. I. Tatarskii, *The Effects of the Turbulent Atmosphere on Wave Propagation* (translated from the Russian and issued by the National Technical Information Office, U. S. Department of Commerce, Springfield, VA 22161, 1971), 231.

[3] R. S. Cole, K. L. Ho and N. D. Mavrokoukoulakis, "The Effect of the Outer Scale of Turbulence and Wavelength on Scintillation Fading at Millimeter Wavelengths," *IEEE Transactions on Antennas and Propagation*, **AP-26**, No. 5, 712–715 (September 1978).

[4] See [2], pages 247–253.

[5] A. Ishimaru, *Wave Propagation and Scattering in Random Media*, vol. 2 (Academic Press, San Diego, CA, 1978), 378–379.

[6] See [1], page 184.

[7] See [5], page 382.

[8] See [1], page 137.

[9] D. L. Fried and J. D. Cloud, "Propagation of an Infinite Plane Wave in a Randomly Inhomogeneous Medium," *Journal of the Optical Society of America*, **56**, No. 12, 1667–1676 (December 1966).

[10] See [2], page 238.

[11] D. L. Fried, "Propagation of a Spherical Wave in a Turbulent Medium," *Journal of the Optical Society of America*, **57**, No. 2, 175–180 (February 1967).

[12] See [5], page 382.

[13] A. Ishimaru, "Fluctuations of a Focused Beam Wave for Atmospheric Turbulence Probing," *Proceedings of the IEEE*, **57**, No. 4, 407–414 (April 1969).

[14] W. B. Miller, J. C. Ricklin and L. C. Andrews, "Log-Amplitude Variance and Wave Structure Function: A New Perspective for Gaussian Beams," *Journal of the Optical Society of America A*, **10**, No. 4, 661–672 (April 1993).

[15] L. C. Andrews and R. L. Phillips, *Laser Beam Propagation through Random Media* (SPIE Optical Engineering Press, Bellingham, Washington, 1998), 137–138.

[16] V. P. Lukin, *Atmospheric Adaptive Optics* (SPIE Optical Engineering Press, Bellingham, Washington, 1995; originally published in Russian in 1986), 64–66. (The quantity described by Equation (3.9) on page 65 of this reference is actually the wave structure function in our terminology. Moreover, in the English version an exponent 5/6 was omitted from the function $C(x)$ which occurs outside the curly brackets of that equation.)

[17] A. I. Kon and V. I. Tatarskii, "Parameter Fluctuations of a Space-Limited Light Beam in a Turbulent Atmosphere," *Izvestiya Vysshikh Uchebnykh Zavedenii Radiofizika (Soviet Radiophysics)*, **8**, No. 5, 617–620 (September–October 1965).

[18] See [2], page 268.

[19] S. F. Clifford, "Temporal-Frequency Spectra for a Spherical Wave Propagating through Atmospheric Turbulence," *Journal of the Optical Society of America*, **61**, No. 10, 1285–1292 (October 1971).

[20] V. P. Lukin and V. V. Pokasov, "Optical Wave Phase Fluctuations," *Applied Optics*, **20**, No. 1, 121–135 (1 January 1981).

[21] A. Ishimaru, "Temporal Frequency Spectra of Multifrequency Waves in Turbulent Atmosphere," *IEEE Transactions on Antennas and Propagation*, **AP-20**, No. 1, 10–19 (January 1972).

[22] See [2], page 231.

[23] See [5], pages 357 and 359.

[24] See [2], page 239.

[25] M. C. Thompson, L. E. Wood, H. B. Janes and D. Smith, "Phase and Amplitude Scintillations in the 10 to 40 GHz Band," *IEEE Transactions on Antennas and Propagation*, **AP-23**, No. 6, 792–797 (November 1975).

[26] E. J. Fremouw, R. C. Livingston and D. A. Miller, "On the Statistics of Scintillating Signals," *Journal of Atmospheric and Terrestrial Physics*, **42**, No. 8, 717–731 (August 1980).

[27] V. P. Lukin, V. V. Pokasov and S. S. Khmelevtsov, "Investigation of the Time Characteristics of Fluctuations of the Phases of Optical Waves Propagating in the Bottom Layer of the Atmosphere," *Izvestiya Vysshikh Uchebnykh Zavedenii, Radiofizika (Soviet Radiophysics)*, **15**, No. 12, 1426–1430 (December 1972).

[28] J. W. Herbstreit and M. C. Thompson, "Measurements of the Phase of Radio Waves Received over Transmission Paths with Electrical Lengths Varying as a Result of Atmospheric Turbulence," *Proceedings of the IRE*, **43**, No. 10, 1391–1401 (October 1955).

[29] See [2], page 269.

8

Double Scattering

The second-order term in the Rytov series plays an influential role in determining moments of the field strength and in demonstrating that energy is conserved. The importance of this term was recognized in early studies in which the Rytov approximation was developed to describe propagation in random media [1][2]. Some treatments have used the expected equivalence of the mean irradiance and its free-space values for plane and spherical waves to relate the second-order Rytov term to the phase and amplitude variances estimated to first order [3]. This is often characterized as a consequence of conservation of energy and we will examine that proposition in Chapter 9.

It soon became clear that the preferred approach was to calculate the second-order solution from first principles and then use the result to predict properties of the signal. This was done first for plane waves [1][4][5] and provided reasonable estimates for important properties of signals. That agreement finally established the Rytov approximation as a trustworthy description for the weak-scattering regime [6]. This work was later extended to describe spherical waves [7]. The development of optical lasers stimulated many applications involving beam waves and the corresponding second-order solution was calculated with improving accuracy over the next two decades [8][9][10][11].

The material presented in this chapter is highly analytical and is intended primarily for specialists in the field. The reader who is more interested in applications may wish to proceed directly to Chapters 9 and 10.

8.1 Basic Expressions

It is important to remember that we have two versions of the second Rytov approximation that are completely equivalent. Tatarskii's version of the second-order solution is given by the following volume integral from (2.56):

$$\psi_2(\mathbf{R}) = -\int_V d^3r \, \frac{G(\mathbf{R}, \mathbf{r})E_0(\mathbf{R})}{E_0(\mathbf{r})}[\nabla\psi_1(\mathbf{r})]^2 \tag{8.1}$$

where

$$\psi_1(\mathbf{R}) = -k^2 \int_V d^3r \, \frac{G(\mathbf{R}, \mathbf{r})E_0(\mathbf{R})}{E_0(\mathbf{r})}\, \delta\varepsilon(\mathbf{r}) \tag{8.2}$$

is the first-order Rytov solution. In Yura's version (2.64) the same solution is expressed as the following combination of single- and double-scattering terms in the Born series:

$$\psi_2(\mathbf{R}) = c - \tfrac{1}{2}a^2 + \tfrac{1}{2}b^2 + i(d - ab) \tag{8.3}$$

Specific expressions for the second-order solution have been generated using both descriptions for various types of wave.

For making signal-strength estimates we will rely primarily on Yura's description. Since we developed first-order expressions for the variances of a and b in Chapter 3, our attention here is focused primarily on the double-scattering terms c and d which are defined by (2.63). The average value of that expression provides the two moments that are needed most often in our work:

$$\langle c + id \rangle = k^4 \int_V d^3r_2 \, G(\mathbf{R}, \mathbf{r}_2)$$
$$\times \int_V d^3r_1 \, G(\mathbf{r}_2, \mathbf{r}_1)\left(\frac{E_0(\mathbf{r}_1, \mathbf{T})}{E_0(\mathbf{R}, \mathbf{T})}\right)\langle\delta\varepsilon(\mathbf{r}_1)\,\delta\varepsilon(\mathbf{r}_2)\rangle \tag{8.4}$$

They depend on the spatial correlation of $\delta\varepsilon$, which we replace by its three-dimensional wavenumber representation. How one does so depends on the type of propagation being examined.

8.1.1 Terrestrial Paths

Many microwave and optical measurements are made on terrestrial paths near the surface. One can often ignore the variation of C_n^2 with altitude in these experiments and use the traditional wavenumber-integral expression for the turbulence spectrum:

$$\langle\delta\varepsilon(\mathbf{r}_1)\,\delta\varepsilon(\mathbf{r}_2)\rangle = 4\int d^3\kappa \, \Phi_n(\kappa)\exp[i\kappa\cdot(\mathbf{r}_1-\mathbf{r}_2)]$$

We have shifted to the spectrum of refractive-index fluctuations because we are primarily concerned with propagation in the lower atmosphere. The electromagnetic and turbulent features of the propagation are completely separated when one combines this with (8.4),

$$\langle c + id \rangle = 4k^4 \int d^3\kappa \, \Phi_n(\kappa) \Theta(\kappa, \mathbf{R}, T) \tag{8.5}$$

where the *double-scattering weighting function* for terrestrial paths is defined by

$$\Theta(\kappa, \mathbf{R}, T) = \int_V d^3r_1 \int_V d^3r_2 \, G(\mathbf{R}, \mathbf{r}_2) G(\mathbf{r}_2, \mathbf{r}_1)$$
$$\times \left(\frac{E_0(\mathbf{r}_1, T)}{E_0(\mathbf{R}, T)} \right) \exp[i\kappa \cdot (\mathbf{r}_1 - \mathbf{r}_2)] \tag{8.6}$$

This formulation will be used to describe spherical-wave and beam-wave transmissions near the surface.

8.1.2 Astronomical Observations

For astronomical observations we must recognize the inhomogeneous nature of the atmosphere since stellar signals travel through it before reaching ground-based telescopes. To treat this case we use the following description of inhomogeneous random media which was established in Section 2.2.8 of Volume 1:

$$\langle \delta\varepsilon(\mathbf{r}_1) \, \delta\varepsilon(\mathbf{r}_2) \rangle = 4 \int d^3\kappa \, \Phi_n\left(\kappa, \frac{\mathbf{r}_1 + \mathbf{r}_2}{2} \right) \exp[i\kappa \cdot (\mathbf{r}_1 - \mathbf{r}_2)] \tag{8.7}$$

The usual procedure is to represent the spectrum as the product of a locally homogeneous wavenumber spectrum $\Omega(\kappa)$ that does not contain C_n^2 and a second function that describes the variation of C_n^2 with location. The review of tropospheric measurements in Section 2.3.3 of Volume 1 showed that the structure constant C_n^2 changes rapidly with altitude but does not vary appreciably in the horizontal plane. This suggests that we can describe the atmospheric spectrum by

$$\Phi_n\left(\kappa, \frac{\mathbf{r}_1 + \mathbf{r}_2}{2} \right) = \Omega(\kappa) C_n^2\left(\frac{z_1 + z_2}{2} \right) \tag{8.8}$$

where the structure-constant profile depends only on the average height. On combining this description with (8.4) we find that the ensemble average of the double-scattering terms can again be expressed as a weighted wavenumber integration of the normalized turbulence spectrum:

$$\langle c + id \rangle = 4k^4 \int d^3\kappa \, \Omega(\kappa) \Pi(\kappa, \mathbf{R}, T) \tag{8.9}$$

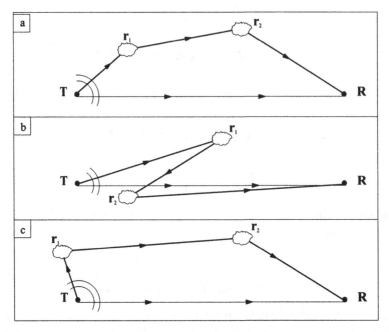

Figure 8.1: Typical propagation paths for double scattering of a spherical wave: (a) two forward-scattering events, (b) two back-scattering events, and (c) a combination of forward- and back-scattering events.

The *double-scattering weighting function* for astronomical observations Π is defined by a pair of volume integrations which are coupled by the profile of C_n^2:

$$\Pi(\boldsymbol{\kappa}, \mathbf{R}, \mathbf{T}) = \int_V d^3 r_1 \int_V d^3 r_2 \, C_n^2 \left(\frac{z_1 + z_2}{2} \right) G(\mathbf{R}, \mathbf{r}_2)$$

$$\times \, G(\mathbf{r}_2, \mathbf{r}_1) \left(\frac{E_0(\mathbf{r}_1, \mathbf{T})}{E_0(\mathbf{R}, \mathbf{T})} \right) \exp[i \boldsymbol{\kappa} \cdot (\mathbf{r}_1 - \mathbf{r}_2)] \qquad (8.10)$$

We have used a new symbol for this weighting function because it contains the structure constant, in contrast to the weighting function for horizontal propagation Θ, which does not. This expression will be used to describe signals from stellar sources that are measured by ground-based telescopes.

8.2 Spherical Waves

Many terrestrial experiments involve the use of omnidirectional transmitters. The field is well represented by a spherical wave for these applications. Three double-scattering sequences are illustrated in Figure 8.1. The first panel shows the most likely sequence in which the spherical wave reaches the receiver by two forward-scattering events. This doubly scattered component supplements the coherent field which travels directly along the line-of-sight path. The second sequence shows how the transmitted signal can be back-scattered twice before reaching the receiver. The third panel illustrates a combination of back-scattering and forward-scattering events.

The volume integrals in (8.6) can be evaluated exactly for spherical waves and our description will not rely on the paraxial approximation. This means that the wavelength need not be small relative to the smallest irregularities. We can there-fore describe wide-angle scattering and treat microwave propagation for which the wavelength is often intermediate between the inner and outer scale lengths of the turbulent irregularities.

8.2.1 A General Description

The radiation field provides an adequate description for the transmitted spherical wave since the vast majority of the scattering elements lies beyond the region where the static and induction fields are important. The field strength is therefore described by

$$E_0(\mathbf{r}, \mathbf{T}) = \frac{\mathcal{E}_0}{4\pi |\mathbf{r} - \mathbf{T}|} \exp(ik|\mathbf{r} - \mathbf{T}|) \qquad (8.11)$$

where the constant $\mathcal{E}_0$ is proportional to the square root of the transmitter power. The electric-field strength for diverging waves thus depends only on the scalar distance from the transmitter.

The first challenge is to decide how to represent the transmitted field and Green's functions in the weighting-function expression (8.6). Notice that they are both spher-ical waves. We can use their explicit description or their equivalent Fourier-integral representation to characterize them analytically. We select the explicit description (8.11) for the term $E_0(\mathbf{R}, \mathbf{T})$ in the denominator. By contrast, we use the Fourier-transform description given in Appendix L for Green's functions, for reasons that will soon become apparent:

$$G(\mathbf{r}, \mathbf{r}') = \frac{1}{(2\pi)^3} \int d^3q \, \frac{\exp[i\mathbf{q} \cdot (\mathbf{r} - \mathbf{r}')]}{k^2 - q^2 + i\epsilon} \qquad (8.12)$$

Since the transmitted wave is simply a scaled version of the Green function, it makes sense to use a similar description for the transmitted field in the volume integrations:

$$E_0(\mathbf{r}_1, \mathbf{T}) = \frac{\mathcal{E}_0}{(2\pi)^3} \int d^3s \, \frac{\exp[i\mathbf{s} \cdot (\mathbf{r}_1 - \mathbf{T})]}{k^2 - s^2 + i\epsilon} \qquad (8.13)$$

With these choices we can express the double-scattering weighting function in the following manner:

$$\Theta_s(\kappa, \mathbf{R}, \mathbf{T}) = \frac{1}{(2\pi)^9} \int_V d^3 r_1 \int_V d^3 r_2 \exp[i\kappa \cdot (\mathbf{r}_1 - \mathbf{r}_2)]$$

$$\times \int d^3 p \frac{\exp[i\mathbf{p} \cdot (\mathbf{R} - \mathbf{r}_2)]}{k^2 - p^2 + i\epsilon} \int d^3 q \frac{\exp[i\mathbf{q} \cdot (\mathbf{r}_2 - \mathbf{r}_1)]}{k^2 - q^2 + i\epsilon}$$

$$\times \int d^3 s \frac{\exp[i\mathbf{s} \cdot (\mathbf{r}_1 - \mathbf{T})]}{k^2 - s^2 + i\epsilon} 4\pi |\mathbf{R} - \mathbf{T}| \exp(-ik|\mathbf{R} - \mathbf{T}|)$$

The random medium is unbounded from the perspective of a terrestrial link, although we will reconsider that assumption later. The volume integrals can be run over all space in this case and each produces a three-dimensional delta function

$$\int d^3 r \exp(i\mathbf{r} \cdot \boldsymbol{\eta}) = (2\pi)^3 \delta(\boldsymbol{\eta})$$

as noted in Appendix F. We choose the coordinate origin to coincide with the transmitter to simplify matters:

$$\Theta_s(\kappa, \mathbf{R}) = \frac{R}{2\pi^2} \exp(-ikR) \int d^3 p \frac{\exp(i\mathbf{p} \cdot \mathbf{R})}{k^2 - p^2 + i\epsilon} \int d^3 q \frac{1}{k^2 - q^2 + i\epsilon}$$

$$\times \int d^3 s \frac{1}{k^2 - s^2 + i\epsilon} \delta(\mathbf{q} - \mathbf{p} - \kappa)\delta(\mathbf{s} - \mathbf{q} + \kappa)$$

The delta functions collapse two of the Fourier-transform integrations:

$$\Theta_s(\kappa, \mathbf{R}) = \frac{R}{2\pi^2} \exp(-ikR) \int d^3 p \frac{\exp(i\mathbf{p} \cdot \mathbf{R})}{[k^2 - p^2 + i\epsilon]^2 [k^2 - (\mathbf{p} + \kappa)^2 + i\epsilon]}$$

This expression can be further reduced when the denominator terms are combined by using the second Feynman formula given in Appendix O. We then make the vector transformation $\mathbf{t} = \mathbf{p} - u\kappa$ to generate the following description:

$$\Theta_s(\kappa, \mathbf{R}) = \frac{R}{\pi^2} \exp(-ikR) \int_0^1 du(1 - u) \exp(iu\kappa \cdot \mathbf{R})$$

$$\times \int d^3 t \frac{\exp(i\mathbf{t} \cdot \mathbf{R})}{[k^2 - \kappa^2 u(1 - u) - t^2 - i\epsilon]^3}$$

This three-dimensional integral can be evaluated with contour integration techniques and the result is found in Appendix N:

$$\Theta_s(\kappa, \mathbf{R}) = -\frac{R}{4} \int_0^1 du (1 - u) \exp(iu\kappa \cdot \mathbf{R})$$

$$\times \exp\left[-ikR + iR\sqrt{k^2 - \kappa^2 u(1 - u)}\right]$$

$$\times \left(\frac{R}{k^2 - \kappa^2 u(1 - u)} + \frac{i}{[k^2 - \kappa^2 u(1 - u)]^{\frac{3}{2}}}\right) \qquad (8.14)$$

It is significant that we have made no assumption so far about the relative values of the turbulent and electromagnetic wavenumbers.

8.2.2 Isotropic Random Media

To proceed further one must begin to describe the turbulence spectrum which appears in the basic expression (8.5). The irregularities are approximately isotropic near the surface, where terrestrial experiments are usually done. In this case it is natural to use spherical coordinates to express the wavenumber integration:

$$\langle c + id \rangle = -Rk^4 \int_0^\infty d\kappa \, \kappa^2 \Phi_n(\kappa) \int_0^\pi d\psi \sin\psi \int_0^{2\pi} d\omega \int_0^1 du \, (1-u)$$

$$\times \exp(i\kappa \, Ru \cos\psi) \exp\left[-ikR + iR\sqrt{k^2 - \kappa^2 u(1-u)}\right]$$

$$\times \left(\frac{R}{k^2 - \kappa^2 u(1-u)} + \frac{i}{[k^2 - \kappa^2 u(1-u)]^{\frac{3}{2}}}\right)$$

The angular integrations are easily performed:

$$\langle c + id \rangle = -2\pi k^4 \int_0^\infty d\kappa \, \kappa \Phi_n(\kappa) \int_0^1 du \, (1-u)\left(\frac{\sin(\kappa Ru)}{u}\right)$$

$$\times \exp\left[-ikR + iR\sqrt{k^2 - \kappa^2 u(1-u)}\right]$$

$$\times \left(\frac{1}{k^2 - \kappa^2 u(1-u)} + \frac{i}{R[k^2 - \kappa^2 u(1-u)]^{\frac{3}{2}}}\right)$$

The path length is usually much greater than the important irregularities and the limiting definition of the delta function discussed in Appendix F is used to collapse the parametric integration:

$$\lim_{\kappa R \to \infty} \left(\frac{\sin(\kappa R u)}{u}\right) = \pi \delta(u)$$

In doing so, one must be careful to note that the integration over u picks up only half the area under the symmetrical delta function. The result is surprisingly simple [7].

$$\langle c + id \rangle = -\frac{1}{2}\left(1 + \frac{i}{kR}\right)\left[4\pi^2 k^2 R \int_0^\infty d\kappa \, \kappa \Phi_n(\kappa)\right] \tag{8.15}$$

The term in square brackets is precisely the phase variance computed with geometrical optics. One can therefore express the result in terms of a quantity that is measured by single-path phase experiments:

$$\langle c + id \rangle = -\frac{1}{2}\langle \varphi_0^2 \rangle \left(1 + \frac{i}{kR}\right) \tag{8.16}$$

We will appeal to the phase measurements discussed in Section 4.1 of Volume 1 when we need to estimate these two average values.

8.2.3 The Relation to Phase and Amplitude Variances

The scattering of spherical waves was described in Section 3.2.2 using the paraxial approximation, which assumes that small-angle scattering applies. We found there that the sum of the phase and amplitude variances is exactly the same as the quantity contained in the square brackets of (8.15). The following relationship is valid for homogeneous isotropic random media when that description is appropriate:

$$\text{Spherical wave:} \quad \langle c + id \rangle = -\frac{1}{2}\left(1 + \frac{i}{kR}\right)(\langle \chi^2 \rangle + \langle \varphi^2 \rangle) \tag{8.17}$$

The imaginary term is extraordinarily small for all situations of practical interest and one has the following result:

$$\text{Spherical wave:} \quad \langle c \rangle = -\frac{1}{2}(\langle \chi^2 \rangle + \langle \varphi^2 \rangle) \quad \text{and} \quad \langle d \rangle = 0 \tag{8.18}$$

These relationships have a profound influence on many estimates of propagation in random media.

8.2.3.1 Wide-angle Scattering

Our next task is to determine whether the relationships (8.18) are valid for the wide-angle-scattering sequences illustrated in Figure 8.1. We have shown that (8.15) is valid for homogeneous isotropic random media when the path length is considerably greater than the outer scale length. The analytical techniques developed to estimate $\langle c + id \rangle$ can be used to calculate the phase and amplitude variances for the same conditions. With the basic description of single scattering

$$\chi + i\varphi = -k^2 \int_V d^3r \, \frac{G(\mathbf{R}, \mathbf{r}) E_0(\mathbf{r})}{E_0(\mathbf{R})} \, \delta\varepsilon(\mathbf{r})$$

the required combination

$$\Sigma_v = \langle \chi^2 \rangle + \langle \varphi^2 \rangle \tag{8.19}$$

is related to the complex weighting function

$$\Lambda(\kappa) = \int_V d^3r \, \frac{G(\mathbf{R}, \mathbf{r}) E_0(\mathbf{r})}{E_0(\mathbf{R})} \, \exp(i\kappa \cdot \mathbf{r})$$

by the following expression:

$$\Sigma_v = 4k^4 \int d^3\kappa \, \Phi_n(\kappa) |\Lambda(\kappa)|^2 \tag{8.20}$$

Before proceeding to evaluate this expression, we must deal with a subtle problem that lies buried in it. We shall find that the spectral weighting function $|\Lambda(\kappa)|^2$ becomes infinite at the resonance wavenumber $\kappa = 2k$ unless one recognizes other atmospheric phenomenon that influence the signal.

Our previous descriptions of phase and amplitude fluctuations were all based on the small-scattering-angle approximation which supposed that all the eddies are large relative to the wavelength. This assumption is clearly frequency-sensitive. It correctly describes optical propagation in the lower atmosphere. By contrast, the wavelengths used for microwave links are often intermediate between the inner and outer scales of irregularities in concentrations of water vapor. Some of the eddies will scatter the incident energy over wide angles because the ratio λ/ℓ need not be small. Microwave scatter propagation beyond the radio horizon gives clear evidence of this wide-angle scattering [12][13].

If wide-angle scattering is also important for line-of-sight propagation, one must consider volume elements that are far removed from the line-of-sight path. That introduces an ever larger number of eddies into consideration and their scattered fields mount rapidly at the receiver. We should expect infinite estimates for the signal variances unless we limit the effective scattering volume in some reasonable way. In the paraxial approximation this limit is provided by the small-scattering-angle constraint that eliminates eddies far from the nominal path. At microwave frequencies we must identify other constraints. There are two limitations to the unbounded addition of scattered components.

8.2.3.2 Geometrical Constraints

The first is geometrical. In the troposphere the decreasing height profile of C_n^2 limits the accumulation of scattered microwaves. We should provide appropriate volume limits to represent these natural boundaries, which we did not do in running the volume integration to infinity earlier. Recall that we used a *slab model* of tropospheric irregularities in describing interferometric observations in Section 5.2 of Volume 1. The level of turbulent activity is assumed to be constant up to a height R and vanish above it in that model.

8.2.3.3 Microwave Attenuation

The more profound influence is that of microwave attenuation [14][15]. We must ask whether it is legitimate to characterize the incident field everywhere in the random medium as the spherical wave that would reach the receiver if there were no atmosphere. The physical reality is that the incident field is attenuated as it passes through the troposphere. By the same token, intensities of waves that are scattered by distant eddies are diminished relative to those of waves scattered by nearby volume elements. We have nowhere accounted for this attenuation. We can do so by adding a small imaginary component to the electromagnetic wavenumber:

$$k \rightarrow k(1 + i\alpha) \tag{8.21}$$

When we make this replacement in the Green function and in the incident field wave, we can be sure that both are suitably attenuated.

Before proceeding with this modification, we must identify the attenuation coefficient α with the relevant physical processes and estimate its magnitude. It is customary to express attenuation in terms of the logarithmic reduction of signal intensity

$$10 \log_{10} \left(\left| \frac{E(R)}{E_0(0)} \right|^2 \right) = -\gamma R \tag{8.22}$$

where γ is measured in decibels per kilometer and the transmission distance R is measured in kilometers [16]. This common engineering parameter is related to the dimensionless constant α by

$$\alpha = 1.15 \times 10^{-6} \left(\frac{\lambda_{cm}}{2\pi} \right) \gamma_{dB/km} \tag{8.23}$$

Two quite different physical processes contribute to microwave attenuation: absorption and out-scattering.

Absorption by oxygen and water-vapor molecules represents a pervasive source of microwave attenuation in the lower atmosphere. A general theory of gaseous absorption valid in the frequency range 100–50 000 MHz was developed by Van Vleck [17][18] and has been confirmed by numerous laboratory experiments. The variation of γ with microwave frequency is plotted in Figure 8.2. Absorption by oxygen is the dominant mechanism below 10 000 MHz and the resonance at $\lambda = 0.5$ cm is due to the permanent magnetic dipole of oxygen. One would expect the following value at L band:

$$\text{Absorption:} \qquad \alpha \simeq 3 \times 10^{-8} \qquad \text{for} \qquad \lambda = 30 \, \text{cm} \tag{8.24}$$

Absorption by water vapor dominates the propagation at 20 000 MHz and the attenuation is approximately

$$\text{Absorption:} \qquad \alpha \simeq 3 \times 10^{-9} \qquad \text{for} \qquad \lambda = 1.5 \, \text{cm} \tag{8.25}$$

Attenuation by water vapor peaks at the molecular resonance $\lambda = 1.35$ cm. It is directly proportional to the absolute humidity and varies substantially with location and season. Climatological maps that have been prepared for the USA allow one to predict the absorption by water vapor with reasonable confidence [19]. Clouds and fog can also cause local changes in absorption due to concentrations of water

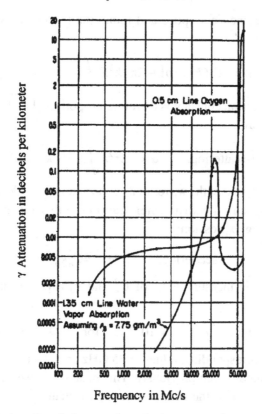

Figure 8.2: Atmospheric absorption of microwaves by molecular oxygen and water vapor summarized by Bean and Dutton [19].

vapor. Microwave absorption has a significant variation with altitude. This means that elevated links and transmissions passing completely through the atmosphere are influenced differently than are those near the surface.

Scattering out of the primary beam is a second source of attenuation. Scattering by rainfall was studied extensively when radar was first developed [20]. Rain drops scatter microwave energy in all directions because the wavelength-to-size ratio is large.[1] The wave's irradiance is reduced by this out-scattering as it progresses into a rainstorm. This attenuation depends strongly on the rate of rainfall and can exceed that due to gaseous absorption – when and where it occurs.

Another pervasive scattering mechanism is provided by out-scattering. This is the effect we have been studying in previous chapters. Energy is constantly scattered out of the incident wave by turbulent irregularities. We analyzed this loss in Chapter 9

---

[1] The raindrops are usually considered to be spherical. Electromagnetic scattering by a dielectric sphere is a solved problem and the results are usually expressed as a series of multipole terms. Microwave attenuation is proportional to the total scattering cross section, which is dominated by the electric-dipole and -quadropole terms plus the magnetic-dipole term. This means that the scattering is not omnidirectional.

of Volume 1 with geometrical optics and found the following expression for the average field strength.

$$\langle E(R) \rangle = E_0(R) \exp\left(-2\pi^2 k^2 R \int_0^\infty d\kappa\, \kappa \Phi_n(\kappa) \right)$$

The equivalent attenuation constant is therefore proportional to the first moment of the turbulence spectrum.

$$\text{Out-scattering:} \qquad \alpha = 2\pi^2 k \int_0^\infty d\kappa\, \kappa \Phi_n(\kappa) \qquad (8.26)$$

Energy is also scattered out of the waves that connect turbulent eddies to the receiver. For a Kolmogorov spectrum with an outer-scale cutoff this universal attenuation is given by the following expression:

$$\text{Out-scattering:} \qquad \alpha = 0.4 C_n^2 \kappa_0^{-\frac{5}{3}} k \qquad (8.27)$$

Using the tropospheric measurements of C_n^2 and κ_0 reviewed in Section 2.3 of Volume 1, we find that

$$\text{Out-scattering:} \qquad \alpha \simeq 4 \times 10^{-9} \qquad \text{for} \qquad \lambda = 30 \text{ cm} \qquad (8.28)$$

This result and the corresponding value for 35 GHz are a factor of ten smaller than the absorption estimates given previously. This means that we should concentrate on microwave absorption as the volume-limiting mechanism.

8.2.3.4 The Influence of Microwave Attenuation

We are now prepared to compute the combination Σ_v defined by (8.19). The small numerical value of α allows one to replace k^2 by $k^2(1 + 2i\alpha)$ in the volume integrals. We use Fourier-integral representations for Green's functions and the field strength to express the complex weighting function:

$$\Lambda(\kappa) = \frac{4\pi R e^{-ikR}}{(2\pi)^6} \int_V d^3r \exp(i\kappa \cdot r) \int d^3q\, \frac{\exp[iq \cdot (R - r)]}{k^2(1 + 2i\alpha) - q^2 + i\epsilon}$$

$$\times \int d^3s\, \frac{\exp(is \cdot r)}{k^2(1 + 2i\alpha) - s^2 + i\epsilon}$$

With this enlargement we are free to run the volume integration to infinity, thereby generating a vector delta function:

$$\Lambda(\kappa) = \frac{R e^{-ikR}}{2\pi^2} \int d^3q\, \frac{\exp(iq \cdot R)}{k^2(1 + 2i\alpha) - q^2 + i\epsilon} \int d^3s\, \frac{\delta(\kappa - q + s)}{k^2(1 + 2i\alpha) - s^2 + i\epsilon}$$

$$= \frac{R e^{-ikR}}{2\pi^2} \int \frac{d^3q \exp(iq \cdot R)}{[k^2(1 + 2i\alpha) - q^2 + i\epsilon][k^2(1 + 2i\alpha) - (q - \kappa)^2 + i\epsilon]}$$

The purely mathematical term $i\varepsilon$ in Green's function is now identified with the absorption term $i\alpha$ which describes important physics. The denominators can be combined using the first result in Appendix O,

$$\Lambda(\kappa) = \frac{Re^{-ikR}}{2\pi^2} \int_0^1 du \int d^3q \frac{\exp(i\mathbf{q}\cdot\mathbf{R})}{[k^2(1+2i\alpha) - \kappa^2 u(1-u) - (\mathbf{q}-u\kappa)^2 + i\epsilon]^2}$$

and the translation $\mathbf{p} = \mathbf{q} - u\kappa$ brings this into the form of a wave integral in Appendix N,

$$\Lambda(\kappa) = -\frac{iR\exp(-ikR)}{2} \int_0^1 du \exp(iu\kappa\cdot\mathbf{R}) f(u,\kappa,k,R,\alpha) \qquad (8.29)$$

where

$$f(u,\kappa,k,R,\alpha) = \frac{\exp\left[iR\sqrt{k^2(1+2i\alpha) - \kappa^2 u(1-u)}\right]}{\sqrt{k^2(1+2i\alpha) - \kappa^2 u(1-u)}} \qquad (8.30)$$

With this result we return to estimate Σ_v using the definition (8.20). As we did in deriving (8.15), we assume that the random medium is isotropic:

$$\Sigma_v = R^2 k^4 \int_0^\infty d\kappa \, \kappa^2 \Phi_n(\kappa) \int_0^\pi d\psi \sin\psi \int_0^{2\pi} d\omega \int_0^1 du \int_0^1 dv$$
$$\times \exp[i(u-v)\kappa R\cos\psi] \, f(u,\kappa,k,R,\alpha)f^*(v,\kappa,k,R,\alpha)$$

The angular integrations are readily performed:

$$\Sigma_v = 4\pi Rk^4 \int_0^\infty d\kappa \, \kappa\Phi_n(\kappa) \int_0^1 du \int_0^1 dv \left[\frac{\sin[\kappa R(u-v)]}{u-v}\right]$$
$$\times f(u,\kappa,k,R,\alpha)f^*(v,\kappa,k,R,\alpha)$$

Since the path length is much greater than the size of the largest eddies, the term in square brackets reduces to a delta function that collapses one of the parametric integrations. The result is expressed as the spectrum of irregularities times a weighting function:

$$\Sigma_v = 4\pi^2 Rk^4 \int_0^\infty d\kappa \, \kappa\Phi_n(\kappa)W(\kappa,k,R,\alpha) \qquad (8.31)$$

where

$$W(\kappa,k,R,\alpha) = \int_0^1 du \, |f(u,\kappa,k,R,\alpha)|^2 \qquad (8.32)$$

One can now combine (8.30) and (8.32) to generate an integral expression for the weighting function:

$$W(\kappa, k, R, \alpha) = \int_0^1 du \, \frac{1}{\rho(u)} \exp\left[-2R \sin\left(\frac{\phi}{2}\right) \sqrt{\rho(u)}\right] \qquad (8.33)$$

where

$$\rho(u) = \sqrt{\left[k^2 - \kappa^2 u(1-u)\right]^2 + 4k^4 \alpha^2} \qquad (8.34)$$

and

$$\tan \phi = \frac{2k^2 \alpha}{k^2 - \kappa^2 u(1-u)} \qquad (8.35)$$

The combination $\rho(u)$ in the denominator would vanish at the midpoint of the integration when $\kappa = 2k$ were it not for the attenuation term $4k^4 \alpha^2$. This is the source of the predicted resonance.

The weighting function is rapidly attenuated for eddies that lie above the resonance wavenumber $\kappa = 2k$. These small eddies generate *evanescent waves* that are exponentially damped and play no important role in the outcome. We focus instead on the region below resonance which plays the dominant role in setting the phase and amplitude variances. Unless one is very close to the resonance, one can approximate the weighting function quite accurately by the following expression:

$$\kappa < 2k: \qquad W(\kappa, k, R, \alpha) = \int_0^1 du \, \frac{\exp(-Rk\alpha)}{k^2 - \kappa^2 u(1-u)} \qquad (8.36)$$

Although the product Rk is usually very large for microwave links, the absorption coefficient is even smaller and the exponential term can be ignored[2] and the parametric integration is easily done:

$$\kappa < 2k: \qquad W(\kappa, k, R, \alpha) = \frac{4k^2}{\kappa \sqrt{4k^2 - \kappa^2}} \arctan\left(\frac{\kappa}{\sqrt{4k^2 - \kappa^2}}\right) \qquad (8.37)$$

This weighting function is plotted in Figure 8.3 as a function of the ratio $\kappa/(2k)$ and rises rapidly as one approaches resonance. Notice that diffraction effects are absent from this expression and that the Fresnel length has completely disappeared. This cancelation also occurred on combining the phase and amplitude variances estimated with the paraxial approximation that led to the sum rule (7.10).

The last step in our exploration is to determine how the weighting function plotted in Figure 8.3 influences the combination Σ_v and its relationship to the double-scattering terms. We consider first its influence on optical signals since we

---

[2] To take a specific example, consider the L-band experiment described in [21], for which a path length $R = 16$ km and $\lambda = 30$ cm were used. With the estimate for α given by (8.24) we see that the exponent is approximately 0.001.

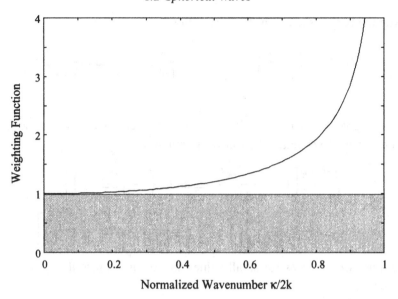

Figure 8.3: The weighting function for the combined phase and amplitude variances predicted by (8.37) and plotted as a function of $\kappa/(2k)$.

have already established that answer in (8.15). We want to use a turbulence spectrum that describes the troposphere. We select the Kolmogorov model with inner- and outer-scale cutoffs which was introduced in Section 2.2.6 of Volume 1:

$$\Sigma_v = 0.033 \times 4\pi^2 Rk^2 C_n^2 \int_{\kappa_0}^{\kappa_m} \frac{d\kappa}{\kappa^{\frac{8}{3}}} \left[\frac{4k^2}{\kappa\sqrt{4k^2 - \kappa^2}} \arctan\left(\frac{\kappa}{\sqrt{4k^2 - \kappa^2}} \right) \right] \quad (8.38)$$

The large-wavenumber cutoff κ_m occurs well below resonance since the inner scale length ℓ_0 is much larger than λ. The weighting function is unity in this region, so

$$\text{Optical:} \qquad \Sigma_v = 0.782 Rk^2 C_n^2 (\kappa_0)^{-\frac{5}{3}} \qquad (8.39)$$

which reproduces the expression for $\langle \varphi_0^2 \rangle$ estimated with geometrical optics. This is gratifying but not our real purpose.

We have gone to considerable trouble to describe microwaves. With the same spectrum model and Figure 8.3 we observe two important features at these frequencies. The inner-scale wavenumber lies beyond the resonance value so the falling spectrum is influenced in some measure by all values of the weighting function. By contrast, the outer-scale wavenumber κ_0 is very small relative to $2k$ and the powerful large eddies fall in the range within which the weighting function is close to unity. As the weighting function is rising towards resonance the spectrum is falling rapidly. Our task is to gauge how that competition is resolved. To do so we rewrite the basic expression (8.38) as follows:

$$\Sigma_v = 0.782 R k^2 C_n^2 \int_{\kappa_0}^{2k} \frac{d\kappa}{\kappa^{\frac{8}{3}}} \left\{ 1 - \left[1 - \frac{4k^2}{\kappa\sqrt{4k^2 - \kappa^2}} \arctan\left(\frac{\kappa}{\sqrt{4k^2 - \kappa^2}}\right) \right] \right\}$$

(8.40)

The first term gives the optical result (8.39) and our attention focuses on the second. We therefore need to evaluate the following integral:

$$\begin{aligned}
\mathcal{J} &= \int_{\kappa_0}^{2k} \frac{d\kappa}{\kappa^{\frac{8}{3}}} \left[1 - \frac{4k^2}{\kappa\sqrt{4k^2 - \kappa^2}} \arctan\left(\frac{\kappa}{\sqrt{4k^2 - \kappa^2}}\right) \right] \\
&= (2k)^{-\frac{5}{3}} \int_{\kappa_0/(2k)}^{1} \frac{du}{u^{\frac{8}{3}}} \left[1 - \frac{1}{u\sqrt{1 - u^2}} \arctan\left(\frac{u}{\sqrt{1 - u^2}}\right) \right]
\end{aligned}$$

The integrand varies as $u^{\frac{1}{3}}$ for small values so we can replace the lower limit by zero to find

$$\mathcal{J} = -4.468(2k)^{-\frac{5}{3}}$$

and the desired combination becomes

Microwave: $$\Sigma_v = 0.782 R k^2 C_n^2 (\kappa_0)^{-\frac{5}{3}} \left[1 - 5.468\left(\frac{\lambda}{2L_0}\right)^{\frac{5}{3}} \right]$$ (8.41)

The microwave values for λ are always much smaller than the outer scale length and the second term is not important. This means that the wavenumber spectrum was falling faster than the resonance weighting function was rising. We would have discovered the same result had we used the complete expression for the weighting function (8.33) which depends on α explicitly. The large eddies evidently dominate the outcome and are little influenced by the weighting function. For microwave links this means that the double-scattering terms are also described by (8.18), which is an important conclusion.

8.3 Plane Waves

Astronomical observations at optical and radio wavelengths are described in terms of plane waves passing through the atmosphere along oblique paths. We will explain how to calculate the double scattering of these signals by examining the simpler case of an overhead source. The propagation geometry for a plane wave falling vertically on the atmosphere is illustrated in Figure 8.4. The random medium begins at ground level and extends indefinitely upward. The level of turbulent activity declines rapidly with altitude according to a profile function $C_n^2(z)$ that can be measured in several

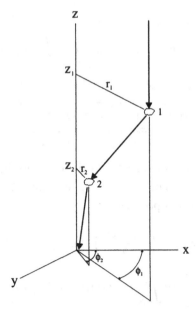

Figure 8.4: Coordinates used to locate typical scattering-volume elements that contribute to double scattering of a plane wave by atmospheric irregularities for an overhead source.

ways.[3] The plane wave is scattered first by an irregularity at $\mathbf{r}_1$. The reradiated field then propagates as a spherical wave from $\mathbf{r}_1$ to a second irregularity at $\mathbf{r}_2$. This wave is scattered again at $\mathbf{r}_2$ and proceeds to the receiver as a spherical wave. The individual contributions must be summed over all possible pairs of volume elements.

8.3.1 General Description

The atmosphere is not homogeneous and we must use the descriptions (8.9) and (8.10) to recognize that influence. In our coordinate system the arriving plane wave is described by

$$E_0(\mathbf{r}) = \mathcal{E}_0 \exp(-ikz) \tag{8.42}$$

Green's functions are expressed in terms of the rectangular coordinates which identify the two scattering points. The receiver is placed at the origin and the basic expression for the weighting function becomes

$$\Pi(\kappa) = \int_0^\infty dz_2 \int_{-\infty}^\infty dx_2 \int_{-\infty}^\infty dy_2 \, Q(x_2, y_2, z_2) \frac{\exp\left(ik\sqrt{x_2^2 + y_2^2 + z_2^2}\right)}{4\pi\sqrt{x_2^2 + y_2^2 + z_2^2}} \tag{8.43}$$

[3] These techniques and typical profiles were discussed in Section 2.3.3 of Volume 1.

where

$$Q(x_2, y_2, z_2) = \int_0^\infty dz_1 \exp(-ikz_1) \int_{-\infty}^\infty dx_1 \int_{-\infty}^\infty dy_1\, C_n^2\left(\frac{z_1 + z_2}{2}\right)$$

$$\times \exp\{i[\kappa_x(x_1 - x_2) + \kappa_y(y_1 - y_2) + \kappa_z(z_1 - z_2)]\}$$

$$\times \frac{\exp\left[ik\sqrt{(x_1 - x_2)^2 + (y_1 - y_2)^2 + (z_1 - z_2)^2}\right]}{4\pi\sqrt{(x_1 - x_2)^2 + (y_1 - y_2)^2 + (z_1 - z_2)^2}} \tag{8.44}$$

In the second expression we transform the horizontal coordinates with $u = x_1 - x_2$ and $v = y_1 - y_2$. The coordinates x_2 and y_2 disappear at the infinite limits and Q depends only on the vertical position:

$$Q(z_2) = \int_0^\infty dz_1\, C_n^2\left(\frac{z_1 + z_2}{2}\right) \exp[-ikz_1 + i\kappa_z(z_1 - z_2)]$$

$$\times \int_{-\infty}^\infty du \int_{-\infty}^\infty dv \exp[i(\kappa_x u + \kappa_y v)]$$

$$\times \frac{\exp\left[ik\sqrt{u^2 + v^2 + (z_1 - z_2)^2}\right]}{4\pi\sqrt{u^2 + v^2 + (z_1 - z_2)^2}}$$

If we transform the last two integrals into the cylindrical coordinates defined in Figure 8.4 the angular integration yields a Bessel function:

$$Q(z_2) = \frac{1}{2}\int_0^\infty dz_1\, C_n^2\left(\frac{z_1 + z_2}{2}\right) \exp[-ikz_1 + i\kappa_z(z_1 - z_2)]$$

$$\times \int_0^\infty dr_1\, r_1\, J_0(r_1\kappa_r) \frac{\exp\left[ik\sqrt{r_1^2 + (z_1 - z_2)^2}\right]}{\sqrt{r_1^2 + (z_1 - z_2)^2}}$$

The radial integration can be done without approximation using Lamb's integral, which is described in Appendix D:

$$Q(z_2) = \frac{1}{2}\int_0^\infty dz_1\, C_n^2\left(\frac{z_1 + z_2}{2}\right) \exp[-ikz_1 + i\kappa_z(z_2 - z_1)]$$

$$\times \begin{cases} \dfrac{i \exp\left(i|z_1 - z_2|\sqrt{k^2 - \kappa_r^2}\right)}{\sqrt{k^2 - \kappa_r^2}} & \text{for } \kappa_r \le k \\[2em] \dfrac{\exp\left(-|z_1 - z_2|\sqrt{\kappa_r^2 - k^2}\right)}{\sqrt{\kappa_r^2 - k^2}} & \text{for } \kappa_r \ge k \end{cases}$$

The first volume integration can be done using the same techniques:

$$\Pi(\kappa) = \int_0^\infty dz_2 \, Q(z_2) \int_0^\infty dr_2 \, r_2 \int_0^{2\pi} d\phi_2 \, \frac{\exp\left(ik\sqrt{r_2^2 + z_2^2}\right)}{4\pi\sqrt{r_2^2 + z_2^2}}$$

$$= \frac{i}{2k} \int_0^R dz_2 \, Q(z_2) \exp(ikz_2)$$

On combining these results we find an expression for the double-scattering weighting function that describes astronomical observations of an overhead source:

$$\Pi(\kappa) = \frac{i}{4k} \int_0^\infty dz_1 \int_0^\infty dz_2 \, C_n^2\left(\frac{z_1 + z_2}{2}\right) \exp[i(z_1 - z_2)(\kappa_z - k)]$$

$$\times \begin{cases} \dfrac{i}{\sqrt{k^2 - \kappa_r^2}} \exp\left(i|z_1 - z_2|\sqrt{k^2 - \kappa_r^2}\right) & \text{for } \kappa_r \le k \\[4mm] \dfrac{1}{\sqrt{\kappa_r^2 - k^2}} \exp\left(-|z_1 - z_2|\sqrt{\kappa_r^2 - k^2}\right) & \text{for } \kappa_r \ge k \end{cases} \tag{8.45}$$

To estimate the moments of c and d we combine this result with the basic expression (8.9). That combination can be simplified by shifting to sum and difference coordinates:

$$u = z_1 - z_2 \qquad \text{and} \qquad z = \tfrac{1}{2}(z_1 + z_2)$$

The average value of the second-order terms is then given by the following expression:

$$\langle c + id \rangle = ik^3 \int d^3\kappa \, \Omega(\kappa) \int_0^\infty dz \, C_n^2(z) \int_{-z}^z du \, \exp[iu(\kappa_z - k)]$$

$$\times \begin{cases} \dfrac{i}{\sqrt{k^2 - \kappa_r^2}} \exp\left(i|u|\sqrt{k^2 - \kappa_r^2}\right) & \text{for } \kappa_r \le k \\[4mm] \dfrac{1}{\sqrt{\kappa_r^2 - k^2}} \exp\left(-|u|\sqrt{\kappa_r^2 - k^2}\right) & \text{for } \kappa_r \ge k \end{cases} \tag{8.46}$$

The exponential decay of evanescent waves scattered by eddies smaller than the wavelength of the radiation is evident in the form for $\kappa_r \ge k$. This term can usually be omitted because it contributes so little to the outcome. From this expression it is also apparent that the plane-wave resonance condition occurs for $\kappa_r = k$, in contrast to the spherical-wave relation $\kappa_r = 2k$ found previously. It is significant that this result

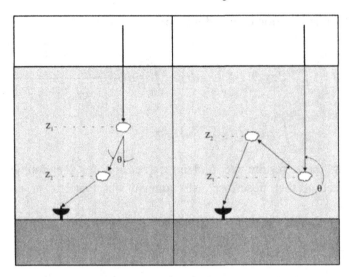

Figure 8.5: Feynman diagrams describing the double scattering of a plane wave falling on the atmosphere from an overhead source. The sequence on the left involves two forward-scattering events, whereas that on the right has two back-scattering events. The scattering angle is identified by θ.

does not depend on the assumption of small-angle forward scattering and therefore describes the quite different double-scattering sequences suggested in Figure 8.5.

8.3.2 Optical and Infrared Observations

One can complete the integrations in (8.46) quite easily for optical and infrared signals. In these bands the smallest tropospheric irregularities are large relative to the wavelength and one can ignore the second form. Only small-angle forward scattering is important and one can exclude the double-back-scattering sequences suggested in Figure 8.5. This means that $u = z_1 - z_2$ assumes only positive values and one establishes the following expression:

$$\langle c + id \rangle = -k^3 \int_{\kappa_r < k} d^3\kappa \, \Omega(\kappa) \int_0^\infty dz \, C_n^2(z)$$

$$\times \int_0^z du \, \frac{\exp\left[iu\left(\kappa_z - k + \sqrt{k^2 - \kappa_r^2}\right)\right]}{\sqrt{k^2 - \kappa_r^2}} \qquad (8.47)$$

One can further simplify this because the small wavelength-to-eddy ratio implies that

$$\kappa_z - k + \sqrt{k^2 - \kappa_r^2} \approx \kappa_z \qquad \text{and} \qquad \sqrt{k^2 - \kappa_r^2} \approx k \qquad (8.48)$$

These relations define the paraxial approximation for small-angle forward scattering and (8.47) now simplifies to

$$\langle c + id \rangle = ik^2 \int_{\kappa < k} d^3\kappa \, \Omega(\kappa) \int_0^\infty dz \, C_n^2(z) \left(\frac{\exp(iz\kappa_z) - 1}{\kappa_z} \right)$$

To proceed further we assume that the irregularities are isotropic. This is apparently a bold assumption in view of the description of tropospheric anisotropy and outer scale lengths presented in Section 2.3.5 of Volume 1. On the other hand, the result is not strongly dependent on that assumption, so we write

$$\langle c + id \rangle = 2\pi ik^2 \int_0^k d\kappa \, \kappa^2\Omega(\kappa) \int_0^\pi d\psi \sin\psi$$
$$\times \int_0^\infty dz \, C_n^2(z) \left(\frac{\exp(iz\kappa \cos\psi) - 1}{\kappa \cos\psi} \right)$$

and, on changing the polar integration to $w = \kappa z \cos\psi$, we find that

$$\langle c + id \rangle = -2\pi k^2 \int_0^k d\kappa \, \kappa^2\Omega(\kappa) \int_0^\infty dz \, C_n^2(z) \int_{-\kappa z}^{\kappa z} \frac{dw}{w} \sin w$$

where we have dropped a term that is odd in w. We next recall two types of tropospheric measurements that were reviewed in Section 2.3 of Volume 1. It was noted there that the profile of C_n^2 has significant values up to altitudes of several kilometers. Since we are considering an overhead source, the eddy dimension in the vertical direction is the relevant scale length. The largest eddy size measured in the vertical direction is a few tens of meters.[4] This means that the product $z\kappa$ is large and the integration over w gives simply π:

Optical and IR: $\qquad \langle c + id \rangle = -2\pi^2 k^2 \int_0^k d\kappa \, \kappa\Omega(\kappa) \int_0^\infty dz \, C_n^2(z) \qquad$ (8.49)

8.3.3 Radio and Microwave Observations

The wavelength is larger than 1 cm for frequencies less than 30 GHz. This is greater than the inner scale length of tropospheric irregularities and one cannot assume that $k > \kappa$. Wide-angle scattering of the signal may therefore be important.

The double-scattering averages can be evaluated if one restricts the scattering angle θ identified in Figure 8.5 to values less than $\pi/2$. That corresponds to ignoring the combination of back-scattering sequences shown in the right-hand panel of Figure 8.5. On the other hand, one can show that these sequences make only a small contribution to the final result. The first scattering point is always higher than

---

[4] For oblique propagation one must consider the much larger horizontal scale lengths and anisotropy.

the second for this class of eddies. We can therefore use (8.47) but must not use the paraxial approximations (8.48) in doing so:

$$\text{Microwave } \theta \leq \pi/2: \quad \langle c + id \rangle = ik^3 \int_{\kappa_r < k} d^3\kappa \, \Omega(\kappa) \int_0^\infty dz \, C_n^2(z) \mathcal{J}(\kappa, z)$$

(8.50)

where

$$\mathcal{J}(\kappa, z) = i \frac{\exp\left[iz\left(\kappa_z - k + \sqrt{k^2 - \kappa_r^2}\right)\right] - 1}{\sqrt{k^2 - \kappa_r^2}\left(\kappa_z - k + \sqrt{k^2 - \kappa_r^2}\right)}$$

(8.51)

To proceed further we will assume that the irregularities are isotropic, as we did in treating optical and infrared transmissions. Using spherical wavenumber coordinates, the double-scattering terms become

$$\langle c + id \rangle = 2\pi ik^2 \int_0^k d\kappa \, \kappa^2 \Omega(\kappa) \int_0^\infty dz \, C_n^2(z) Q(\kappa, z)$$

where

$$Q(\kappa, z) = \int_0^\pi d\psi \sin \psi \, \frac{1 - \exp\left[z\left(\kappa \cos \psi - k + \sqrt{k^2 - \kappa^2 \sin^2 \psi}\right)\right]}{\left(k^2 - \kappa^2 \sin^2 \psi\right)^{\frac{1}{2}}\left(\kappa \cos \psi - k + \sqrt{k^2 - \kappa^2 \sin^2 \psi}\right)}$$

This integral can be evaluated with the following substitution:

$$w = z\left(\kappa \cos \psi - k + \sqrt{k^2 - \kappa^2 \sin^2 \psi}\right)$$

so that

$$Q(\kappa, z) = -\frac{z}{\kappa} \int_{-\kappa z}^{\kappa z} \frac{dw}{w(w + kz)} (1 - \cos w - i \sin w)$$

The product kz is very large and $w + kz$ can be replaced by kz because the important contributions to the integration over w come from small values. We then apply the previous arguments about the large value of the combination κz and drop the odd term:

$$Q(\kappa, z) = \frac{i}{k\kappa} \int_{-\kappa z}^{\kappa z} \frac{dw}{w} \sin w$$

This gives the following expression for double scattering of plane waves that pass vertically through the atmosphere but are not back-scattered:

$$\text{Microwave } \theta \leq \pi/2: \qquad \langle c + id \rangle = -2\pi^2 k^2 \int_0^k d\kappa\, \kappa\Omega(\kappa) \int_0^\infty dz\, C_n^2(z)$$

$$(8.52)$$

This is the same result as that established above for optical and infrared signals. We suggest how the term corresponding to large scattering angles can be evaluated in Problem 1.

8.3.4 The Relation to Phase and Amplitude Variances

The variance of phase and amplitude fluctuations imposed on astronomical signals was addressed in Chapters 3 and 7 using the paraxial approximation. Their sum is independent of diffraction effects. By comparing that description with the double-scattering terms in (8.49) we confirm that the following result is valid for optical and infrared signals passing through the atmosphere:

$$\text{Optical and IR astronomy:} \qquad \langle c \rangle = -\tfrac{1}{2}[\langle \chi^2 \rangle + \langle \varphi^2 \rangle] \quad \text{and} \quad \langle d \rangle = 0 \quad (8.53)$$

This relationship must be explored independently for radio astronomy observations. Atmospheric irregularities can generate wide-angle scattering of these microwave signals, as we noted in our discussion of spherical waves. We can describe this case using (8.20) if we are careful to use the spectrum for inhomogeneous media provided by (8.8):

$$\Sigma_v = 4k^4 \int d^3\kappa\, \Omega(\kappa) \int_V d^3 r_1 \exp(i\boldsymbol{\kappa}\cdot\mathbf{r}_1) \left(\frac{G(\mathbf{R}, \mathbf{r}_1)E_0(\mathbf{r}_1)}{E_0(\mathbf{R})} \right)$$

$$\times \int_V d^3 r_2 \exp(-i\boldsymbol{\kappa}\cdot\mathbf{r}_2) \left(\frac{G(\mathbf{R}, \mathbf{r}_2)E_0(\mathbf{r}_2)}{E_0(\mathbf{R})} \right)^* C_n^2\left(\frac{z_1 + z_2}{2} \right) \qquad (8.54)$$

If we write the volume integrations as

$$\int_V d^3 r = \int_0^\infty dz \int_0^\infty dr\, r \int_0^{2\pi} d\phi$$

we can express

$$\Sigma_v = 4k^4 \int d^3\kappa\, \Omega(\kappa) \int_0^\infty dz_1 \int_0^\infty dz_2\, C_n^2\left(\frac{z_1 + z_2}{2} \right) \exp[i\kappa_z(z_1 - z_2)]\, Q_1 Q_2$$

where

$$Q_1 = \int_0^\infty dr_1\, r_1 \int_0^{2\pi} d\phi_1 \left(\frac{G(\mathbf{R}, \mathbf{r}_1)E_0(\mathbf{r}_1)}{E_0(\mathbf{R})} \right) \exp[i(\kappa_x x_1 + \kappa_y y_1)]$$

and Q_2 is the complex conjugate expressed as a function of z_2. We use cylindrical coordinates for $\mathbf{r}$ and $\boldsymbol{\kappa}$ to describe the terms in this expression:

$$Q_1 = \frac{e^{ik(R-z_1)}}{4\pi} \int_0^\infty dr_1\, r_1 \int_0^{2\pi} d\phi_1 \frac{e^{ik\sqrt{r_1^2+z_1^2}}}{\sqrt{r_1^2+z_1^2}} e^{i\kappa_r r_1 \cos(\phi_1-\omega)}$$

These integrations can be performed analytically using Lamb's integral:

$$Q_1 = \frac{e^{ik(R-z_1)}}{2}
\begin{cases}
\dfrac{i}{\sqrt{k^2-\kappa_r^2}} \exp\left(iz_1\sqrt{k^2-\kappa_r^2}\right) & \text{for } \kappa_r \leq k \\[2ex]
\dfrac{1}{\sqrt{\kappa_r^2-k^2}} \exp\left(-z_1\sqrt{\kappa_r^2-k^2}\right) & \text{for } \kappa_r \geq k
\end{cases}$$

One can calculate the other factor from this result and their product becomes

$$Q_1 Q_2 = \frac{e^{ik(z_2-z_1)}}{4}
\begin{cases}
\dfrac{i}{k^2-\kappa_r^2} \exp\left[i(z_1-z_2)\sqrt{k^2-\kappa_r^2}\right] & \text{for } \kappa_r \leq k \\[2ex]
\dfrac{1}{\kappa_r^2-k^2} \exp\left[-(z_1+z_2)\sqrt{\kappa_r^2-k^2}\right] & \text{for } \kappa_r \geq k
\end{cases}$$

The second term is attenuated exponentially with height and can be dropped. If one does not work too close to the resonance wavenumber $\kappa = k$, one can ignore atmospheric attenuation in treating the first term. We then change to sum and difference coordinates:

$$\Sigma_v = k^4 \int_{\kappa_r < k} d^3\kappa\, \Omega(\kappa) \int_0^\infty dz\, C_n^2(z) \int_{-z}^z du\, \frac{\exp\left[iu\left(\kappa_z - k + \sqrt{k^2-\kappa_r^2}\right)\right]}{k^2-\kappa_r^2} \tag{8.55}$$

This is the same as the original expression (8.50) except that the denominator is modified. For isotropic irregularities the techniques described previously can be used to evaluate this result and one finds that the following relationship is also valid for microwave and radio frequencies:

Radio astronomy: $\quad \langle c \rangle = -\frac{1}{2}[\langle \chi^2 \rangle + \langle \varphi^2 \rangle] \quad$ and $\quad \langle d \rangle = 0 \quad$ (8.56)

8.4 Beam Waves

The combination of laser sources and optical telescopes has led to important applications of beam waves. Double scattering of a collimated optical signal formed in this way is illustrated in Figure 8.6. This is a special type of beam wave and we need to consider the general class of such signals.

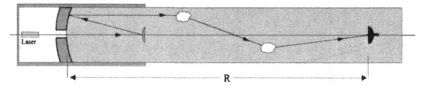

Figure 8.6: Double scattering of a collimated beam formed by a laser and a Cassegrain telescope.

Accurate descriptions for the double scattering of beam waves are a good deal more difficult to establish than the results for plane and spherical waves. An exact solution for this important type of signal has not yet been developed and we resort to a familiar approximation in order to obtain useful results. The free-space field strength for a type of beam wave is given by (3.80).

To describe the double scattering of a beam wave we assume that the signal travels near the surface over a relatively short distance. We can therefore assume that the turbulence conditions are constant along the path and that the irregularities are nearly isotropic. This means that we can use the description for terrestrial paths given by (8.5) and (8.6). Beam waves are usually formed by optical or infrared lasers. Since their wavelengths are much smaller than the inner scale length, we can use the paraxial approximation to simplify the computational task. Green's function that connects the receiver to the second scattering point can therefore be simplified as follows:

$$G(\mathbf{R}, \mathbf{r}_2) = \frac{1}{4\pi(R - x_2)} \exp\left[ik\left((R - x_2) - \frac{r_2^2}{2(R - x_2)}\right)\right] \qquad (8.57)$$

Green's function that connects the first and second scattering points is simplified in the same way:

$$G(\mathbf{r}_2, \mathbf{r}_1) = \frac{1}{4\pi|x_2 - x_1|} \exp\left[ik\left(|x_2 - x_1| - \frac{r_1^2 + r_2^2 - 2r_1r_2\cos(\phi_1 - \phi_2)}{2|x_2 - x_1|}\right)\right]$$

$$(8.58)$$

where the cylindrical coordinates are defined by Figure 8.7.

8.4.1 The Expression for $\langle c + id \rangle$

The small-scattering-angle condition requires that the distances of the receiver and scattering points are sequential so that $R > x_2 > x_1$ and the integration over x_1 stops at x_2. This means that one can drop the absolute-magnitude notation. It is natural to use cylindrical coordinates to define the wavenumber vector:

$$\kappa = \mathbf{i}_z \kappa_r \cos\omega + \mathbf{i}_y \kappa_r \sin\omega + \mathbf{i}_x \kappa_x$$

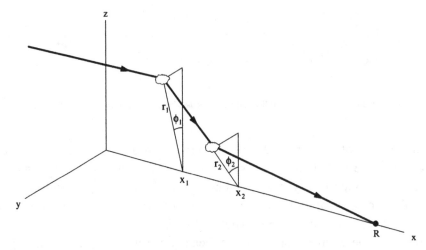

Figure 8.7: The propagation geometry used to calculate the double-scattering weighting function for a beam wave.

With these observations the double-scattering weighting function (8.6) becomes

$$\Theta_b(\kappa, R) = \frac{1}{16\pi^2} \int_0^R dx_2 \int_0^{x_2} dx_1 \frac{\exp[i\kappa_z(x_1 - x_2) + ik(R - x_1)]}{(R - x_2)(x_2 - x_1)}$$

$$\times \int_0^\infty dr_1\, r_1 \int_0^{2\pi} d\phi_1 \left(\frac{E_0(r_1, x_1)}{E_0(0, R)} \right) \exp\left(\frac{ikr_1^2}{2(x_2 - x_1)} \right)$$

$$\times \exp[i\kappa_r r_1 \cos(\phi_1 - \omega)] Q(r_1, \phi_1)$$

We have delayed introducing the explicit description for the free-space field strength for notational simplicity. The substitute function is defined by

$$Q(r_1, \phi_1) = \int_0^\infty dr_2\, r_2 \int_0^{2\pi} d\phi_2 \exp[-i\kappa_r r_2 \cos(\phi_2 - \omega)]$$

$$\times \exp\left(\frac{ikr_2^2}{2(R - xz_2)} \right) \exp\left[ik\left(\frac{r_2^2 - 2r_1 r_2 \cos(\phi_1 - \phi_2)}{2(x_2 - x_1)} \right) \right]$$

and the required integrations can be found in Appendix D:

$$\int_0^{2\pi} d\phi\, \exp[i(a\cos\phi + b\sin\phi)] = 2\pi J_0\left(\sqrt{a^2 + b^2} \right)$$

and

$$\int_0^\infty dr\, r\, J_0(r\eta) \exp(-r^2 p^2) = \frac{1}{2\,p^2} \exp\left(-\frac{\eta^2}{4p^2} \right)$$

so that

$$Q(r_1, \phi_1) = 2\pi i \frac{(R - x_2)(z_2 - x_1)}{k(R - x_1)} \exp\left(\frac{-i(R - x_2)}{2k(R - x_1)} \Xi(r_1, \phi_1) \right)$$

where

$$\Xi(r_1, \phi_1) = \kappa_r^2(x_2 - x_1) + r_1^2 \frac{k^2}{x_2 - x_1} + 2r_1 k \kappa_r \cos(\phi_1 - \omega)$$

With the beam-wave field-strength expression the weighting function is

$$\Theta_b(\kappa, R) = \frac{i}{4k} \int_0^R dx_2 \int_0^{x_2} dx_1 \frac{\exp[i\kappa_x(x_1 - x_2)]}{R - x_1} \frac{1 + i\alpha R}{1 + i\alpha x_1}$$
$$\times \mathcal{J}(x_1, x_2) \exp\left(-i \frac{\kappa_r^2}{2k} \frac{(R - x_2)(x_2 - x_1)}{R - x_1}\right)$$

where the function $\mathcal{J}(x_1, x_2)$ represents the integrations over r_1 and ϕ_1. The azimuth integral can again be expressed in terms of a Bessel function:

$$\mathcal{J}(x_1, x_2) = \int_0^\infty dr_1 \, r_1 \exp\left[-r_1^2\left(\gamma - \frac{ik}{2(R - x_1)}\right)\right] J_0\left[r_1 \kappa_r \left(\frac{x_2 - x_1}{R - x_1}\right)\right]$$

and the remaining radial integration can be done using the previous result:

$$\Theta_b(\kappa, R) = -\frac{1}{4k^2} \int_0^R dx_2 \int_0^{x_2} dx_1 \exp[i\kappa_x(x_1 - x_2)]$$
$$\times \exp\left[-i \frac{\kappa_r^2}{2k} \frac{x_2 - x_1}{R - x_1}\left(R - x_2 + (x_2 - x_1)\frac{1 + i\alpha x_1}{1 + i\alpha R}\right)\right]$$

$$(8.59)$$

This expression provides a complete description of the double-scattering term within the context of small-angle forward scattering. On the other hand, that basic assumption means that one can simplify the result still further. Since all the turbulent eddy are of sizes large relative to the wavelength in this approximation the following inequality is valid:

$$\kappa_x \gg \frac{\kappa_r^2}{2k}$$

This suggests that we can neglect the second exponential in the weighting function:

$$\Theta_b(\kappa, R) \simeq -\frac{1}{4k^2} \int_0^R dx_2 \int_0^{x_2} dx_1 \exp[i\kappa_x(x_1 - x_2)]$$

The double integration along the path was evaluated for our analysis of phase fluctuations using geometrical optics:[5]

$$\Theta_b(\kappa, R) \simeq -\frac{\pi R}{4k^2} \delta(\kappa_x)$$

---

[5] Equation (5.27) in Volume 1 addresses this double integral for the complete square and gives twice the result needed here.

which leads to the same result as that found for plane and spherical waves [7][11]:

Beam waves: $\langle c \rangle = -2\pi^2 k^2 R \displaystyle\int_0^\infty d\kappa\, \kappa \Phi_n(\kappa)$ and $\langle d \rangle = 0$ (8.60)

The expressions for the average values of c and d are important for estimating the average field strength.

8.4.2 The Expression for $\langle \psi_2 \rangle$

By contrast, the irradiance depends on the real part of the entire second-order Rytov solution, which we have denoted by ψ_2. One needs the average value of this quantity in order to estimate the mean irradiance. Manning established an explicit expression for the average value of ψ_2 using Tatarskii's formulation of the second-order solution and the paraxial approximation [10]. It is defined by a hypergeometric function that depends on the complex parameter α which describes the beam wave:

$$\langle \psi_2 \rangle = -1.119 R^{\frac{11}{6}} k^{\frac{7}{6}} C_n^2 \exp\left(i\frac{5\pi}{6} \right) \,{}_2F_1\left(-\frac{5}{6}, \frac{11}{6}, \frac{17}{6}; \frac{i\alpha R}{1 + i\alpha R} \right) \qquad (8.61)$$

8.4.3 The Relation to Phase and Amplitude Variances

For plane and spherical waves we established the important relationship (8.18) between the average values of the double-scattering terms and the first-order phase and amplitude variances. We cannot assume that this connection is valid for beam waves. To explore this question analytically we use the generic beam-wave expressions which describe the on-axis variances in terms of wavenumber and propagation-path integrations:

$$\langle \chi^2 \rangle = 2\pi^2 k^2 \int_0^R dx \int_0^\infty d\kappa\, \kappa \Phi_n(\kappa)[\mathfrak{M}(\kappa, x) + \mathfrak{N}(\kappa, x)]$$

$$\langle \varphi^2 \rangle = 2\pi^2 k^2 \int_0^R dx \int_0^\infty d\kappa\, \kappa \Phi_n(\kappa)[\mathfrak{M}(\kappa, x) - \mathfrak{N}(\kappa, x)]$$

Their sum is thus given by

$$\frac{1}{2}\Sigma_v = 2\pi^2 k^2 \int_0^R dx \int_0^\infty d\kappa\, \kappa \Phi_n(\kappa) \exp\left(-\frac{\kappa^2 \gamma_2 (R - x)}{k} \right)$$

where we have introduced the explicit form of $\mathfrak{M}(\kappa, x)$ suggested by (3.92) with $\rho = 0$. We can rewrite this in a manner that emphasizes the relationship to $\langle c \rangle$ by adding and subtracting unity in the integrand:

$$-\frac{1}{2}\Sigma_v = \langle c \rangle + 2\pi^2 k^2 \int_0^R dx \int_0^\infty d\kappa\, \kappa \Phi_n(\kappa)\left[1 - \exp\left(-\frac{\kappa^2 \gamma_2 (R - x)}{k} \right) \right]$$

Using the Kolmogorov spectrum we see that the upper and lower wave-number limits can be ignored:

$$-\frac{1}{2}\Sigma_v = \langle c \rangle + 2.176\, k^{\frac{7}{6}} C_n^2 \int_0^R dx\, [\gamma_2(R-x)]^{\frac{5}{6}}$$

Recalling the definition of the imaginary part of γ from (3.89) and noting that it also depends on the distance from the receiver, we find

$$-\frac{1}{2}\Sigma_v = \langle c \rangle + 0.816\, k^{\frac{7}{6}} C_n^2 R^{\frac{11}{6}} \left(\frac{\alpha_1 R}{(1-\alpha_2 R)^2 + (\alpha_1 R)^2} \right)^{\frac{5}{6}}$$

Using the definitions of α_1 and α_2 given by (3.81), we see that the combination in large parentheses can be expressed in terms of the transmission distance and beam width at the receiver:

$$\frac{\alpha_1 R}{(1-\alpha_2 R)^2 + (\alpha_1 R)^2} = \frac{R\lambda}{\pi w^2(R)}$$

This ratio is large for most practical applications and one therefore cannot rely on the familiar connection (8.18) for beam waves:

$$\text{Beam waves:} \qquad \langle c \rangle \neq -\frac{1}{2}[\langle \chi^2 \rangle + \langle \varphi^2 \rangle] \qquad (8.62)$$

8.5 Problems

Problem 1

Calculate the correction to the expression (8.52) that corresponds to the double-back-scattering sequence for microwave transmission through the atmosphere illustrated in Figure 8.5.

Problem 2

Can you see how to generalize the expressions for astronomical observations to include sources that are not at zenith?

Problem 3

For a beam wave show that the real part of Yura's second-order solution

$$\langle c \rangle - \frac{1}{2}\langle a^2 \rangle + \frac{1}{2}\langle b^2 \rangle = \langle c(R) \rangle - \frac{1}{2}\langle \chi^2(\rho, R) \rangle + \frac{1}{2}\langle \varphi^2(\rho, R) \rangle$$

is independent of the radial distance ρ from the centerline of the beam and described by

$$\Re\left[-1.119\, R^{\frac{11}{6}} k^{\frac{7}{6}} C_n^2 \exp\left(i\frac{5\pi}{6} \right) \, _2F_1\left(-\frac{5}{6}, \frac{11}{6}, \frac{17}{6}; \frac{i\alpha R}{1+i\alpha R} \right) \right]$$

Notice that this agrees with (8.61).

References

[1] V. I. Tatarskii, "Second Approximation to the Problem of Propagation of Waves in a Medium with Random Inhomogeneities," *Izvestiya Vysshikh Uchebnykh Zavedenii, Radiofizika (Soviet Radiophysics)*, **5**, No. 3, 164–202 (1962).

[2] R. A. Schmeltzer, "Means, Variances, and Covariances for Laser Beam Propagation Through a Random Medium," *Quarterly Journal of Applied Mathematics*, **24**, No. 4, 339–354 (January 1967).

[3] S. M. Rytov, Yu. A. Kravtsov and V. I. Tatarskii, *Principles of Statistical Radiophysics 4, Wave Propagation Through Random Media* (Springer-Verlag, Berlin, 1989), 68.

[4] H. T. Yura, "The Second-Order Rytov Approximation," The RAND Corporation, Memorandum RM-5787-PR (February 1969).

[5] T. A. Shirokova, "Second Approximation in the Method of Smooth Perturbations," *Akusticheskii Zhurnal (Soviet Physics – Acoustics)*, **5**, No. 4, 498–503 (April–June 1960).

[6] H. T. Yura, "Optical Propagation through a Turbulent Medium," *Journal of the Optical Society of America*, **59**, No. 1, 111–112 (January 1969).

[7] H. T. Yura and S. G. Hanson, "Second-order Statistics for Wave Propagation through Complex Optical Systems," *Journal of the Optical Society of America A*, **6**, No. 4, 564–575 (April 1989).

[8] F. G. Gebhardt and S. A. Collins, "Log-Amplitude Mean for Laser-Beam Propagation in the Atmosphere," *Journal of the Optical Society of America*, **59**, No. 9, 1139–1148 (September 1969).

[9] Z. I. Feyzulin, "Amplitude and Phase Fluctuations of a Confined Wave Beam Propagating in a Randomly Inhomogeneous Medium," *Radiotekhnika i Electronika (Radio Engineering and Electronic Physics)*, **15**, No. 7, 1189–1195 (1970).

[10] R. M. Manning, "Beam Wave Propagation within the Second Rytov Perturbation Approximation," *Izvestiya Vysshikh Uchebnykh Zavedenii, Radiofizika (Radiophysics and Quantum Electronics)*, **39**, No. 4, 423–436 (August 1996).

[11] L. C. Andrews and R. L. Phillips, *Laser Beam Propagation through Random Media* (SPIE Optical Engineering Press, Bellingham, Washington, 1998), 105–109.

[12] J. H. Chisholm, P. A. Portmann, J. T. deBettencourt and J. F. Roche, "Investigations of Angular Scattering and Multipath Properties of Tropospheric Propagation of Short Radio Waves Beyond the Horizon," *Proceedings of the IRE*, **43**, No. 10, 1317–1335 (October 1955).

[13] K. Bullington, "Characteristics of Beyond-the-Horizon Radio Transmission," *Proceedings of the IRE*, **43**, No. 10, 1175–1180 (October 1955).

[14] S. F. Clifford and J. W. Strohbehn, "The Theory of Microwave Line-of-Sight Propagation Through a Turbulent Atmosphere," *IEEE Transactions on Antennas and Propagation*, **AP-18**, No. 2, 264–274 (March 1970).

[15] S. F. Clifford, "Wave Propagation in a Turbulent Medium," Ph.D. dissertation, Thayer School of Engineering, Dartmouth College, Hanover, NH (June 1969).

[16] J. H. Van Vleck, E. M. Purcell and H. Goldstein, "Atmospheric Attenuation," in *Propagation of Short Radio Waves*, vol. 13, edited by D. E. Kerr (MIT Radiation Laboratory Series, McGraw-Hill, New York, 1951), 641–692.

[17] J. H. Van Vleck, "Absorption of Microwaves by Oxygen," *Physical Review*, **71**, No. 7, 413–424 (April 1947).

[18] J. H. Van Vleck, "The Absorption of Microwaves by Uncondensed Water Vapor," *Physical Review*, **71**, No. 7, 425–433 (April 1947).

[19] B. R. Bean and E. J. Dutton, *Radio Meteorology* (National Bureau of Standards Monograph 92, U. S. Government Printing Office, March 1966), 270 *et seq.*

[20] R. K. Crane, *Electromagnetic Wave Propagation Through Rain* (John Wiley & Sons, New York, 1996).

[21] J. W. Herbstreit and M. C. Thompson, "Measurements of the Phase of Radio Waves Received over Transmission Paths with Electrical Lengths Varying as a Result of Atmospheric Turbulence," *Proceedings of the IRE*, **43**, No. 10, 1391–1401 (October 1955).

9

Field-strength Moments

All of the development so far has concentrated on the variances of phase and amplitude – or their correlations with respect to separation, time delay and frequency separation. We turn now to moments of the field strength itself. These are the quantities that are often measured in astronomical observations and terrestrial experiments. They are usually calculated with theories that characterize strong scintillation. It is significant that the Rytov approximation gives identical results for many of these quantities. This approach requires only the tools we have already developed.

The average field strength and mean irradiance set important reference levels for measurement programs. They are also needed in order to describe other features of propagation in random media. Accurate estimates for these quantities require both first- and second-order solutions in the Rytov expansion. These depend on the double-scattering expressions which were established for plane, spherical and beam waves in Chapter 8. Field-strength moments are more difficult to estimate because they contain the Born terms in exponential form. We shall learn that the mean field is attenuated very rapidly with distance. By contrast, the mean irradiance for plane and spherical waves is everywhere equal to its free-space value.

These calculations set the stage for analyzing two important features of the electromagnetic field. With the second Rytov approximation one can demonstrate that energy is conserved in a nonabsorbing medium. The mutual coherence function is calculated in the same way and provides an expression identical to that predicted both by geometrical optics and by strong-scattering theories. This result is confirmed by astronomical and terrestrial experiments.

Irradiance fluctuations are often quite large at optical and millimeter-wave frequencies and their data is usually recorded with logarithmic amplifiers to compress the dynamic range. Descriptions of the mean value and variance of logarithmic irradiance also require the second-order solution and their results are confirmed in the weak-scattering range by numerous experiments. However, the predictions and measurements diverge dramatically above $\langle \chi^2 \rangle > 1$. Strong scattering causes saturation in this range and then a gradual decline in scintillation level. These experiments establish a limit for the Rytov approximation that agrees with the theoretical estimate that will be established in Chapter 12.

By contrast, intensity fluctuations are usually quite modest for optical and radio-astronomical observations unless the source is close to the horizon or a small aperture is employed. The scintillation imposed on most satellite signals is modest at the frequencies used for civilian communications.

9.1 The Mean Field

The average field strength or mean field provides an important reference level for describing propagation in random media. We will calculate this quantity in two ways. In the first approach we rely on an analytical technique. This yields a universal expression that describes attenuation of the coherent component of the field strength. It shows that the coherent field rapidly disappears as the wave travels through a random medium, even when the irregularities are very weak. This result is confirmed using an intuitive physical model, which also provides a basis for understanding conservation of energy and other moments of the field.

9.1.1 An Analytical Description

We begin with an analytical description for the mean field. To estimate this properly one must use Yura's second-order Rytov solution (2.64). Taking the ensemble average of that expression, we can write

$$\langle E_2(\mathbf{R}) \rangle = E_0(\mathbf{R}) \langle \exp[a + c - \tfrac{1}{2}a^2 + \tfrac{1}{2}b^2 + i(b + d - a\,b)] \rangle \qquad (9.1)$$

The techniques of cumulant analysis are needed for the evaluation of this average because we are not confident that the Born terms are all Gaussian random variables. It is convenient to represent the exponential by a single complex variable,

$$Z = a + c - \tfrac{1}{2}a^2 + \tfrac{1}{2}b^2 + i(b + d - a\,b)$$

so that

$$\langle E_2(\mathbf{R}) \rangle = E_0(\mathbf{R}) \langle \exp Z \rangle$$

We use the basic result of Appendix M, keeping only the first two cumulants because we are working to second order:

$$\langle \exp(iqZ) \rangle = \exp\left[iq\langle Z \rangle - \tfrac{1}{2}q^2(\langle Z \rangle^2 - \langle Z \rangle^2)\right] \qquad (9.2)$$

By setting $q = -i$ in this expression we establish a fundamental expression that will be used repeatedly in our work:

$$\langle \exp Z \rangle = \exp\left[\langle Z \rangle + \tfrac{1}{2}(\langle Z \rangle^2 - \langle Z \rangle^2) \right] \tag{9.3}$$

The required moments of Z are represented by moments of the first- and second-order terms in the Born series:

$$\langle Z \rangle = \langle c - \tfrac{1}{2}a^2 + \tfrac{1}{2}b^2 + i(d - ab) \rangle$$
$$= \langle c \rangle - \tfrac{1}{2}\langle \chi^2 \rangle + \tfrac{1}{2}\langle \varphi^2 \rangle + i[\langle d \rangle - \langle \varphi \chi \rangle]$$
$$\langle Z^2 \rangle = \langle (a + ib)^2 \rangle = \langle \chi^2 \rangle - \langle \varphi^2 \rangle + 2i \langle \varphi \chi \rangle$$

On combining these results with (9.3) we see that the field strength depends on the double-scattering terms averaged over the ensemble of atmospheric irregularities:

$$\langle E_2(\mathbf{R}) \rangle = E_0(\mathbf{R}) \exp[\langle c \rangle + i \langle d \rangle] \tag{9.4}$$

We are primarily interested in the field strength measured at the receiver. The corresponding average values of the double-scattering components for terrestrial paths were calculated in Chapter 8. The real part is identical for plane waves, spherical waves and beam waves when the irregularities are isotropic and homogeneous:

$$\langle c \rangle = -2\pi^2 R k^2 \int_0^\infty d\kappa \, \kappa \, \Phi_n(\kappa, z)$$

The imaginary term is small relative to this component,

$$\langle d \rangle = \frac{1}{kR} \langle c \rangle$$

and can be discarded in making field-strength estimates. The mean field therefore depends on frequency, distance and the spectrum's first moment [1][2][3]:

$$\langle E_2(\mathbf{R}) \rangle = E_0(\mathbf{R}) \exp\left(-2\pi^2 R k^2 \int_0^\infty d\kappa \, \kappa \, \Phi_n(\kappa)\right) \tag{9.5}$$

This predicts a rapid attenuation of the average field strength as the wave passes through the random medium. The description of astronomical signals traveling along an inclined path through the entire atmosphere is

$$\langle E_2(\mathbf{R}) \rangle = E_0(\mathbf{R}) \exp\left(-2\pi^2 k^2 \sec \vartheta \int_0^R dz \int_0^\infty d\kappa \, \kappa \, \Phi_n(\kappa, z)\right) \tag{9.6}$$

where ϑ is the zenith angle of the source. In both cases the exponent is proportional to the single-path phase variance calculated with geometrical optics:

$$\langle E_2(\mathbf{R})\rangle = E_0(\mathbf{R})\exp\left(-\tfrac{1}{2}\langle\varphi_0^2\rangle\right) \tag{9.7}$$

Two things are significant about this prediction. The first point to note is that the expression for the average field strength is exactly the same for plane, spherical and beam waves. The second observation is that the Rytov prediction provides a universal expression. The same result was established with geometrical optics[1] and is predicted by theories that describe strong scattering.[2]

9.1.2 The Influence on the Received Signal

We need to evaluate the average field strength for various frequencies and path lengths. The single-path phase measurements reviewed in Chapter 4 of Volume 1 allow one to estimate $\langle\varphi_0^2\rangle$, which occurs in the basic expression (9.7). Data recorded at 1046 MHz on a 16-km path in Colorado showed that the rms phase fluctuation was approximately 3° [4]. The average field strength at the receiver is virtually the same as the free-space value in that case. Similar measurements on a 25-km path in Hawaii at 9414 MHz gave an rms phase excursion of 50° [5], which leads to

$$\text{X band:} \qquad \langle E_2(\mathbf{R})\rangle = 0.7E_0(\mathbf{R})$$

and the coherent field is essentially unchanged. The mean field should be largely eliminated along similar path lengths at 30 GHz because the attenuation increases rapidly with frequency.

The reduction of the average field strength is much greater at optical and infrared wavelengths because of the frequency scaling. The distribution of single-path phase fluctuations was measured with a He–Ne laser using path lengths of 95 and 200 m [6]. The actual data is reproduced in Figure 8.2 of Volume 1 and indicates that the rms phase fluctuation is about 20 radians for those conditions. The corresponding mean field is therefore vanishingly small:

$$\text{Optical:} \qquad \langle E_2(\mathbf{R})\rangle = E_0(\mathbf{R})\exp(-100)$$

Since the phase variance is proportional to distance, this means that the field strength is reduced to 2% of its free-space value during the first few meters of travel.

One should wonder what is happening to the electromagnetic field as this rapid attenuation proceeds. Understanding the physics depends on what we mean by

---

[1] As described in Section 9.1 of Volume 1.
[2] Strong-scattering descriptions of the mean field and other moments are addressed in Volume 3.

the average field strength. A coherent component of the field reaches the receiver traveling along the line-of-sight path, just as the free-space field would. It is progressively diminished because energy is being scattered out of the main beam by irregularities located along this path. These *out-scattering* losses to the coherent field are compensated by an increasingly strong incoherent field component. That field is composed of all the waves scattered into the receiver by irregularities located throughout the random medium. We will demonstrate in Section 9.3 that this compensation takes place in a way that conserves energy.

Let us take this description one step further by studying how these field components are sensed in the receiver. The coherent mean field $\langle E \rangle$ has the same phase as the free-space field would and its magnitude is given by (9.7). This field establishes a voltage component in the receiver that sets the phase reference for the combined signal, as suggested in Figure 9.1. The incoherent field E_s creates voltage components that are in phase and out of phase with respect to that reference. The instantaneous terminus of the incoherent voltage is shown there together with its preceding values, which have traced out the irregular trajectory suggested in Figure 9.1. The top panel is a typical voltage diagram for a microwave link for which the mean field is appreciable. In that case the coherent field provides a robust phase and voltage reference. The second panel describes a millimeter-wave signal when the mean field is comparable to the signal generated by the incoherent field.

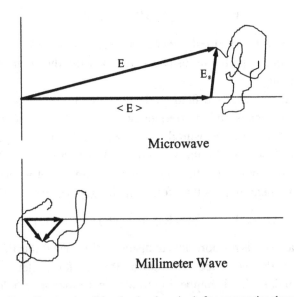

Figure 9.1: Vector voltage diagrams describing the signal received after propagation through a random medium as suggested by Uscinski [7]. The phase reference is established by the coherent field whose mean value is $\langle E \rangle$. The instantaneous voltage induced by the incoherent field E_s has both in-phase and out-of-phase components. The first diagram represents microwave propagation, for which the mean field is substantial. The second describes a millimeter-wave link and the mean field is greatly reduced.

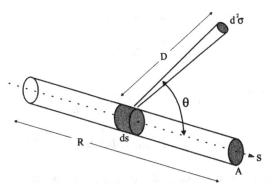

Figure 9.2: The geometry used to analyze scattering out of the coherent field as it travels through the volume defined by the receiver aperture. The scattered wave component is identified by the radial distance D and the angle θ.

The mean field disappears completely at optical wavelengths and the measured voltage is due entirely to the incoherent components.

9.1.3 A Physical Model of Out-scattering

One can construct a physical description of the out-scattering by using the *far-field* or *cross-section approximation* [8][9]. This approach relates the power scattered from a volume element to the turbulence spectrum evaluated at the *Bragg wavenumber*. This uniquely important wavenumber is defined by the scattering angle and wavelength:

$$\kappa_{\mathrm{B}} = \frac{4\pi}{\lambda} \sin\left(\frac{\theta}{2}\right) \tag{9.8}$$

In this approximation the average power that flows outward to a distant surface element is described by

$$\frac{d^5 P}{d^2\sigma \, dV} = |E_0(s)|^2 \frac{2\pi k^4}{D^2} \Phi_n\left[2k \sin\left(\frac{\theta}{2}\right)\right] \tag{9.9}$$

where D is the scalar distance to $d^2\sigma$ from the volume element dV. Notice that, for $\theta = \pi/2$, this reduces to the back-scattering relationship[3] which was used to describe profile measurements of C_n^2 with sounding radars.

To apply this result to the problem at hand, we imagine a plane wave traveling horizontally in the troposphere as illustrated in Figure 9.2. We identify the volume element with a thin slice of the cylinder $dV = A \, ds$ which is defined by the receiver aperture. The power lost per unit volume by out-scattering is estimated by

---

[3] See equation (2.108) in Volume 1.

integrating (9.9) over a distant spherical surface:

$$\frac{dP}{dV} = -\iint d^2\sigma \, |E_0(s)|^2 \frac{2\pi k^4}{D^2} \Phi_n\left[2k \sin\left(\frac{\theta}{2}\right)\right]$$

$$= -D^2 \int_0^\pi d\theta \sin\theta \int_0^{2\pi} d\phi \, |E_0(s)|^2 \frac{2\pi k^4}{D^2} \Phi_n\left[2k \sin\left(\frac{\theta}{2}\right)\right]$$

$$= -4\pi^2 k^4 |E_0(s)|^2 \int_0^\pi d\theta \sin\theta \, \Phi_n\left[2k \sin\left(\frac{\theta}{2}\right)\right]$$

We next change the integration variable to an equivalent wavenumber by setting $\kappa = 2k \sin(\theta/2)$. The power lost per unit length along the line-of-sight path is just this expression multiplied by the area of the aperture:

$$\frac{dP}{ds} = -A|E_0(s)|^2 4\pi^2 k^2 \int_0^{2k} d\kappa \, \kappa \Phi_n(\kappa, s)$$

In writing this expression we have kept available the possibility that the spectrum of irregularities changes along the path, as it does for signals that transit the atmosphere. The electromagnetic wavenumber k is much greater than any of the turbulence wavenumbers in the spectrum of irregularities for optical propagation and we can let $2k \to \infty$. The situation is more complicated for microwave signals but we will not consider that case here. The power flowing in the cylinder is $A|E_0(s)|^2$ and we can rewrite the previous expression as a differential equation:

$$\frac{dP}{ds} = P(s)\left(-4\pi^2 k \int_0^\infty d\kappa \, \kappa \Phi_n(\kappa, s)\right)$$

This is easily solved to give

$$P(R) = P(0) \exp\left(-4\pi^2 k^2 \int_0^R ds \int_0^\infty d\kappa \, \kappa \Phi_n(\kappa, s)\right) \qquad (9.10)$$

If the turbulent conditions are constant along the path,

$$P(R) = P(0) \exp\left(-4\pi^2 R k^2 \int_0^\infty d\kappa \, \kappa \Phi_n(\kappa)\right) \qquad (9.11)$$

These expressions describe the progressive attenuation of power density as the wave travels through the random medium. They are consistent with the estimate of the average field strength (9.5).

9.2 The Mean Irradiance

The *average intensity* or *mean irradiance* provides an important reference level for an electromagnetic field traveling through a random medium. Using the second-order field-strength expression (2.64), the instantaneous intensity is expressed in terms of the Born components as

$$I_2(\mathbf{R}) = |E_0(\mathbf{R})|^2 \exp(2a + 2c - a^2 + b^2) \qquad (9.12)$$

The mean irradiance is therefore given in terms of the free-space value $I_0(\mathbf{R})$ and the ensemble average of a new combination of terms:

$$\langle I_2(\mathbf{R}) \rangle = I_0(\mathbf{R})\langle \exp(2a + 2c - a^2 + b^2) \rangle$$

The cumulant-analysis result (9.3) can again be exploited if we define the exponential for this estimate as follows:

$$Z = 2a + 2c - a^2 + b^2$$

To second order the required moments are

$$\langle Z \rangle = \langle 2c - a^2 + b^2 \rangle = 2\langle c \rangle - \langle \chi^2 \rangle + \langle \varphi^2 \rangle$$
$$\langle Z^2 \rangle = \langle (2a)^2 \rangle = 4\langle \chi^2 \rangle$$

and the ensemble average becomes

$$\langle I_2(\mathbf{R}) \rangle = I_0(\mathbf{R}) \exp(2\langle c \rangle + \langle \chi^2 \rangle + \langle \varphi^2 \rangle) \qquad (9.13)$$

To proceed further one must specify the type of wave.

9.2.1 Plane and Spherical Waves

The average value of the real component of the double-scattering term is proportional to the sum of the phase and amplitude variances for plane and spherical waves, as we demonstrated in Chapter 8:

$$\langle c \rangle = -\tfrac{1}{2}(\langle \chi^2 \rangle + \langle \varphi^2 \rangle)$$

The mean irradiance is therefore equal to the free-space value everywhere in the random medium [3][10][11]:

$$\text{Plane and spherical waves:} \qquad \langle I_2(\mathbf{R}) \rangle = I_0(\mathbf{R}) \qquad (9.14)$$

The mean irradiance is invariant for plane waves and decreases as the inverse square of the distance from the transmitter for spherical waves. These results are consistent with the observation that there are no preferred points in the random medium for these types of wave. Most optical measurements of the mean irradiance are made with logarithmic amplifiers, which compress the wide swings that are observed, and we will describe those experiments in Section 9.5.1.

In early studies of such propagation, the mean irradiance was estimated using only the basic Rytov solution:

$$\langle I_1(\mathbf{R})\rangle = I_0(\mathbf{R})\exp\left(2\langle \chi^2\rangle\right)$$

This suggested that $\langle I\rangle$ should increase exponentially as the wave travels through a random medium since the amplitude variance itself increases with distance as $R^{\frac{11}{6}}$. That prediction contradicts one's intuition and experimental results. This posed a problem for the Rytov method during the first decade of its use and was often described as a violation of conservation of energy. We now know that the mean irradiance and conservation of energy are closely linked. Both are properly described when one includes the second Rytov approximation for the electric-field strength.

9.2.2 Beam Waves

The irradiance of a beam wave in vacuum changes with distance along the path and with radial distance from the centerline of the beam. This joint variability is defined for a Gaussian beam wave in (3.94). We want to examine how that behavior is modified by the random medium. To do so we describe the field strength using the first and second terms of the Rytov series:

$$E_2(\mathbf{R}) = E_0(\mathbf{R})\exp(\psi_1 + \psi_2) \tag{9.15}$$

The mean irradiance can be expressed in terms of the moments of the real parts of these functions using (9.3):

$$\langle I_2(\rho, x)\rangle = I_0(\rho, x)\exp\left[2\langle\Re(\psi_2)\rangle + 2\langle(\Re(\psi_1))^2\rangle\right]$$

The second term in the exponent is the logarithmic amplitude variance $\langle\chi^2(\rho, x)\rangle$ defined by (3.94). An explicit expression for $\langle\psi_2\rangle$ was presented in (8.64) and depends only on the downrange distance from the transmitter:

$$\langle\psi_2(x)\rangle = -2.176 i^{\frac{5}{6}} C_n^2 R^{\frac{11}{6}} k^{\frac{7}{6}} \frac{6}{11}\, {}_2F_1\left(-\frac{5}{6}, \frac{11}{6}, \frac{17}{6}; \frac{i\alpha x}{1+i\alpha x}\right) \tag{9.16}$$

The real part of this result cancels out a similar term in the variance expression and their combination depends on the first Kummer function [12][13]:

$$\langle I_2(\rho, x)\rangle = I_0(\rho, x)\exp\left[-1.632 C_n^2 x^{\frac{8}{3}} k^{\frac{7}{6}} (\alpha_1)^{\frac{5}{6}}\left(\frac{w_0}{w(x)}\right)^{\frac{5}{3}} M\left(-\frac{5}{6}, 1; \frac{2\rho^2}{w^2(x)}\right)\right]$$

$$\tag{9.17}$$

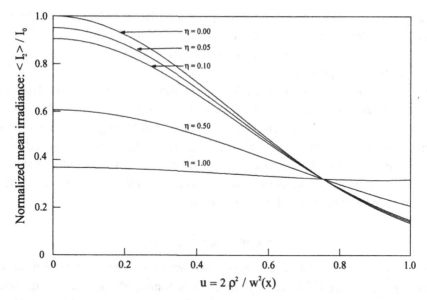

Figure 9.3: The mean irradiance for a beam wave predicted by the second-order Rytov solution. The mean irradiance is normalized by $I_0 = |\mathcal{E}_0|^2 w_0^2/w^2(x)$ and the horizontal axis measures the off-centerline radial distance divided by the local beam radius. The parameter η is defined by (9.19).

To illustrate the beam broadening induced by refractive-index irregularities it is helpful to combine this result with the expression (3.82) for the vacuum intensity:

$$\langle I_2(\rho, x)\rangle = |\mathcal{E}_0|^2 \left(\frac{w_0}{w(x)}\right)^2 \exp\left[-\frac{2\rho^2}{w^2(x)} - \eta(x)M\left(-\frac{5}{6}, 1; \frac{2\rho^2}{w^2(x)}\right)\right] \quad (9.18)$$

The parameters which describe the random medium and beam-wave features are combined to give a single parameter that depends on the downrange distance:

$$\eta(x) = 1.632 C_n^2 x^{\frac{8}{3}} k^{\frac{7}{6}}(\alpha_1)^{\frac{5}{6}}\left(\frac{w_0}{w(x)}\right)^{\frac{5}{3}} \quad (9.19)$$

The normalized mean irradiance defined by (9.18) is plotted in Figure 9.3 as a function of the dimensionless ratio

$$u = 2\rho^2/w^2(x)$$

for four values of η. The unperturbed Gaussian beam corresponds to $\eta = 0$ and provides a useful reference.[4] Along the beam axis refractive-index irregularities cause the beam to shrink relative to this reference by a factor of $\exp(-\eta)$. This centerline reduction increases rapidly with distance because η does. Scattering in the random medium evidently spreads the beam energy in the radial direction. This

---

[4] From Appendix G we have $M(a, b, 0) = 1$ for all a and b.

spreading becomes more pronounced as η takes on larger values, either through stronger turbulence conditions or as the downrange distance increases. These curves merge at $\rho = 0.754\omega$ because the Kummer function vanishes there. Beyond that point their relative importance reverses.

We have not plotted the mean irradiance beyond $\rho = \omega$ because the Kummer function turns negative at 1.14, which rapidly drives $\langle I_2(\rho, x) \rangle$ to values larger than $I_0(\rho, x)$. That unrealistic behavior is caused by our having violated the small-scattering-angle assumption that we used in deriving the first- and second-order Rytov components which were used to establish (9.17). Large radial distances correspond to large scattering angles and they require a more complete description than is provided by the paraxial approximation.

9.3 Conservation of Energy

We turn now to the important question of conservation of energy. There is no doubt that energy is conserved if the random medium is nonabsorbing. The only question is whether the Rytov approximation reflects this fundamental requirement. Using both qualitative and quantitative arguments we shall find that it does.

The flow of energy is closely related to the mean irradiance, but they are not quite the same thing. We have just found that the mean irradiance is everywhere equal to the unperturbed values for plane and spherical waves. This is not true for beam waves, yet we are convinced that energy is conserved. We need to approach conservation of energy as an independent feature of the propagation. Only with the second Rytov solution can one demonstrate that energy is conserved. This was not recognized in the first applications of the Rytov approximation and it appeared that it did not conserve energy. Yura first clarified this problem when he developed a second-order solution [1][14].

9.3.1 A Diagrammatic Demonstration

Most of our attention has been focused on waves that are scattered into the receiver by refractive-index irregularities. In examining the mean field we studied the energy loss suffered by the coherent field component that is caused by out-scattering along the line-of-sight path. To understand conservation of energy we must combine these two features in a balanced description. These scattering mechanisms are illustrated in Figure 9.4 for a plane wave incident on a random medium.

A typical eddy in the medium scatters some of the incident plane wave in all directions. The component of this scattered wave that reaches the receiver is specified by its scattering angle θ and by the physical properties of the medium at depth d. Now consider a comparable eddy on the line-of-sight path at the same depth. This eddy scatters some fraction of the incident wave in each direction. One of those

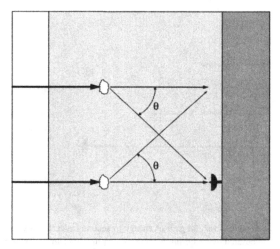

Figure 9.4: Single scattering of a plane wave, indicating the diagrammatic equivalence of losses from the coherent field due to out-scattering and projection of incoherent energy into the receiver by eddies lying at the same depth in the random medium.

out-scattering components makes the same angle θ with respect to the propagation direction. Since the level of turbulence is presumably a function only of depth in the medium, the two scattered waves shown in Figure 9.4 should be the same – at least in the mean-square sense. The receiver should lose and gain the same amount of energy on average from this pair of eddies. For every volume element that scatters energy into the receiver there is a compensating volume element on the direct path that scatters energy out of the initial plane wave at precisely the same angle. If the receiver is omnidirectional the net decrease caused by out-scattering is exactly matched by the buildup of the incoherent field generated by eddies throughout the turbulent medium. From this diagrammatic argument we conclude that energy is conserved [15].

Two questions arise in this discussion. If the receiver is directional, its preferential sensitivity will discriminate against some of the off-axis volume elements. Out-scattering by irregularities on the direct path can exceed the energy provided by the incoherent field in that case. This is an antenna/lens discrimination effect and does not violate conservation of energy.

One might also ask about attenuation of the primary wave that arrives at the off-axis scattering eddy. The energy which reaches a typical volume element is less than that in the initial plane wave because of out-scattering along the line-of-sight path that leads to it. The diagram of Figure 9.5 reminds us that double-scattering sequences project energy into the scattering volume element at $\mathbf{r}_1$. For every volume element along the path that scatters energy out of the plane wave reaching $\mathbf{r}_1$, there is a compensating volume element that scatters the same amount into $\mathbf{r}_1$. Attenuation of the incident plane wave prior to its scattering is therefore compensated by the

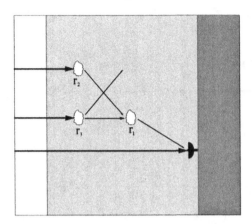

Figure 9.5: Diagramatic balancing of loss and gain of energy to second order. Energy lost from the plane wave by out-scattering at r_3 is compensated by scattering from r_2 to r_1 at the same angle. Energy reaching the receiver is thus balanced by double-scattering.

incoherent waves scattered into the volume element by surrounding eddies. To second order the wave that reaches the receiver exactly balances the energy that is lost from the line-of-sight path. In this way one reasons that energy should be conserved at each level of scattering.

9.3.2 A Description with the Poynting Vector

We need to support the foregoing discussion with an analytical demonstration of conservation of energy. The *Poynting vector* is the traditional way to describe the flow of energy in an electromagnetic field [16][17]. It defines the direction and magnitude of the flow of energy across a unit area and is expressed by the cross product of the electric- and magnetic-field vectors:

$$\mathbf{S} = \frac{c}{4\pi}\mathbf{E} \times \mathbf{H} \qquad (9.20)$$

We will show in Chapter 11 that the polarization of a line-of-sight signal does not change in a measurable way as a wave passes through a random medium. This means that one can replace the vector fields by their scalar equivalents. The energy flux density averaged over one cycle of the carrier frequency can be written as

$$\mathbf{S} = \frac{c}{8\pi k}\Im(E^* \, \boldsymbol{\nabla} E) \qquad (9.21)$$

This vector has the important property that its divergence must vanish everywhere in the medium – except at the transmitter. To show this we use the vector identity

$$\boldsymbol{\nabla} \cdot (\phi \mathbf{A}) = \mathbf{A} \cdot \boldsymbol{\nabla}\phi - \phi \, \boldsymbol{\nabla} \cdot \mathbf{A}$$

and change to the explicit form of the energy-flow vector:

$$\nabla \cdot \mathbf{S} = \frac{c}{16\pi i k} \nabla \cdot (E^* \nabla E - E \nabla E^*)$$

$$= \frac{c}{16\pi i k} (E^* \nabla^2 E - E \nabla^2 E^*)$$

The wave equation which defines the scalar field,

$$\nabla^2 E + k^2 (1 + \delta\varepsilon) E = -i 4\pi k j(\mathbf{r})$$

allows the expression above to be written as follows:

$$\nabla \cdot \mathbf{S} = \frac{c}{2} [E(\mathbf{r}) j^*(\mathbf{r}) - E^*(\mathbf{r}) j(\mathbf{r})]$$

The current density vanishes everywhere except at the transmitting antenna, so that everywhere else in the propagation medium the divergence of the Poynting vector must be zero:

$$\nabla \cdot \mathbf{S} = 0 \qquad (9.22)$$

To show that energy is conserved it is necessary to use the second-order Rytov solution:

$$\mathbf{S}(\mathbf{r}) = \frac{c}{8\pi k} \Im \left[E_2^*(\mathbf{r}) \, \nabla E_2(\mathbf{r}) \right] \qquad (9.23)$$

The field strength depends on the first two terms in the Rytov expansion. We break the sum of these terms into their real and imaginary parts

$$\psi_1 + \psi_2 = u + iv$$

so as to express the electric-field strength concisely:

$$E_2(\mathbf{r}) = E_0(\mathbf{r}) \exp(u + iv) \qquad (9.24)$$

The Poynting vector must be averaged over the ensemble of atmospheric conditions that generate the random field-strength components:

$$\langle \mathbf{S}(\mathbf{r}) \rangle = \frac{c}{8\pi k} \Im \langle E_0^*(\mathbf{r}) e^{u-iv} \, \nabla [E_0(\mathbf{r}) e^{u+iv}] \rangle$$

$$= \frac{c}{8\pi k} \Im [E_0^*(\mathbf{r}) \nabla E_0(\mathbf{r})] \langle e^{2u} \rangle + \frac{c}{8\pi k} |E_0(\mathbf{r})|^2 \langle e^{2u} \nabla v \rangle$$

The first term is the Poynting vector for the unperturbed field except for the ensemble average $\langle e^{2u} \rangle$.

The second term is quite different and is best evaluated by raising the gradient to exponential form:

$$\langle e^{2u}\,\nabla v\rangle = \frac{\partial}{\partial\zeta}\langle\exp(2u + \zeta\,\nabla v)\rangle|_{\zeta=0}$$

We can evaluate the ensemble average by setting

$$Z = 2u + \zeta\,\nabla v$$

and, with the help of (9.3), we find that

$$\langle e^{2u}\,\nabla v\rangle = \frac{\partial}{\partial\zeta}\exp\left(\langle 2u + \zeta\,\nabla v\rangle + \frac{1}{2}\langle(2u + \zeta\,\nabla v)^2\rangle\right)\Bigg|_{\zeta=0}$$
$$= \langle(1 + 2u)\,\nabla v\rangle\exp(\langle 2u\rangle + 2\langle u^2\rangle)$$

With this result the Poynting vector is expressed in terms of its free-space value and ensemble averages of u and v:

$$\langle \mathbf{S}(\mathbf{r})\rangle = \mathbf{S}_0(\mathbf{r})\langle e^{2u}\rangle + \frac{c}{8\pi k}|E_0(\mathbf{r})|^2\langle(1 + 2u)\,\nabla v\rangle\langle e^{2u}\rangle \qquad (9.25)$$

This is a general result which does not depend on the type of signal that is radiated by the transmitter.

9.3.3 Plane and Spherical Waves

We can demonstrate that conservation of energy is maintained for plane and spherical waves with a reasonable calculation. In these two common cases it is convenient to express the components u and v in terms of the first- and second-order Born terms:

$$u = a + c - \tfrac{1}{2}a^2 + \tfrac{1}{2}b^2 \qquad \text{and} \qquad v = b + d - ab \qquad (9.26)$$

From our examination of the mean irradiance we know that

$$\langle e^{2u}\rangle = 1$$

for these types of waves. Using the definitions given above for u and v, the Poynting vector is given by the following equation:

$$\langle \mathbf{S}(\mathbf{r})\rangle = \mathbf{S}_0(\mathbf{r}) + \frac{c}{8\pi k}|E_0(\mathbf{r})|^2\langle a\,\nabla b - b\,\nabla a\rangle \qquad (9.27)$$

We will have proven that energy is conserved if we can show that the second term vanishes.

The ensemble average of the gradient differences can be expressed as the sum of two terms by using the vector relations in Appendix J:

$$\langle a\,\nabla b - b\,\nabla a\rangle = -i\langle(a - ib)\,\nabla(a + ib)\rangle + \tfrac{1}{2}i\,\nabla\langle a^2 + b^2\rangle$$

The first can be written out explicitly using the single-scattering expression for $a + ib$ given by (2.27):

$$\langle(a - ib)\,\nabla(a + ib)\rangle = \left\langle (a - ib)\,\nabla_R \int d^3r \left(\frac{G(\mathbf{R}, \mathbf{r})E_0(\mathbf{r})}{E_0(\mathbf{R})}\right)\delta\varepsilon(\mathbf{r})\right\rangle$$

$$= \left\langle (a - ib)\int d^3r\, E_0(\mathbf{r})\,\delta\varepsilon(\mathbf{r})\,Q(\mathbf{R}, \mathbf{r})\right\rangle \qquad (9.28)$$

In the second formulation we have summarized several gradient terms by

$$Q(\mathbf{R}, \mathbf{r}) = \frac{\nabla_R G(\mathbf{R}, \mathbf{r})}{E_0(\mathbf{R})} - \frac{\nabla_R E_0(\mathbf{R})}{E_0(\mathbf{R})}\frac{G(\mathbf{R}, \mathbf{r})}{E_0(\mathbf{R})}$$

The following relationship is evidently valid for plane waves:

$$\text{Plane wave:} \qquad \frac{\nabla E_0}{E_0} = ik\mathbf{i}_x \qquad (9.29)$$

The exact expression for spherical waves is

$$\nabla\left(\frac{e^{ikr}}{r}\right) = ik\left(1 + \frac{i}{kr}\right)\left(\frac{e^{ikr}}{r}\right)\mathbf{i}_r$$

but the imaginary correction is trivial for paths of practical interest:

$$\text{Spherical wave:} \qquad \nabla\left(\frac{e^{ikr}}{r}\right) \simeq ik\mathbf{i}_r \qquad (9.30)$$

Since Green's function depends only on the scalar distance $|\mathbf{R} - \mathbf{r}|$ the following useful property is evident:

$$\nabla_R G(\mathbf{R}, \mathbf{r}) = -\nabla_r G(\mathbf{R}, \mathbf{r})$$

Integrating the first term in $Q(\mathbf{R}, \mathbf{r})$ by parts gives the following relationship:

$$\langle(a - ib)\,\nabla(a + ib)\rangle = \left\langle (a - ib)\int d^3r \left(\frac{G(\mathbf{R}, \mathbf{r})E_0(\mathbf{r})}{E_0(\mathbf{R})}\right)\nabla_r[\delta\varepsilon(\mathbf{r})]\right\rangle$$

The term $a - ib$ can be represented by the complex conjugate of the same single-scattering integral. We substitute the wavenumber-spectrum representation for the spatial correlation and change to the index of refraction using $\delta\varepsilon = 2\,\delta n$.

$$\langle(a - ib)\,\nabla(a + ib)\rangle = 4i\int d^3\kappa\,\Phi_n(\kappa)|\Lambda(\kappa, R)|^2\kappa \qquad (9.31)$$

The paraxial approximation for the complex weighting function (3.16) gives

$$|\Lambda(\boldsymbol{\kappa}, \boldsymbol{R})|^2 = 2\pi R \delta(\kappa_z)$$

The result (9.31) is conveniently written in spherical wavenumber coordinates if the random medium is isotropic:

$$\langle (a - ib)\,\nabla(a + ib)\rangle = 2\pi R \int_0^\infty d\kappa\,\kappa^3 \Phi_n(\kappa) \int_0^\pi d\psi\,\sin\psi \int_0^{2\pi} d\omega$$

$$\times\,\delta(\kappa\cos\psi)(\mathbf{i}_x\,\cos\psi + \mathbf{i}_y\,\sin\psi\cos\omega + \mathbf{i}_z\,\sin\psi\sin\omega)$$

The delta function kills the x component and the other terms disappear when the azimuth integration is performed. With these observations we demonstrate the following useful property:

$$\langle (a - ib)\,\nabla(a + ib)\rangle = \int d^3\kappa\,\Phi_n(\kappa)|\Lambda(\boldsymbol{\kappa}, r)|^2 \boldsymbol{\kappa} = 0 \qquad (9.32)$$

and the average Poynting vector becomes

$$\langle \mathbf{S}(\mathbf{r})\rangle = \mathbf{S}_0(\mathbf{r}) + i\,\frac{c}{16\pi k}|E_0(\mathbf{r})|^2\,\nabla\langle a^2 + b^2\rangle \qquad (9.33)$$

The remaining ensemble average is just the geometrical-optics expression for the phase variance, according to the sum rule (7.10). We can write its gradients in several ways:

$$\nabla\langle a^2 + b^2\rangle = \mathbf{i}_R\,\frac{\partial}{\partial R}\left(4\pi R k^2 \int_0^\infty d\kappa\,\kappa\Phi_n(\kappa)\right)$$

$$= \mathbf{i}_R\left(4\pi k^2 \int_0^\infty d\kappa\,\kappa\Phi_n(\kappa)\right)$$

$$= \mathbf{i}_R\left(\frac{1}{R}\,\langle\varphi_0^2\rangle\right) \qquad (9.34)$$

On combining the last result with (9.33) we have

$$\langle \mathbf{S}(\mathbf{r})\rangle = \mathbf{S}_0(\mathbf{r})\left(1 + \frac{i}{2kR}\langle\varphi_0^2\rangle\right) \qquad (9.35)$$

To estimate the imaginary term we can use the numerical values for the phase variance that were reviewed in Section 4.1.5 of Volume 1. The measured phase shifts for a 1046-MHz signal on a 16-km path give a correction of one part in a billion. The correction is 10^{-8} or less for an optical signal on a 500-m path. To high accuracy, we have thus demonstrated that energy is conserved for plane

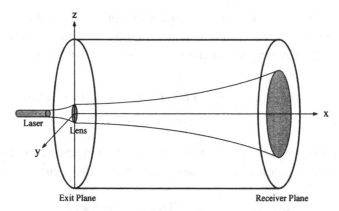

Figure 9.6: The geometry used to describe the flow of energy for a beam wave propagating in a random medium. The cylindrical volume is bounded by end planes that are normal to the line-of-sight direction. The first boundary coincides with the exit plane of the transmitter optics. The receiver plane is located at an arbitrary distance from the transmitter.

and spherical waves if one uses both the first-order and the second-order Rytov approximation.

9.3.4 Beam Waves

The broadening of beam waves by refractive-index irregularities was examined first using the mean irradiance as a yardstick. It seems unlikely that the average Poynting vector is everywhere equal to its vacuum value in view of this broadening. On the other hand, one can establish a valuable expression for the Poynting vector integrated over successive planes that are normal to the centerline of the beam. To establish this relationship we exploit Gauss' theorem and consider the cylindrical volume defined in Figure 9.6. We will eventually let the radius of this cylinder go to infinity. The closing ends of this cylinder are two planes that are normal to the beam axis. The first coincides with the exit plane of the transmitter at $x = 0$. The second closing plane is located an arbitrary distance from the transmitter.

Gauss' theorem relates the volume integral of the divergence of any vector to its normal components integrated over all the surfaces which bound the volume. We apply this general theorem to the Poynting vector to describe the flow of energy:

$$\iiint_V d^3r \, \boldsymbol{\nabla} \cdot \mathbf{S} = \underbrace{\iint d^2\sigma \, (\mathbf{i}_r \cdot \mathbf{S})}_{\text{cylinder}} + \underbrace{\iint d^2\sigma \, \langle -\mathbf{i}_x \cdot \mathbf{S} \rangle}_{\substack{\text{exit} \\ \text{plane}}} + \underbrace{\iint d^2\sigma \, \langle \mathbf{i}_x \cdot \mathbf{S} \rangle}_{\substack{\text{receiver} \\ \text{plane}}}$$

$$(9.36)$$

The divergence of $\mathbf{S}$ vanishes everywhere inside the volume according to (9.22) because the source is to the left of the exit plane. The integral over the large cylindrical surface vanishes because the beam-wave field strength falls off as $\exp(-\eta r^2)$. By taking the ensemble average of the remaining terms we establish a fundamental relationship for the Poynting vector integrated over successive planes:

$$\iint_{\text{receiver plane}} d^2\sigma \, \langle \mathbf{i}_x \cdot \mathbf{S}(r, x) \rangle = \iint_{\text{exit plane}} d^2\sigma \, \langle \mathbf{i}_x \cdot \mathbf{S}(r, 0) \rangle = \text{constant} \qquad (9.37)$$

The Poynting vector at the exit plane is not influenced by the random medium in the paraxial approximation and it can be estimated with the vacuum field strength:

$$\mathbf{i}_x \cdot \mathbf{S}(r, 0) = \frac{c}{8\pi k} \Im \left[E_0^*(r, x) \frac{\partial}{\partial x} E_0(r, x) \right]_{x=0}$$

By introducing the description (3.80) for a Gaussian beam wave one establishes the following expression:

$$S_x(r, 0) = \frac{c}{8\pi} |\mathcal{E}_0|^2 \left\{ 1 - \left(\frac{\lambda}{2\pi\omega_0} \right)^2 + \frac{1}{2} r^2 \left[\left(\frac{\lambda}{\pi\omega_0^2} \right)^2 - \frac{1}{R_0^2} \right] \right\} \exp\left(-2\frac{r^2}{\omega_0^2} \right)$$

This provides a general expression for the horizontal component of the flow of energy integrated over any plane that is normal to the beam axis,

$$\int_0^\infty dr \, r \int_0^{2\pi} d\phi \, \langle \mathbf{i}_x \cdot \mathbf{S}(r, x) \rangle = \frac{c}{16} |\mathcal{E}_0|^2 \omega_0^2 N(\lambda, \omega_0, \mathcal{R}_0) \qquad (9.38)$$

where

$$N(\lambda, \omega_0, \mathcal{R}_0) = \left[1 - \left(\frac{\lambda}{2\pi\omega_0} \right)^2 \right] - \frac{1}{4} \left[\left(\frac{\lambda}{\pi\omega_0} \right)^2 - \left(\frac{\omega_0}{\mathcal{R}_0} \right)^2 \right] \qquad (9.39)$$

The correction terms are small for the optical systems used to create beam waves and one can usually replace the combination $N(\lambda, \omega_0, \mathcal{R}_0)$ by unity:

$$\iint_{\text{receiver plane}} d^2\sigma \, \langle \mathbf{i}_x \cdot \mathbf{S}(r, x) \rangle = \frac{c}{16} |\mathcal{E}_0|^2 \omega_0^2 \qquad (9.40)$$

It is significant that this basic relationship does not depend on the distance traveled in the random medium. Refractive-index irregularities can redistribute the flow of energy in the radial direction but they can do so only in a way that keeps the surface integral constant.

This conservation requirement is often expressed in terms of the mean irradiance [18][19]:

$$\int_0^\infty dr \, r \int_0^{2\pi} d\phi \, \langle I(r, x) \rangle = \text{constant} \qquad (9.41)$$

This would follow from (9.40) if $\langle I \rangle$ were proportional to the average value of the Poynting vector's gradient in the x direction. Using the field-strength expression (9.24), we can describe the mean irradiance in terms of its vacuum value and a single ensemble average:

$$\langle I \rangle = |E_0|^2 \langle e^{2u} \rangle$$

The term $\langle e^{2u} \rangle$ occurs in the general description of the Poynting vector (9.25), which we can write in the following way:

$$\langle \mathbf{S} \rangle = \langle I \rangle \frac{c}{8\pi k} \left(\frac{\Im(E_0^* \, \boldsymbol{\nabla} E_0)}{|E_0|^2} + \langle (1 + 2u) \, \boldsymbol{\nabla} v \rangle \right)$$

With the standard expression for the vacuum field strength the x component of this vector can be written as follows:

$$\langle S_x \rangle = \langle I \rangle \frac{c}{8\pi k} \left\{ \left[k - \Im \left(\frac{\alpha}{1 + i\alpha x} \right) + \frac{1}{2} k r^2 \Im \left(\frac{\alpha^2}{[1 + i\alpha x]^3} \right) \right] \right.$$
$$\left. + \left\langle a \frac{\partial}{\partial x} b - b \frac{\partial}{\partial x} a \right\rangle \right\} \tag{9.42}$$

where we have replaced u and v by their definitions in terms of the Born terms a, b, c and d using (9.26). It is not hard to show that the electromagnetic wavenumber is the largest term in the square brackets. This also means that one can make the following approximation when one is evaluating the ensemble average in (9.42):

$$\text{Beam wave:} \qquad \frac{\boldsymbol{\nabla} E_0}{E_0} \approx ik \mathbf{i}_x \tag{9.43}$$

That was the essential step in proving (9.32) for plane and spherical waves. Without further calculation we arrive at the following representation.

$$\text{Beam wave:} \qquad \langle S_x \rangle = \langle I \rangle \frac{c}{8\pi} \left[1 + O \left(\frac{1}{kR} \right) \right] \tag{9.44}$$

This means that we can justify (9.41) on the basis of conservation of energy alone. In using that relationship it is important to recall the fairly crude approximation (9.43) that we have used twice to connect $\langle S_x \rangle$ and $\langle I \rangle$.

It is important to see whether the suggested relationship (9.41) is satisfied by the beam-wave expression (9.17) established with the second-order Rytov solution and the paraxial approximation. The constant in that expression for $\langle I \rangle$ is found by integrating the vacuum field intensity over the exit plane of the transmitter. This is possible because we are dealing with small-angle forward scattering and the random medium plays virtually no role in configuring the field strength at the exit plane:

$$\int_0^\infty dr\, r \int_0^{2\pi} d\phi\, I_0(r, 0) = \frac{\pi}{2}\omega_0^2 |\mathcal{E}_0|^2 \qquad (9.45)$$

With this result we want to see whether the following equation is valid for all distances x from the transmitter:

$$\frac{\pi}{2}\omega_0^2 |\mathcal{E}_0|^2 \overset{?}{=} \int_0^\infty dr\, r \int_0^{2\pi} d\phi\, |\mathcal{E}_0|^2 \left(\frac{\omega_0}{\omega(x)}\right)^2$$

$$\times \exp\left[-2\frac{r^2}{\omega^2(x)} - \eta(x)M\left(-\frac{5}{6}, 1; \frac{2r^2}{w^2(x)}\right)\right]$$

where $\eta(x)$ is defined by (9.19). The substitution $u = 2r^2/\omega^2(x)$ eliminates the variable-beam-radius terms and one is left with the following question:

$$\int_0^\infty du\, \exp(-u)\exp\left[-\eta(x)M\left(-\tfrac{5}{6}, 1; u\right)\right] \overset{?}{=} 1$$

The downrange distance survives in the combination $\eta(x)$ and it would appear that this relationship is not valid for all values of x. On the other hand, the second-order solution is valid only for small values of the Rytov variance. We should therefore ask whether the first-order expansion of the second exponential provides agreement:

$$\int_0^\infty du\, \exp(-u)\left[1 - \eta(x)M\left(-\tfrac{5}{6}, 1; u\right)\right] \overset{?}{=} 1$$

The integral of the first term yields unity and we are left with the following question:

$$\int_0^\infty du\, \exp(-pu)M\left(-\tfrac{5}{6}, 1; u\right)|_{p=1} \overset{?}{=} 0$$

The Laplace transform of the Kummer function is given in Appendix G and one can verify that it vanishes for $p = 1$. The relationship (9.41) is therefore satisfied to first order in the propagation parameter $\eta(x)$ as was first noted by Manning [20].

9.4 The Mutual Coherence Function

We noted previously that the average field strength and mean irradiance set important reference levels for an electromagnetic signal passing though a random medium. The *mutual coherence function* takes this description to the next level and provides a fundamental measure for the field. It describes the similarity or coherence of the field at adjacent points in the medium. This function plays an important

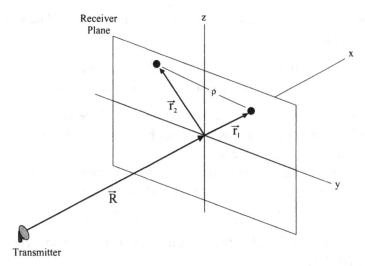

Figure 9.7: The propagation geometry used to define the mutual coherence function. The measurement points r_1 and r_2 lie in a plane that is normal to the line-of-sight vector $R = i_x R$. The scalar distance between the receivers is denoted by ρ.

role in a number of practical applications [21]. It describes the broadening of a laser beam by refractive-index irregularities and sets the signal-to-noise ratio for an optical heterodyne detector. It defines the limiting resolution which can be achieved along an atmospheric path.

The mutual coherence function is usually described with theories that incorporate multiple-scattering effects. It is significant that the same expression emerges from the second-order Rytov solution [22]. This function is usually measured for points located in a plane that is normal to the line of sight, as illustrated in Figure 9.7. It is defined analytically as the ensemble average of the field strength measured at one point multiplied by its complex conjugate measured at a second point:

$$\Gamma(\mathbf{R}, \mathbf{r}_1, \mathbf{r}_2) = \left\langle E(\mathbf{R} + \mathbf{r}_1) E^*(\mathbf{R} + \mathbf{r}_2) \right\rangle \qquad (9.46)$$

The same combination is sometimes used to describe the coherence of the field strength measured at different times [23] or at different frequencies [24].

9.4.1 The Expression for Plane and Spherical Waves

Calculating the mutual coherence function is straightforward when the transmitted signal is either a plane or a spherical wave. In these cases it is convenient to use Yura's version of the second-order solution:

$$\Gamma(\mathbf{R}, \mathbf{r}_1, \mathbf{r}_2) = \Gamma_0(\mathbf{R}, \mathbf{r}_1, \mathbf{r}_2) \langle \exp Z \rangle \qquad (9.47)$$

The complex exponent Z depends on Born terms that must be evaluated at the two measurement points:

$$Z = a_1 + c_1 - \tfrac{1}{2}a_1^2 + \tfrac{1}{2}b_1^2 + i(b_1 + d_1 - a_1b_1)$$
$$+ a_2 + c_2 - \tfrac{1}{2}a_2^2 + \tfrac{1}{2}b_2^2 - i(b_2 + d_2 - a_2b_2) \tag{9.48}$$

Subscripts on the components refer to the locations of the receivers $\mathbf{r}_1$ and $\mathbf{r}_2$. When these basic expressions are used to describe time or frequency coherence the subscripts refer to those parameters.

Before we address the ensemble average $\langle \exp Z \rangle$ we must identify the normalization which is defined in terms of the vacuum field strength:

$$\Gamma_0(\mathbf{R}, \mathbf{r}_1, \mathbf{r}_2) = E_0(\mathbf{R} + \mathbf{r}_1)E_0^*(\mathbf{R} + \mathbf{r}_2) \tag{9.49}$$

For a plane wave this is equal to the mean irradiance and does not depend on the locations of receivers:

$$\text{Plane Wave:} \qquad \Gamma_0(\mathbf{R}, \mathbf{r}_1, \mathbf{r}_2) = |\mathcal{E}_0 \exp(ikx)|^2 = I_0 \tag{9.50}$$

The situation is more complicated for spherical waves:

$$\text{Spherical Wave:} \qquad \Gamma_0(\mathbf{R}, \mathbf{r}_1, \mathbf{r}_2) = |\mathcal{E}_0|^2 \frac{\exp[ik(|\mathbf{R} + \mathbf{r}_1| - |\mathbf{R} + \mathbf{r}_2|)]}{|\mathbf{R} + \mathbf{r}_1||\mathbf{R} + \mathbf{r}_2|} \tag{9.51}$$

The path length is large relative to the displacements of receivers in the usual measurement situation and this expression simplifies to

$$\text{Spherical Wave:} \qquad \Gamma_0(\mathbf{R}, \mathbf{r}_1, \mathbf{r}_2) = \frac{|\mathcal{E}_0|^2}{R^2} \exp\left(\frac{ik}{2R}(r_1^2 - r_2^2) \right) \tag{9.52}$$

With these preliminaries completed, we turn to evaluating the influence of turbulent irregularities on the mutual coherence function. The ensemble average required in (9.47) depends on the first two moments of Z. The average value becomes

$$\langle Z \rangle = -\langle a_1^2 \rangle - \langle a_2^2 \rangle - i\langle a_1b_1 - a_2b_2 \rangle$$

where we have used the double-scattering relationships in (8.18) to express $\langle c \rangle$ in terms of the first-order variances at the location of each receiver and ignored $\langle d \rangle$. In the same way the second moment is given by the following expression:

$$\langle Z^2 \rangle = \langle a_1^2 + 2a_1a_2 + a_2^2 \rangle - \langle b_1^2 - 2b_1b_2 + b_2^2 \rangle$$
$$+ 2i[\langle a_1b_1 - a_2b_2 \rangle - \langle a_1b_2 - a_2b_1 \rangle]$$

If the transmitted signal is a plane wave, the variances of a and b are the same everywhere in the measurement plane since they depend only on the depth to which the wave has penetrated. The situation is slightly different for spherical waves. The variances then depend on the radial distances from the transmitter to the locations of the receivers. The differential distance is small since the path length is usually large relative to the distance between receivers. Moreover, the variances change slowly with distance. It is therefore a good approximation to assume that $\langle a^2 \rangle$ and $\langle b^2 \rangle$ are the same at both locations.[5] With these observations we can write the mutual coherence function in the following form:

$$\Gamma(\mathbf{R}, \mathbf{r}_1, \mathbf{r}_2) = \Gamma_0(\mathbf{R}, \mathbf{r}_1, \mathbf{r}_2) \exp\left(-\langle a^2 - a_1 a_2 \rangle - \langle b^2 - b_1 b_2 \rangle - i\langle a_1 b_2 - a_2 b_1 \rangle\right)$$

The imaginary term in the exponent vanishes for the two types of wave we are considering. To show this we write the difference using vector notations to describe the locations of receivers:

$$\langle a_1 b_2 - a_2 b_1 \rangle = \langle a(\mathbf{R} + \mathbf{r}_1) b(\mathbf{R} + \mathbf{r}_2) - a(\mathbf{R} + \mathbf{r}_2) b(\mathbf{R} + \mathbf{r}_1) \rangle$$

We can expand each term in a Taylor series since the displacements of receivers are small relative to the path length. Keeping only terms that are first order in the inter-receiver displacements, we find that

$$\langle a_1 b_2 - a_2 b_1 \rangle = (\mathbf{r}_1 - \mathbf{r}_2) \cdot \langle b(\mathbf{R}) \nabla a(\mathbf{R}) - a(\mathbf{R}) \nabla b(\mathbf{R}) \rangle$$

The ensemble average which multiplies the difference vector was evaluated for plane and spherical waves in demonstrating that conservation of energy holds. We showed there that it is of order $1/(kR)$ and thus negligible. The mutual coherence function therefore depends on familiar quantities, which we write in terms of the logarithmic amplitude and phase fluctuations:

$$\Gamma(\mathbf{R}, \mathbf{r}_1, \mathbf{r}_2) = \Gamma_0(\mathbf{R}, \mathbf{r}_1, \mathbf{r}_2) \exp\left[-\langle \chi^2 - \chi_1 \chi_2 \rangle - \langle \varphi^2 - \varphi_1 \varphi_2 \rangle\right] \quad (9.53)$$

Let us evaluate this expression for a plane wave passing through the atmosphere from a source with zenith angle ϑ. In examining the spatial correlation of amplitude and phase in Chapters 4 and 7, respectively, we found that they depend only on the scalar distance between the receivers $\rho = |\mathbf{r}_1 - \mathbf{r}_2|$. With the usual decomposition for inhomogeneous irregularities described by a vertical profile $C_n^2(z)$ one can establish the following explicit representations for the terms in (9.53) by remembering that $a = \chi$ and $b = \varphi$:

$$\langle a^2 \rangle - \langle a_1 a_2 \rangle = 4\pi^2 k^2 \sec \vartheta \int_0^\infty dz\, C_n^2(z) \int_0^\infty d\kappa\, \kappa \Omega(\kappa) [1 - J_0(\kappa\rho)]$$

$$\times \sin^2\left(\frac{\kappa^2 z \sec \vartheta}{2k}\right)$$

$$\langle b^2 \rangle - \langle b_1 b_2 \rangle = 4\pi^2 k^2 \sec\vartheta \int_0^\infty dz\, C_n^2(z) \int_0^\infty d\kappa\, \kappa \Omega(\kappa)[1 - J_0(\kappa\rho)]$$

$$\times \cos^2\!\left(\frac{\kappa^2 z \sec\vartheta}{2k}\right)$$

The diffraction terms disappear when these expression are added:

$$\Gamma(\rho, \vartheta) = I_0 \exp\!\left(-4\pi^2 k^2 \sec\vartheta \int_0^\infty dz\, C_n^2(z) \int_0^\infty d\kappa\, \kappa \Omega(\kappa)[1 - J_0(\kappa\rho)]\right)$$

$$(9.54)$$

The exponent is the transmission expression for the wave structure function $\mathcal{D}(\rho)$ of a plane wave that was introduced in Section 7.2.1:

$$\text{Plane wave:} \qquad \Gamma(\rho, \vartheta) = I_0 \exp\!\left[-\tfrac{1}{2}\mathcal{D}(\rho, \vartheta)\right] \qquad (9.55)$$

This result suggests that the received field strength is correlated only over a distance for which the wave structure function is small. Since the mean-square phase variation increases with the transmission distance, the field should become uncorrelated at smaller and smaller distances as the wave proceeds through a random medium. This effect is observed for starlight propagating through the atmosphere. The optical fields intercepted by different surface elements of large telescopes are uncorrelated if they are separated by more than 10 or 20 cm. That reality has important consequences for astronomical observations of faint sources.

Next let us consider the relation (9.53) for a spherical wave propagating along a terrestrial path of length R. The level of turbulence is relatively constant and we can use the appropriate spatial-correlation results from (7.44) to write

$$\langle a^2 - a_1 a_2 \rangle + \langle b^2 - b_1 b_2 \rangle = 4\pi^2 k^2 R \int_0^\infty d\kappa\, \kappa \Phi_n(\kappa) \int_0^1 du\, [1 - J_0(\kappa\rho u)]$$

This combination is again proportional to the wave structure function:

$$\text{Spherical wave:} \qquad \Gamma(\mathbf{R}, \mathbf{r}_1, \mathbf{r}_2) = \Gamma_0(\mathbf{R}, \mathbf{r}_1, \mathbf{r}_2) \exp\!\left[-\tfrac{1}{2}\mathcal{D}(\rho)\right] \qquad (9.56)$$

We can estimate the coherence of the arriving field by using the results of microwave and optical phase-difference experiments that were reviewed in Chapter 5 of Volume 1.

The result (9.56) is remarkable for several reasons. It contains no terms that depend on diffraction effects. Moreover, it is identical to the mutual coherence function calculated with geometrical optics.[6] The same expression will be generated

---

[6] See Section 9.2 of Volume 1.

in Volume 3 with the Markov approximation, the method of path integrals and diagrammatic techniques. It is a universal result in this respect – as is the mean-field expression (9.7). It is significant that this generic result emerges from the second-order Rytov solution.

9.4.2 Experimental Confirmations

To measure the mutual coherence function one must measure the amplitudes of both signals and their difference in phase:

$$\Gamma(\mathbf{R}, \mathbf{r}_1, \mathbf{r}_2) = \langle A_1 A_2 \exp[i(\phi_1 - \phi_2)] \rangle \tag{9.57}$$

It is easy to do so at microwave frequencies since both components of the field strength are readily available. One must augment intensity measurements of optical signals with interferometric techniques to sense the phase difference.

This was done in a pioneering experiment in which an interferometer was operated at the focal plane of an astronomical telescope [25]. Two bright stars with different zenith angles were observed for several nights and the mutual coherence function was measured for each star. Experimental data for a typical observation is reproduced in Figure 9.8. The measurements for each separation were made approximately 5 min apart. These plots agree generally with the exponential decrease

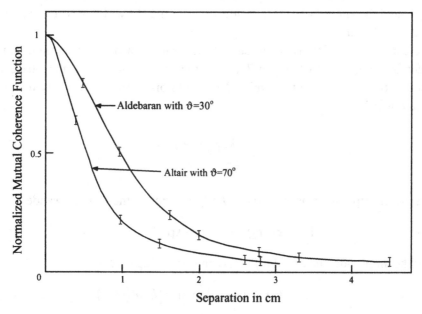

Figure 9.8: Astronomical observations of the mutual coherence function by Roddier and Roddier [25], who made interferometric measurements at the focal plane of a telescope. Two sources with different zenith angles were used to measure this function sequentially at each separation.

predicted by (9.56). However, comparisons with the commonly assumed wave structure function

$$\mathcal{D}(\rho) = \text{constant} \times \rho^{\frac{5}{3}} \qquad (9.58)$$

gave agreement with the measurements only for separations less than $\rho = 1$ cm. That is consistent with the discussion of Section 5.1.2 of Volume 1 in which it was found that one should use the exact solution plotted in Figure 5.8, rather than the simple 5/3 relationship.

Similar experiments were performed on terrestrial links using He–Ne lasers and Mach–Zehnder interferometers [26][27]. Densitometer measurements of fringe patterns indicated the behavior of the wave structure function. They confirmed the 5/3 relationship for separations less than 4 cm but a significant departure was noted for larger spacings. That is consistent with the exact theory which includes outer-scale effects [21].

A different approach is based on the observation that the average illumination of each point on the focal plane of a telescope is proportional to the two-dimensional Fourier transform of the mutual coherence function [28][29]. By scanning the focal plane with a narrow slit it is possible to relate the measured signal to the one-dimensional Fourier transform of $\Gamma(\rho, x)$ calculated in the direction orthogonal to the slit. Terrestrial experiments based on this connection were performed using He–Ne laser signals and several path lengths. The dissipation range of the spectrum complicated the analysis of these results because the slit width was comparable to the inner scale length.

The third method involves measuring the wander of a focused beam and its broadening [30][31]. In Chapter 7 of Volume 1 we found that the phase difference between two points is related to angle-of-arrival fluctuations for small separations ρ:

$$\delta\theta = \frac{\lambda}{2\pi}\left(\frac{\varphi_1 - \varphi_2}{\rho}\right)$$

This relationship allows us to rewrite (9.57) in terms of the angular wander:

$$\Gamma(\mathbf{R}, \mathbf{r}_1, \mathbf{r}_2) = \langle A_1 A_2 \exp(ik\rho\,\delta\theta)\rangle \qquad (9.59)$$

When the amplitude and angular fluctuations are independent

$$\Gamma(\mathbf{R}, \mathbf{r}_1, \mathbf{r}_2) = \langle A_1 A_2\rangle \exp\left(-\tfrac{1}{2}k^2\rho^2\langle\delta\theta^2\rangle\right) \qquad (9.60)$$

Other expressions relate the broadening of a focused beam to the mutual coherence function.

9.4.3 The Beam-wave Expression

Optical beam waves are readily formed with coherent laser sources using combinations of mirrors and lenses. This type of signal is widely used in civilian and military applications. Refractive-index irregularities broaden these beam waves in important ways. There is a strong incentive to calculate the mutual coherence function, since it best describes the variation of field strength within the beam. The analytical challenge presented by this calculation is considerably greater than those encountered so far in our work. Attempts to find a reliable description for these cases were marked by frustrations and inconsistencies for more than a decade.

The first step is to establish the mutual coherence function for the vacuum field. It can be expressed succinctly in terms of sum and difference vector coordinates of the receivers, namely

$$\mathbf{r} = \tfrac{1}{2}(\mathbf{r}_1 + \mathbf{r}_2) \qquad \text{and} \qquad \rho = \mathbf{r}_1 - \mathbf{r}_2 \tag{9.61}$$

as follows:

$$\text{Beam wave:} \quad \Gamma_0(\mathbf{R}, \mathbf{r}_1, \mathbf{r}_2) = |\mathcal{E}_0|^2 \frac{w_0^2}{w^2(R)} \exp\left(-\frac{4\mathbf{r}^2 + \rho^2}{2w^2(R)} - i\frac{k}{\mathcal{R}}\rho \cdot \mathbf{r} \right) \tag{9.62}$$

Here w_0 is the beam radius and $\mathcal{R}$ the radius of curvature at the receiver plane.[7] The beam radius $w(R)$ at the receiver is defined by (3.83). This expression reduces to the mean irradiance for the vacuum field when the measurement points coincide.

Calculation of the mutual coherence function for a random medium requires considerable care [32]. To do so we must return to the basic expression (9.48) and adapt it to the special conditions of a beam wave. The average value of the complex quantity is given by

$$\langle Z \rangle = \langle c_1 + c_2 + id_1 - id_2 \rangle - \tfrac{1}{2}\langle b_1^2 + b_2^2 - a_1^2 - a_2^2 \rangle - i\langle a_1 b_1 - a_2 b_2 \rangle$$

and the mean-square value to second order becomes

$$\langle Z^2 \rangle = \langle a_1^2 + 2a_1 a_2 + a_2^2 \rangle - \langle b_1^2 - 2b_1 b_2 + b_2^2 \rangle$$
$$+ 2i[\langle a_1 b_1 - a_2 b_2 \rangle - \langle a_1 b_2 - a_2 b_1 \rangle]$$

The average values of c and d for homogeneous isotropic random media were established in (8.60) and allow us to write the term that modifies the vacuum mutual-coherence-function expression for a beam wave as follows:

$$\langle \exp Z \rangle = \exp\left(-4\pi^2 k^2 R \int_0^\infty d\kappa \, \kappa \Phi_n(\kappa) + \langle a_1 a_2 \rangle - \langle b_1 b_2 \rangle - i\langle a_1 b_2 - a_2 b_1 \rangle \right)$$
$$\tag{9.63}$$

---

[7] These quantities were defined in Section 3.2.5.

The imaginary term vanishes, as it did for plane and spherical waves. The spatial covariance of a was established in (4.29) and (4.30). For the spatial covariance of b we simply change the minus sign in the definition (4.30) into a plus sign. This allows one to express the mutual coherence function as

$$\Gamma(\mathbf{R}, \mathbf{r}_1, \mathbf{r}_2) = \Gamma_0(\mathbf{R}, \mathbf{r}_1, \mathbf{r}_2) \exp\left[-\tfrac{1}{2}\Delta(\mathbf{R}, \mathbf{r}_1, \mathbf{r}_2)\right] \qquad (9.64)$$

where

$$\Delta(\mathbf{R}, \mathbf{r}_1, \mathbf{r}_2) = 8\pi^2 k^2 R \int_0^1 du \int_0^\infty d\kappa\, \kappa \Phi_n(\kappa)$$
$$\times \left[1 - J_0(\kappa P) \exp\left(-\kappa^2 \frac{2R^2(1-u)^2}{k^2 w^2(R)}\right)\right] \qquad (9.65)$$

The effective separation P is given in (4.31), which is more conveniently expressed in terms of the sum and difference coordinates of the receivers (9.61) as follows [33]:

$$P = |\rho\gamma_1 - 2i\mathbf{r}\gamma_2| \qquad (9.66)$$

where the absolute-magnitude notation applies only to the vectors.

With these expressions one can proceed to estimate the mutual coherence function for various beam-wave configurations encountered in practice. That involves a good deal of analysis and computation for cases of practical interest. Approximations that have been developed for these applications are presented by Andrews and Phillips [34]. The reader is encouraged to consult that rich source before attempting calculations based on (9.65).

The general solution for the mutual coherence function simplifies for several receiver geometries. As it should, it reduces to the mean-irradiance expression (9.17) when the two receivers are co-located. The reader is encouraged to verify this connection in Problem 2, where several helpful suggestions are provided.

A common experimental arrangement employs two receivers deployed along a radial line at equal and opposite distances from the beam axis so that $\mathbf{r}_1 = -\mathbf{r}_2$. In this case the mutual coherence function depends only on the scalar distance between the receivers and on the downrange distance. Andrews and Phillips investigated this configuration for three limiting forms of a generalized beam wave [34]. Their computations for a plane wave, a spherical wave and a collimated beam are reproduced in Figure 9.9. The mutual coherence function is plotted as a function of the Fresnel ratio $k\rho^2/R$ and demonstrates the considerable difference between these common types of signal.

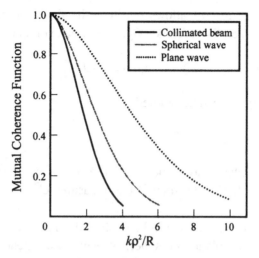

Figure 9.9: The mutual coherence function for three common types of wave calculated by Andrews and Phillips [34]. The results are plotted versus the squared ratio of the inter-receiver separation and the Fresnel length. They are normalized by dividing each by its on-axis value.

9.5 Logarithmic Intensity Fluctuations

Large variations of optical signal intensity encouraged the use of *logarithmic amplifiers*, which compress the dynamic range of the measured fluctuations. To relate the Rytov approximation to this type of measurement we introduce a new symbol to identify the logarithmic irradiance:

$$\mathbb{H} = \log\left(\frac{I(\mathbf{R})}{I_0}\right) \tag{9.67}$$

This quantity is inherently second order and we must include the second-order solution to describe it accurately. Recalling the definition of intensity (9.12), we can describe it as follows:

$$\mathbb{H} = 2a + 2c - a^2 + b^2 \tag{9.68}$$

The moments of $\mathbb{H}$ are expressed directly in terms of the Born moments and we need not use cumulant-analysis techniques.

9.5.1 The Mean Logarithmic Irradiance

The mean value of the logarithmic intensity sets an important reference level for measurements of irradiance fluctuations:

$$\langle \mathbb{H} \rangle = 2\langle c \rangle - \langle a^2 \rangle + \langle b^2 \rangle \tag{9.69}$$

The expression (8.18) for $\langle c \rangle$ allows one to write this average in terms of the amplitude variance when the reference signal is a plane or spherical wave:

$$\langle \mathbb{H} \rangle = -2\langle a^2 \rangle = -2\langle \chi^2 \rangle \tag{9.70}$$

When aperture averaging can be ignored

$$\langle \mathbb{H} \rangle = -2\eta C_n^2 R^{\frac{11}{6}} k^{\frac{7}{6}} \tag{9.71}$$

where $\eta = 0.307$ for plane waves and $\eta = 0.124$ for spherical waves. This prediction can be tested on horizontal links if one measures C_n^2 since R is then known accurately.

An experiment to confirm the relationship (9.71) was performed using a He–Ne laser [35]. Both plane and spherical waves were generated by changing the transmitter optics. The diameter of the receiver was 0.3 mm so that aperture averaging was not important. Propagation paths close to the surface with $R = 250$ m and $R = 1750$ m were used. It was determined that the effects of aerosol scattering and deviative absorption would be much less than the influence of refractive-index irregularities on these paths. Values of C_n^2 inferred from temperature-profile gradients were used to estimate χ_{rms}. The data from this experiment is reproduced in Figure 9.10, where the measured values of $\sqrt{-2\langle \mathbb{H} \rangle}$ are plotted versus the inferred values of $2\chi_{\mathrm{rms}}$. These results support the linear

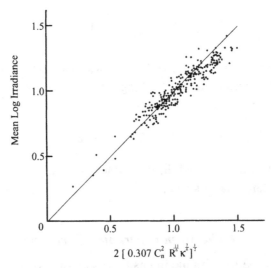

Figure 9.10: Optical measurements of the mean logarithmic irradiance made by Gracheva, Gurvich and Kallistra-tova [35]. They used a He–Ne laser signal on paths of lengths 250 and 1750 m near the surface. The experimental values are plotted versus the estimated rms amplitude fluctuation computed from simultaneous measurements of the structure constant.

prediction (9.71) in the following range:

$$\chi_{rms} = 0.307 C_n^2 R^{\frac{11}{6}} k^{\frac{7}{6}} < 1 \tag{9.72}$$

A similar measurement was made with a He–Ne laser on a mixed terrestrial path [36]. The signal was first transmitted close to the surface before being reflected to a receiver mounted on a 70-m high tower that was 900 m from the mirror. The initial horizontal paths were 100 and 500 m long in two versions of the experiment. These measurements also confirmed the linear relationship (9.71).

The condition $\chi_{rms} < 1$ defines the weak-scattering regime in which the Rytov approximation is considered to be valid. The mean logarithmic-irradiance measurements cited above confirm that understanding. The data departed dramatically from the linear relationship above this limit, thereby indicating the onset of strong scattering. That change is reflected more clearly in the next type of measurement.

9.5.2 The Variance of the Logarithmic Irradiance

The next step is to establish an expression for the *variance of the logarithmic irradiance*. Using the basic relationship (9.12), we can describe it as follows:

$$\langle \mathbb{H}^2 \rangle - \langle \mathbb{H} \rangle^2 = \left\langle (2a + 2c - a^2 + b^2)^2 \right\rangle - \left(2c - a^2 + b^2 \right)^2 \tag{9.73}$$

When the transmitted signal is a plane or spherical wave, the relationships among the Born moments provided in (8.18) show that it is proportional to the Rytov variance:

$$\langle \delta \mathbb{H}^2 \rangle = \langle \mathbb{H}^2 \rangle - \langle \mathbb{H} \rangle^2 = 4 \langle a^2 \rangle = 4 \langle \chi^2 \rangle \tag{9.74}$$

In the experiments to test this relationship small receivers were used so that aperture averaging could be ignored. The rms value of $\delta \mathbb{H}$ is expressed in a simple form in these cases:

$$\delta \mathbb{H}_{rms} = 2\chi_{rms} = 2\sqrt{\eta C_n^2 R^{\frac{11}{6}} k^{\frac{7}{6}}} \tag{9.75}$$

This prediction can be tested on terrestrial paths close to the surface by measuring the structure constant accurately.

Measuring the variance of the logarithmic irradiance was an important goal of several early programs. The basic features of those experiments are summarized in Table 9.1. In the first tests high-pressure mercury lamps were used to generate incoherent light waves. Plane and spherical waves were formed by different optical trains. The signals traveled close to the ground for a few hundred meters. The structure constant C_n^2 was calculated from measurements of the temperature-profile gradient using its assumed height dependence. These experiments seemed

Table 9.1: *A summary of early measurements of the variance of the logarithmic irradiance using optical signals on terrestrial paths*

Year	Source	Diameter (mm)	Type of Wave Plane	Type of Wave Spherical	Distance (m)	Maximum $2\chi_{rms}$	Ref.
1965	Hg lamp	2	x		125–1750	10.5	39
1967	Hg lamp	2–5		x	250–1750	10.5	40
1968	He–Ne laser		x	x	50–6500	9	41
1969	He–Ne laser	1		x	990	3.4	42
1969	Ruby laser	0.4×3		x	200–1500	1.7	43
1970	He–Ne laser		x		1000 and 1400	8.5	36
1970	He–Ne laser	0.3	x	x	250 and 1750	10	44

to confirm the predicted relationship (9.75) during observations made at night when turbulent activity is at a minimum [37][38].

The development of coherent laser light sources provided greater accuracy and allowed much longer path lengths to be investigated. Logarithmic amplifiers were used to compress the large dynamic range of intensity fluctuations that were encountered during conditions of strong turbulence. High-speed platinum-wire resistance thermometers made it possible to obtain accurate estimates of C_n^2 from closely spaced sensors. These advances permitted careful checking of the prediction (9.75) over a wide range of propagation and turbulence conditions.

A comprehensive optical experiment was then performed in the USSR [44]. A He–Ne laser signal was transmitted along horizontal paths of 250 and 1750 m. Both plane and spherical waveforms were used with a point-like receiver. Measured values of $\delta\mathbb{H}_{rms}$ were compared with values of χ_{rms} calculated from simultaneous measurements of C_n^2. A wide variety of conditions was encountered on these paths and the actual data points are reproduced in Figure 9.11. These measurements are both reassuring and surprising. They confirm the linear relationship (9.75) predicted by the Rytov approximation in the weak-scattering regime,

$$2\chi_{rms} < 1 \tag{9.76}$$

as did the measurements of the mean logarithmic intensity. This experimental boundary confirms the analytical estimate for the range of validity for the Rytov approximation discussed in Chapter 12.

The surprising features of the data in Figure 9.11 occur beyond $2\chi_{rms} = 1$. The experimental data does not follow the predicted straight line for larger values. Instead, the rms logarithmic intensity reaches a maximum value for values of $2\chi_{rms}$ between 1.0 and 1.5. It then decreases gradually out to the largest values that were measured. This behavior cannot be explained by the second-order Rytov approximation and represents a different kind of physics than that which we have

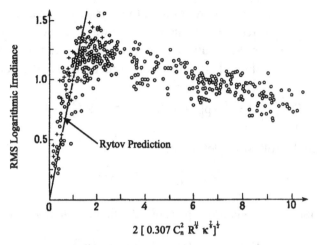

$$2 \, [\, 0.307 \, C_n^2 \, R^{\frac{11}{6}} \, \kappa^{\frac{7}{6}}]^{\frac{1}{2}}$$

Figure 9.11: Plane-wave measurements of the rms logarithmic irradiance made by Gracheva, Gurvich and Kallistratova [44] with a He–Ne laser. Data obtained with a path length of 250 m is plotted as crosses and data for 1750 m as open circles. The horizontal values were calculated from (9.75) using simultaneous measurements of C_n^2 from closely spaced temperature sensors.

been discussing. The observed saturation and decline reflects strong scattering, which will be addressed in Volume 3.

9.6 The Variance of Intensity

Some types of signals do not experience large fluctuations and hence in experiments one often elects to measure the irradiance directly, rather than its logarithmic counterpart. Astronomical observations usually fit this pattern. Radio-astronomy observations do also unless the source is close to the horizon or the frequency is in the low-VHF band. The scintillation experienced on microwave signals from communication satellites also falls in this class.

Each of these applications is characterized by a plane wave passing through the atmosphere and the intensity fluctuations are usually normalized by the mean irradiance:

$$\frac{\delta I}{\langle I \rangle} = \frac{I - \langle I \rangle}{\langle I \rangle} \tag{9.77}$$

The variance of this ratio is called the *scintillation index* S_4 and is the most common measurement in such observations:

$$\left\langle \left(\frac{\delta I}{\langle I \rangle} \right)^2 \right\rangle = (S_4)^2 = \frac{\langle I^2 \rangle - \langle I \rangle^2}{\langle I \rangle^2} \tag{9.78}$$

One needs the second-order Rytov solution to estimate this quantity. The mean irradiance is equal to the free-space value for plane waves as noted in (9.14). The basic irradiance expression (9.12) allows one to write the variance of intensity as the ensemble average involving first- and second-order terms in the Born series:

$$\left\langle \left(\frac{\delta I}{\langle I \rangle} \right)^2 \right\rangle = \langle \exp[4a + 4c - 2a^2 + 2b^2] \rangle - 1 \tag{9.79}$$

We use (9.3) with the following definition:

$$Z = 4a + 4c - 2a^2 + 2b^2 \tag{9.80}$$

Exploiting the plane-wave relationship (8.18), the first two moments become

$$\langle Z \rangle = \langle 4c - 2a^2 + 2b^2 \rangle = -4\langle \chi^2 \rangle$$

$$\langle Z^2 \rangle = \langle (4a)^2 \rangle = 16\langle \chi^2 \rangle$$

The normalized variance of intensity is thus given by the following [45]:

$$\left\langle \left(\frac{\delta I}{\langle I \rangle} \right)^2 \right\rangle = \exp(4\langle \chi^2 \rangle) - 1 \tag{9.81}$$

For plane wave traveling near the surface the variance of irradiance becomes

Terrestrial link: $$\left\langle \left(\frac{\delta I}{\langle I \rangle} \right)^2 \right\rangle = \exp\left(1.228 C_n^2 R^{\frac{11}{6}} k^{\frac{7}{6}} \right) - 1 \tag{9.82}$$

This result suggests that the variance of intensity should increase rapidly as the wave travels through a random medium. That is contradicted by the measurements reproduced in Figure 9.11 which show that the irradiance fluctuations first saturate and then decline. This issue is resolved when we remember that the Rytov approximation is valid only for $\langle \chi^2 \rangle$ less than unity. In that case we can expand the exponential:

Horizontal: $$\left\langle \left(\frac{\delta I}{\langle I \rangle} \right)^2 \right\rangle \simeq 1.228 C_n^2 R^{\frac{11}{6}} k^{\frac{7}{6}} \tag{9.83}$$

The corresponding result for a signal that passes completely through the atmosphere from a source with zenith angle ϑ is given by

Transmission: $$\left\langle \left(\frac{\delta I}{\langle I \rangle} \right)^2 \right\rangle \simeq 2.252 k^{\frac{7}{6}} \sec \vartheta \int_0^\infty dz \, z^{\frac{5}{6}} C_n^2(z) \tag{9.84}$$

9.7 Problems

Problem 1

In calculating the mutual coherence function for plane and spherical waves we assumed that the displacements between receiver were small relative to the path length. Show that this assumption is unnecessary for an unbounded plane wave and that the result (9.55) is valid for all locations of receivers in a plane normal to the line of sight. What relaxation of that assumption is possible for spherical waves?

Problem 2

For beam-wave signals, verify that the mean-irradiance expression (9.17) emerges from the mutual-coherence-function description (9.65) when the measurement points r_1 and r_2 coincide. To complete this calculation you will need Weber's second integral involving Bessel functions in Appendix D and the recurrence relations among Kummer functions in Appendix G.

References

[1] H. T. Yura, C. C. Sung, S. F. Clifford and R. J. Hill, "Second-order Rytov Approximation," *Journal of the Optical Society of America*, **73**, No. 4, 500–502 (April 1983).

[2] L. C. Andrews and R. L. Phillips, *Laser Beam Propagation through Random Media* (SPIE Optical Engineering Press, Bellingham, Washington, 1998), 129.

[3] H. T. Yura and S. G. Hanson, "Second-order Statistics for Wave Propagation through Complex Optical Systems," *Journal of the Optical Society of America A*, **6**, No. 4, 564–575 (April 1989).

[4] J. W. Herbstreit and M. C. Thompson, "Measurements of the Phase of Radio Waves Received over Transmission Paths with Electrical Lengths Varying as a Result of Atmospheric Turbulence," *Proceedings of the IRE*, **43**, No. 10, 1391–1401 (October 1955).

[5] K. A. Norton, J. W. Herbstreit, H. B. Janes, K. O. Hornberg, C. F. Peterson, A. F Barghausen, W. E. Johnson, P. I. Wells, M. C. Thompson, M. J. Vetter and A. W. Kirkpatrick, *An Experimental Study of Phase Variations in Line-of-Sight Microwave Transmissions* (National Bureau of Standards Monograph 33, U. S. Government Printing Office, 1 November 1961).

[6] V. P. Lukin, V. V. Pokasov and S. S. Khmelevtsov, "Investigation of the Time Characteristics of Fluctuations of the Phases of Optical Waves Propagating in the Bottom Layer of the Atmosphere," *Izvestiya Vysshikh Uchebnykh Zavedenii, Radiofizika (Radiophysics and Quantum Electronics)*, **15**, No. 12, 1426–1430 (December 1972).

[7] B. J. Uscinski, *The Elements of Wave Propagation in Random Media* (McGraw-Hill, New York, 1977), 27–29.

[8] V. I. Tatarskii, *The Effects of the Turbulent Atmosphere on Wave Propagation* (translated from the Russian and issued by the National Technical Information Office, U. S. Department of Commerce, Springfield, VA 22161, 1971), 107–123.

[9] S. M. Rytov, Yu. A. Kravtsov and V. I. Tatarskii, *Principles of Statistical Radiophysics 3, Elements of Random Fields* (Springer-Verlag, Berlin, 1989), 189–195.

[10] H. T. Yura, "Electromagnetic Field and Intensity Fluctuations in a Weakly Inhomogeneous Medium," The RAND Corporation, Memorandum RM-5697-PR (July 1968), 17.

[11] S. M. Rytov, Yu. A. Kravtsov and V. I. Tatarskii, *Principles of Statistical Radiophysics 4, Wave Propagation Through Random Media* (Springer-Verlag, Berlin, 1989), 68–69.

[12] R. M. Manning, "Beam Wave Propagation Within the Second Rytov Perturbation Approximation," *Izvestiya Vysshikh Uchebnykh Zavedenii, Radiofizika (Radiophysics and Quantum Electronics)*, **39**, No. 4, 423–436 (August 1996). (In English.)

[13] See [2], pages 133–135.

[14] H. T. Yura, "The Second-Order Rytov Approximation," The RAND Corporation, Memorandum RM-5787-PR (February 1969).

[15] A. D. Wheelon, "Radiowave Scattering by Tropospheric Irregularities," *Journal of Research of the NBS – D. Radio Propagation*, **63D**, No. 2, 205–233 (September–October 1959).

[16] J. A. Stratton, *Electromagnetic Theory* (McGraw-Hill, New York, 1941), 131–137.

[17] See [9], page 189.

[18] See [11], page 95.

[19] R. F. Lutomirski and H. T. Yura, "Propagation of a Finite Optical Beam in an Inhomogeneous Medium," *Applied Optics*, **10**, No. 7, 1652–1658 (July 1971).

[20] R. M. Manning, private communication on 23 January 2001.

[21] R. F. Lutomirski and H. T. Yura, "Wave Structure Function and Mutual Coherence Function of an Optical Wave in a Turbulent Atmosphere," *Journal of the Optical Society of America*, **61**, No. 4, 482–487 (April 1971).

[22] J. W. Strohbehn, "Line-of-Sight Wave Propagation Through the Turbulent Atmosphere," *Proceedings of the IEEE*, **56**, No. 8, 1301–1318 (August 1968).

[23] A. Ishimaru, *Wave Propagation and Scattering in Random Media*, vol. 2 (Academic Press, San Diego, 1978), 279–285 and 422–424.

[24] See [23], pages 314–325 and 425–427.

[25] C. Roddier and F. Roddier, "Correlation Measurements on the Complex Amplitude of Stellar Plane Waves Perturbed by Atmospheric Turbulence," *Journal of the Optical Society of America*, **63**, No. 6, 661–663 (June 1973).

[26] M. Bertolotti, L. Muzii and D. Sette, "Correlation Measurements on Partially Coherent Beams by Means of an Integration Technique," *Journal of the Optical Society of America*, **60**, No. 12, 1603–1607 (December 1970).

[27] R. G. Buser, "Interferometric Determination of the Distance Dependence of the Phase Structure Function for Near-Ground Horizontal Propagation at 6328 Å," *Journal of the Optical Society of America*, **61**, No. 4, 488–491 (April 1971).

[28] A. V. Artem'ev and A. S. Gurvich, "Experimental Study of Coherence-Function Spectra," *Izvestiya Vysshikh Uchebnykh Zavedenii Radiofizika (Radiophysics and Quantum Electronics)*, **14**, No. 5, 580–583 (May 1971).

[29] A. S. Gurvich and V. I. Tatarskii, "Coherence and Intensity Fluctuations of Light in the Turbulent Atmosphere," *Radio Science*, **10**, No. 1, 3–14 (January 1975).

[30] M. A. Kallistratova and V. V. Pokasov, "Defocusing and Fluctuations of the Displacement of a Focused Laser Beam in the Atmosphere," *Izvestiya Vysshikh Uchebnykh Zavedenii Radiofizika (Radiophysics and Quantum Electronics)*, **14**, No. 8, 940–945 (August 1971).

[31] I. A. Starobinets, "The Average Illumination and the Intensity Fluctuations at the Focus of a Light Beam Focused in a Turbulent Atmosphere," *Izvestiya Vysshikh Uchebnykh Zavedenii Radiofizika (Radiophysics and Quantum Electronics)*, **15**, No. 5, 563–566 (May 1972).

[32] H. T. Yura, "Mutual Coherence Function of a Finite Cross Section Optical Beam Propagating in a Turbulent Medium," *Applied Optics*, **11**, No. 6, 1399–1406 (June 1972).

[33] L. C. Andrews, W. B. Miller and J. C. Ricklin, "Spatial Coherence of a Gaussian-beam Wave in Weak and Strong Optical Turbulence," *Journal of the Optical Society of America A*, **11**, No. 5, 1653–1660 (May 1994).

[34] See [2], pages 135–141.

[35] M. E. Gracheva, A. S. Gurvich and M. A. Kallistratova, "Measurement of the Average Amplitude of a Light Wave Propagating in a Turbulent Atmosphere," *Izvestiya Vysshikh Uchebnykh Zavedenii Radiofizika (Radiophysics and Quantum Electronics)*, **13**, No. 1, 36–39 (January 1970).

[36] M. I. Mordukhovich, "Dispersion of the Intensity Fluctuations and Average Amplitude Level of Laser Light During Propagation Along a Strongly Inhomogeneous Route," *Izvestiya Vysshikh Uchebnykh Zavedenii Radiofizika (Radiophysics and Quantum Electronics)*, **13**, No. 2, 212–215 (February 1970).

[37] A. S. Gurvich, V. I. Tatarskii and L. R. Tsvang, "Experimental Investigation of the Statistical Character of the Twinkling of Light Originating at the Earth's Surface," *Doklady Akademii Nauk SSSR (Soviet Physics – Doklady)*, **123**, No. 4, 655–658 (1958). (This paper is available only in Russian.)

[38] V. I. Tatarskii, A. S. Gurvich, M. A. Kallistratova and L. V. Terent'eva, "The Influence of Meteorological Conditions on the Intensity of Light Scintillation Near the Earth's Surface," *Astronomicheskii Zhurnal (Soviet Astronomy)*, **2**, No. 4, 578–580 (July–August 1958).

[39] M. E. Gracheva and A. S. Gurvich, "Strong Fluctuations in the Intensity of Light Propagated Through the Atmosphere Close to the Earth," *Izvestiya Vysshikh Uchebnykh Zavedenii Radiofizika (Soviet Radiophysics)*, **8**, No. 4, 511–515 (July–August 1965).

[40] M. E. Gracheva, "Investigation of the Statistical Properties of Strong Fluctuations in the Intensity of Light Propagated Through the Atmosphere Near the Earth," *Izvestiya Vysshikh Uchebnykh Zavedenii Radiofizika (Radiophysics and Quantum Electronics)*, **10**, No. 6, 424–433 (June 1967).

[41] A. S. Gurvich, M. A. Kallistratova and N. S. Time, "Fluctuations in the Parameters of a Light Wave from a Laser during Propagation in the Atmosphere," *Izvestiya Vysshikh Uchebnykh Zavedenii Radiofizika (Radiophysics and Quantum Electronics)*, **11**, No. 9, 771–776 (September 1968).

[42] G. R. Ochs and R. S. Lawrence, "Saturation of Laser-Beam Scintillation under Conditions of Strong Atmospheric Turbulence," *Journal of the Optical Society of America*, **59**, No. 2, 226–227 (February 1969).

[43] P. H. Deitz and N. J. Wright, "Saturation of Scintillation Magnitude in Near-Earth Optical Propagation," *Journal of the Optical Society of America*, **59**, No. 5, 527–535 (May 1969).

[44] M. E. Gracheva, A. S. Gurvich and M. A. Kallistratova, "Dispersion of 'Strong' Atmospheric Fluctuations in the Intensity of Laser Radiation," *Izvestiya Vysshikh Uchebnykh Zavedenii Radiofizika (Radiophysics and Quantum Electronics)*, **13**, No. 1, 40–42 (January 1970).

[45] See [11], page 69.

10

Amplitude Distributions

Predicting the distribution of measured fluctuations in intensity is one of the great challenges in describing electromagnetic propagation through random media. The amplitude variance addressed in Chapter 3 describes the width of this distribution but tells one nothing about the likelihood that very large or very small fluctuations will occur. By contrast, the probability density function for intensity variations provides a complete portrait of the signal's behavior. This wider perspective is important for engineering applications in which one must predict the complete range of signal values. The same description provides an important insight into the physics of scattering of waves by turbulent irregularities.

The Rytov approximation predicts that the distribution is log-normal. That forecast is confirmed by measurements made over a wide range of conditions on terrestrial links. It is also confirmed by astronomical observations. The agreement is independent of the model of turbulent irregularities employed. It therefore provides a test for the basic theoretical approach to electromagnetic propagation. Second-order refinements to the log-normal distribution predict a skewed log-normal distribution and agree with numerical simulations. This success represents a significant achievement for the Rytov approach.[1]

The distribution provided by Rytov theory needs to be enlarged in some situations. It describes the short-term fluctuations observed over periods of a few minutes to a few hours. One must also consider variations of signal strength measured over weeks, months or years. Long-term statistics are important for communication services employing terrestrial or satellite relays. Amplitude fluctuations imposed on microwave signals by the ionosphere are apparently influenced by additional considerations and one needs to augment the Rytov approximation in this arena. The intermittent behavior of the random medium has an important influence on the

---

[1] Recall that geometrical optics cannot describe intensity fluctuations as noted in Section 9.4 of Volume 1. The Rice distribution predicted by the Born approximation does not agree with experimental propagation data.

signal distribution for short paths and for small data samples. This phenomenon is only beginning to be properly understood and is an area of active research.

10.1 The Log-normal Distribution

The distribution of amplitude and intensity fluctuations is an immediate consequence of the first-order Rytov solution defined by (2.19). Recall that the amplitude
can be measured directly on microwave links if the source is coherent. In optical
and radio-astronomy experiments one can measure only the intensity of the arriving
signal because their sources are incoherent. We shall develop the probability density functions for amplitude and intensity simultaneously by exploiting the weak-
scattering connection (3.2). In this approach, the logarithmic amplitude of the signal
is defined by the volume integral (2.26) taken over the random dielectric variations:

$$\chi = \log\left(\frac{A}{A_0}\right) = -k^2 \int d^3r \, \delta\varepsilon(r) \, \Re\left(\frac{G(R, r)E_0(r)}{E_0(R)}\right) \tag{10.1}$$

where A_0 is proportional to the field strength $E_0(R)$ that would be measured if
there were no irregularities.

If we knew the statistical behavior of $\delta\varepsilon$ we could probably deduce the distribution
of χ. In this endeavor, it is important to begin with the measured distributions
of short-term temperature fluctuations that were reviewed in Section 2.3.2.2 of
Volume 1. We found there that δT and C_n^2 are often distributed as Gaussian random
variables (GRVs). In those cases, one can be sure that χ is also a GRV because a
linear combination of GRVs is itself a GRV. That linear combination is represented
by the volume integral of (10.1) in our application. The average value of χ vanishes
because $\langle\delta\varepsilon\rangle = 0$ and the probability density function (PDF) for the logarithmic
amplitude becomes

$$P(\chi)d\chi = \frac{d\chi}{\tau\sqrt{2\pi}} \exp\left(-\frac{\chi^2}{2\tau^2}\right) \tag{10.2}$$

The amplitude variance was discussed at length in Chapter 3 and is abbreviated as

$$\langle\chi^2\rangle = \tau^2 \tag{10.3}$$

We can exploit the relationship (10.1) between A and χ to express the *probability
density function for the signal amplitude* as follows:

$$P(A)dA = \frac{1}{\tau\sqrt{2\pi}} \frac{dA}{A} \exp\left\{-\frac{1}{2\tau^2}\left[\log\left(\frac{A}{A_0}\right)\right]^2\right\} \tag{10.4}$$

This defines a *log-normal distribution*, which will play a prominent role in our
work. It is discussed further in Appendix E. A similar expression describes the

distribution of fluctuations in intensity or irradiance:

$$P(I)\,dI = \frac{1}{2\tau\sqrt{2\pi}}\,\frac{dI}{I}\,\exp\left\{-\frac{1}{8\tau^2}\left[\log\left(\frac{I}{I_0}\right)\right]^2\right\} \qquad (10.5)$$

These results are remarkable for several reasons. They depend only on the logarithmic amplitude variance τ^2 and not on the type of transmitted wave employed.[2] The turbulence spectrum influences the PDF only through its role in setting the value of τ^2 but does not affect the functional form of the distribution.

Before proceeding any further, we must also consider the situation in which $\delta\varepsilon$ is not distributed as a GRV. Temperature measurements show that the distributions of ΔT and C_n^2 are sometimes log-normal. The volume integration of non-Gaussian dielectric variations, however, should be distributed as a GRV. To demonstrate this we use a mathematical argument exploited by Obukhov and Tatarskii [1]. They break the random medium into volume elements larger than the outer scale length. The $\delta\varepsilon$ are uncorrelated between these volume elements, except for the border regions, which are relatively unimportant. The scattered field components from these volumes are therefore independent. If the path length is large relative to the outer scale length, the received field should be the sum of a large number of independent random field components received from the individual volume elements:

$$E = E_s(1) + E_s(2) + E_s(3) + E_s(4) + \cdots$$

The central-limit theorem tells us that the sum of independent statistical quantities should be distributed as a GRV. This means that (10.4) also describes the distribution of the scattered signal components.

The *cumulative probability* is often measured in microwave and optical experiments. For the signal amplitude it can be established by integrating from $-\infty$ up to a reference level L:

$$\mathcal{P}(A \leq L) = \frac{1}{\tau\sqrt{2\pi}}\int_{-\infty}^{L}\frac{dA}{A}\,\exp\left\{-\frac{1}{2\tau^2}\left[\log\left(\frac{A}{A_0}\right)\right]^2\right\}$$

This integral can be expressed in terms of the error function defined in Appendix C, which is widely available in tabulated form [2]:

$$\mathcal{P}(A \leq L) = \frac{1}{2}\left\{1 + \mathrm{erf}\left[\frac{1}{\tau\sqrt{2}}\log\left(\frac{L}{A_0}\right)\right]\right\} \qquad (10.6)$$

This expression is plotted on *probability coordinates* in Figure 10.1 for several values of τ less than unity that satisfy the Rytov condition (3.1). Each case is

---

[2] We shall learn later that this conclusion changes for strong scattering. Numerical simulations show that the PDF is different for plane, spherical and beam waves. Moreover, it also seems to depend on the inner scale length.

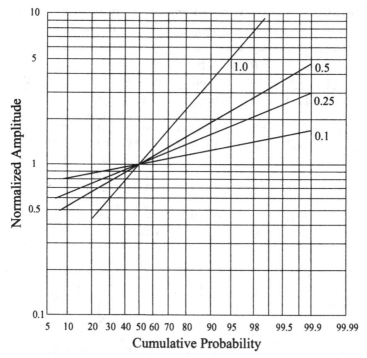

Figure 10.1: The cumulative probability for the logarithmic amplitude relative to the mean signal level that is predicted by the first-order Rytov solution. The results are plotted on probability coordinates for several values of the amplitude variance τ^2.

represented by a straight line on these coordinates and the lines all pass through a common point at the mean value. Their slopes depend on τ^2 and, through the amplitude-variance expressions, on C_n^2, ℓ_0, L_0, path length, size of receiver and frequency. A similar expression describes intensity fluctuations.

The log-normal distribution is quite different than the familiar Gaussian distribution which is often used to describe noise in electronic circuits. We need to compare this prediction carefully with electromagnetic measurements to see whether it accurately describes propagation though the real atmosphere.

10.1.1 Microwave Confirmations

Microwave measurements of amplitude fluctuations are the natural place to begin. They are typically described by weak scattering and should provide a valid test of the prediction. The cumulative probability has been measured for a wide variety of conditions and results of these experiments are summarized in Table 10.1. The remarkable feature of these measurements is that they each confirmed the log-normal distribution – with varying precision. It is worthwhile to review the results of these experiments.

Table 10.1:*A summary of microwave measurements of the cumulative probability for propagation near the surface*

Location	Year	R (km)	h (m)	f (GHz)	$\sqrt{\lambda R}$ (m)	a_r (m)	Ref.
Georgia	1971	43	40–50	4	56.8		3
				6	46.4		
Hawaii	1974	120	0–3000	9.6	61.2	2.4	4
Hawaii	1975	64	0–3000	9.55	44.8	1.35	5
				19.1	31.7	1.35	
				22.2	29.4	1.35	
				25.4	27.5	1.35	
				33.3	24.0	1.00	
UK	1977	4.1	15–50	36	5.8	0.29	6
				110	3.3	0.23	

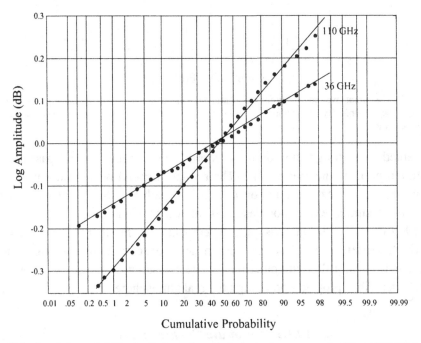

Figure 10.2: Cumulative probability distributions of amplitude fluctuations measured at 36 and 110 GHz by Ho, Mavroukoukoulakis and Cole [6].

The cleanest experiment was performed using signals at 36 and 110 GHz propagating over a 4.1-km path across London [6]. The cumulative probabilities for both signals from a typical data set are reproduced in Figure 10.2. The data shows that the distribution was log-normal. That finding was consistently verified on other days. The slopes of the straight lines are different, as they should be in view of the

Table 10.2:*A summary of optical experiments that measured the cumulative probability for signals propagating near the surface*

Location	Year	R	Source	$2a_r$ (mm)	Ref.
USSR	1965	125–1750 m	Hg lamp	2	7
USSR	1967	250–1750 m	Hg lamp	2–5	8
West Germany	1966	4.5 and 14.5 km	He–Ne laser	5–80	9
USSR	1968	50–6500 m	He–Ne laser	Point	10
Maryland	1969	1200 m	CO_2 laser	250	11, 12
			He–Ne laser	62	
Colorado	1969	990 m	He–Ne laser	1	13
Colorado	1969	5.5 and 15 km	He–Ne laser	1	14
		45 and 145 km	He–Ne laser	2.5	
USSR	1974	1750 m	He–Ne laser	0.3	15

difference in frequency. The slopes of the straight lines should scale as $k^{\frac{7}{6}}$ because the receivers were smaller than the Fresnel lengths, and they do.

In an early experiment a microwave communication link operating at 4 and 6 GHz was used to measure the distribution of amplitude fluctuations [3]. The cumulative amplitude data could be plotted as straight lines on probability coordinates, thus confirming the log-normal model. A series of long-distance microwave experiments was conducted in the Hawaiian Islands on paths of 64 and 120 km using frequencies indicated in Table 10.1 [4][5]. Again the cumulative probability measurements of amplitude fluctuations suggested a log-normal distribution. These measurements may have exceeded the condition $\tau^2 < 1$ and yet provided agreement with the Rytov prediction. We will encounter this feature again in distributions measured with laser signals on long paths.

10.1.2 Optical Confirmations

This straight-line behavior is observed at optical wavelengths and a series of early experiments is summarized in Table 10.2. In the first tests, Gracheva and Gurvich used incoherent light from a mercury lamp to measure the cumulative probability at short distances [7][8]. They found that the fluctuations follow a straight line on probability coordinates.

When coherent laser sources later became available, one could extend the distance and improve the accuracy of these measurements. The first to do so used a He–Ne laser with a variety of path lengths [10]. The log-normal model was always confirmed for path lengths less than 100 m. Many measurements were then made at 650 m. The distribution was log-normal when C_n^2 was small enough to ensure that weak-scattering conditions pertained. On the other hand, departures were noted when large values of C_n^2 generated strong scattering, thereby giving a hint of new physics.

A more ambitious experiment was then done using both He–Ne and CO_2 laser signals transmitted over a 1200-m path [11][12]. These measurements confirmed the log-normal distribution but gave a clear indication of saturation for a cumulative probability greater than 99.8% corresponding to large but unusual signals.

As the fundamental difference between weak and strong scattering became apparent from other experiments, a renewed effort to measure the PDF at great distances was made. The first authors to do so used path lengths of 4.5 and 14.5 km with a He–Ne laser, giving the result that 51 of 68 data sets agreed with the log-normal distribution [9]. An even wider range of transmission distances was then investigated using paths ranging from 5.5 to 145 km [14]. Experimental data from this experiment is reproduced in Figure 10.3. Although there is an indication of saturation in two cases, the data falls remarkably well on straight lines, suggesting that the log-normal distribution describes even these long but

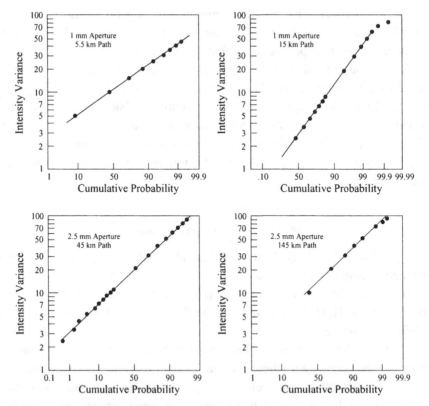

Figure 10.3: Cumulative intensity distributions measured on four long elevated laser links by Ochs, Bergman and Snyder [14].

elevated paths. These results were a surprise and caused a flurry of theoretical speculation.

A comprehensive campaign was then mounted in the USSR using a He–Ne laser on a path of 1750 m [15]. The data showed that the PDF is exactly log-normal for weak scattering and is approximately so for strong scattering. The simple model is deficient for deep fading, corresponding to the saturation noted in earlier experiments. Although these results are not yet fully understood, they are quite useful for engineering applications.

10.1.3 The Bivariate Distribution

The joint distribution for the signal strength measured at adjacent receivers is useful for many applications. It is also interesting scientifically and provides a good test of the basic description. Churnside found the following expression for the *bivariate probability density function* of irradiance for weak scattering [16]:

$$P(I_1, I_2) dI_1 dI_2 = \frac{dI_1 dI_2}{I_1 I_2 8\pi \tau^2 \sqrt{1 - C_\chi^2(\rho)}} \exp\left(-\frac{F(I_1, I_2, C_\chi(\rho))}{8\tau^2(1 - C_\chi^2(\rho))}\right) \quad (10.7)$$

where the exponent is

$$F(I_1, I_2, C_\chi(\rho)) = \left[\log\left(\frac{I_1}{I_0}\right) + 2\tau^2\right]^2 + \left[\log\left(\frac{I_2}{I_0}\right) + 2\tau^2\right]^2$$
$$- 2C_\chi(\rho)\left[\log\left(\frac{I_1}{I_0}\right) + 2\tau^2\right]\left[\log\left(\frac{I_2}{I_0}\right) + 2\tau^2\right] \quad (10.8)$$

and $C_\chi(\rho)$ is the spatial correlation between two receivers separated by ρ. This prediction was tested experimentally using a He–Ne laser on a 100-m path in Colorado. The receiving array consisted of five silicon photodiodes, which provided ten spacings ranging from 0.77 to 12.19 mm. The parameters $C_\chi(\rho)$ and τ were measured independently and used to estimate $W(I_1, I_2)$. The bivariate probability density was measured directly as a function of I_1 and I_2 and agreed with the predicted distributions.

10.1.4 The Influence of Aperture Averaging

One should ask how aperture averaging influences the cumulative probability. If weak scattering is the operative mechanism, one would expect the distribution to remain log-normal as the size of aperture is changed. Aperture smoothing should reduce all signals by the same fraction and the functional form of the distribution should remained unchanged. To be sure, the slope of the straight lines will vary in response to the smaller or larger values of τ implied by the receiver gain factors plotted in Figure 3.9.

This question was investigated experimentally with a He–Ne laser signal on an 8-km path [17]. The distribution of receiver noise was measured by occasionally blanking out the transmission and the results were used to correct fluctuations of the received signal. The adjusted results were approximately log-normal for aperture diameters from 1 mm to 89 cm. Höhn made the same observation on 4.5- and 14.5-km links using apertures between 5 and 80 cm [9]. This agreement masks a potential problem that we should not ignore.

Both links were quite long and it would be surprising if weak scattering were in force. We are somewhat surprised that the log-normal distribution itself seems to fit much of the data. The familiar saturation for large signals was observed on both links. One expects aperture smoothing to reduce the occurrence of signal spikes in strong scattering and that may be what was observed. What is evidently required is a fresh start on all aspects of strong scattering and that will be begun in Volume 3.

10.1.5 The Influence of the Sample Length

Let us return to the apparent suppression of large irradiance values that is often evident in laser data. This bend-over in the cumulative probability for large values can be caused by several effects. Detector saturation can suppress large signals and its impact on estimates of the moments of irradiance has been addressed several times [18][19][20].

The observed saturation can also be produced by a finite data sample. All values in the log-normal distribution would be encountered if a very large data set were used – including exceptionally large values. When a finite sample is used, the maximum necessarily corresponds to the largest irradiance in the actual set that is drawn. This value is finite and the cumulative probability curve must bend over to meet it at the 100% level. If a longer sample is then chosen, it is likely that a larger maximum value will be found and this would redefine the 100% level upward. The influence of sample length on PDF estimates based on moment measurements was investigated with numerical simulations [21].

10.2 The Second-order Correction

It is important to remember that the log-normal distribution is a unique consequence of the basic Rytov solution. This solution is simply the first-order term in the expansion of the surrogate function given by (2.13). In Chapter 9 we found that the second-order solution must be introduced in order to establish proper estimates for the field-strength moments. It would be surprising if the same term did not influence the probability distribution itself. We will find that the correction generated by the second-order solution introduces a fundamental skewness into the log-normal distribution that is apparent in measurements and simulations [22].

10.2.1 The Predicted Distribution

The first step in this exploration is to express the irradiance in terms of the second-order Rytov solution. The field-strength expression (2.64) gives the following description for the instantaneous irradiance:

$$I = I_0 \exp(2a + 2c - a^2 + b^2) \tag{10.9}$$

The dimensionless random variables a, b and c are defined in terms of the single- and double-scattering expressions by (2.27) and (2.63). The received signal is usually passed through a logarithmic amplifier to reduce the dynamic range of the instantaneous fluctuations and the logarithmic irradiance will be the focus of our analysis:

$$\mathbb{H} = \log\left(\frac{I}{I_0}\right) = 2a + 2c - a^2 + b^2 \tag{10.10}$$

Tatarskii suggested a method for calculating the distribution when one is not confident that $\mathbb{H}$ is a Gaussian random variable [23]. He notes that the PDF for a random variable is related to its characteristic function by a Fourier transformation:

$$\langle \exp(iq\mathbb{H}) \rangle = \int_{-\infty}^{\infty} d\mathbb{H} \exp(iq\mathbb{H}) \, P(\mathbb{H}) \tag{10.11}$$

If one knew the characteristic function from other sources, one could calculate the probability distribution from its inverse Fourier transform:

$$P(\mathbb{H}) = \frac{1}{2\pi} \int_{-\infty}^{\infty} dq \exp(-iq\mathbb{H}) \, \langle \exp(iq\mathbb{H}) \rangle \tag{10.12}$$

The method of cumulant analysis allows one to estimate the characteristic function in a straightforward manner. This mathematical technique expresses the ensemble average of a Fourier term as the exponential of a power series in the transform variable:

$$\langle \exp(iq\mathbb{H}) \rangle = \exp\left(\sum_{1}^{\infty} K_n \frac{(iq)^n}{n!}\right) \tag{10.13}$$

The coefficients K_n in this series are called cumulants and they are related to the moments of $\mathbb{H}$ by a system of relations explained in Appendix M. We need keep only the first three terms to estimate the distribution's skewness:

$$\langle \exp(iq\mathbb{H}) \rangle = \exp\left(iqK_1 - \frac{1}{2}q^2K_2 - \frac{i}{6}q^3K_3 \right) \tag{10.14}$$

The distribution would be log-normal if the characteristic function were only quadratic in q. A nonvanishing third cumulant means that the distribution must depart from log-normal. The extent of this departure depends on the magnitude of K_3 relative to K_2. The individual cumulants are defined in terms of the moments of $\mathbb{H}$ as follows:

$$K_1 = \langle \mathbb{H} \rangle$$
$$K_2 = \langle \mathbb{H}^2 \rangle - \langle \mathbb{H} \rangle^2 \tag{10.15}$$
$$K_3 = \langle \mathbb{H}^3 \rangle - 3\langle \mathbb{H} \rangle \langle \mathbb{H}^2 \rangle + 2\langle \mathbb{H} \rangle^3$$

These expressions do not assume that $\mathbb{H}$ is a Gaussian random variable.[3]

To evaluate the moments of $\mathbb{H}$ we use the definition for the logarithmic intensity in terms of the dimensionless random variables a, b and c given by (10.10). In this way one relates the moments of $\mathbb{H}$ to those of a, b and c. We show in Section 8.3.2 that the following relationship connects certain moments of these terms for plane and spherical waves:

$$\langle c \rangle = -\frac{1}{2}[\langle a^2 \rangle + \langle b^2 \rangle] \tag{10.16}$$

The quantity $\langle a^2 \rangle$ is identical to the logarithmic amplitude variance that we have denoted by τ^2. The terms in these moments that contain an odd number of $\delta\varepsilon$ in their definitions can be ignored if the $\delta\varepsilon$ are GRVs with zero mean. With these conventions we find that the first two cumulants can be expressed in terms of τ^2 alone:

$$K_1 = \langle 2a + 2c - a^2 + b^2 \rangle = -2\tau^2$$
$$K_2 = \langle (2a + 2c - a^2 + b^2)^2 \rangle - \langle 2a + 2c - a^2 + b^2 \rangle^2 = 4\tau^2 + O(\tau^4) \tag{10.17}$$

Notice that K_2 is just the *Rytov variance* defined by $\beta_0^2 = 4\tau^2$ that is often employed in scintillation measurements and simulations.

The third cumulant is fourth order in $\delta\varepsilon$ and is uniquely important for estimating the skewness of the distribution. It is related to moments of a, b and c by the following expression:

$$K_3 = 12\left(2\langle a^2c \rangle + \langle a^2b^2 \rangle - \langle a^4 \rangle + 2\langle a^2 \rangle^2 \right) \tag{10.18}$$

---

[3] Recall Problem 3 in Chapter 2 of Volume 1, where the distribution of two GRVs was found to be decidedly non-Gaussian.

Because a and b are GRVs we can express

$$\langle a^4 \rangle = 3\langle a^2 \rangle^2 = 3\tau^4 \qquad \text{and} \qquad \langle a^2 b^2 \rangle = \tau^2 \sigma^2 (1 + 2v^2)$$

as noted in Appendix E. Here v is the normalized correlation of phase and amplitude. If one ignores inner-scale effects, K_3 can be expressed in terms of the appropriate amplitude variances for plane and spherical waves [22]:

$$\text{Plane wave:} \qquad K_3 = -11.142\tau_{pl}^4$$
$$\text{Spherical wave:} \quad K_3 = -5.685\tau_{sph}^4 \tag{10.19}$$

The third cumulant therefore depends on the level of turbulent activity C_n^2, the path length and the electromagnetic frequency. In using these relations one must remember that τ is different for the two types of wave.

With these results in hand, we return to the Fourier-integral expression (10.12) for the PDF and substitute the characteristic function defined by (10.14) into it:

$$P(\mathbb{H}) = \frac{1}{2\pi} \int_{-\infty}^{\infty} dq \, \exp\left(-iq(\mathbb{H} - K_1) - \frac{1}{2}q^2 K_2 - \frac{i}{6}q^3 K_3 \right)$$

Although the integration over q can be expressed in terms of Airy functions, it is simpler to cast it in a form suitable for numerical integration. It is natural to shift and rescale the logarithmic irradiance as follows:

$$u = \frac{\mathbb{H} - K_1}{\sqrt{K_2}} = \frac{\mathbb{H} + 2\tau^2}{2\tau} = \frac{1}{\beta_0}\left[\log\left(\frac{I}{I_0} \right) + \frac{1}{2}\beta_0^2 \right] \tag{10.20}$$

We change the integration variable to

$$x = q\sqrt{K_2} = 2\tau q = \beta_0 q$$

and observe that the integration over x needs to run only over positive values:

$$P(u, \gamma) = \frac{1}{\pi \beta_0} \int_0^{\infty} dx \, \cos(xu + \gamma x^3) \exp\left(-\tfrac{1}{2}x^2\right) \tag{10.21}$$

The distribution of $\mathbb{H}$ would be Gaussian if γ were zero since

$$P(u, 0) = \frac{1}{\pi \beta_0} \int_0^{\infty} dx \, \cos(xu) \exp\left(-\tfrac{1}{2}x^2\right) = \frac{1}{\beta_0 \sqrt{\pi}} \exp\left(-\tfrac{1}{2}u^2\right)$$

which represents a log-normal distribution of irradiance.

The *skewness parameter* γ in (10.21) is clearly an indicator of how far the distribution departs from log-normal. One finds that even small values of γ can generate significant modifications of the PDF. The following relationship defines it:

$$\gamma = \frac{K_3}{6(K_2)^{\frac{3}{2}}} = \frac{K_3}{48\tau^3} \tag{10.22}$$

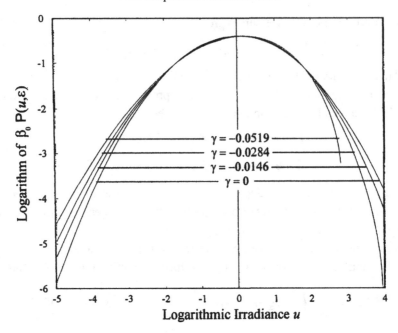

Figure 10.4: The PDF of the logarithmic irradiance predicted by the second Rytov approximation for three negative values of the skewness parameter. The PDF is multiplied by the Rytov deviation β_0 and the independent variable u is identified in (10.20). The log-normal distribution corresponds to $\gamma = 0$.

and, with the expressions (10.19) for K_3, it becomes

$$\text{Plane wave:} \qquad \gamma = -0.232\tau_{\text{pl}}$$
$$\text{Spherical wave:} \qquad \gamma = -0.118\tau_{\text{sph}} \qquad (10.23)$$

The expected values of γ are thus negative. They are also quite small because τ is less than unity for weak scattering. Notice that the initial type of wave influences the resulting distribution both through the appropriate expressions for τ and by virtue of the numerical coefficients that multiply them.

With these preparations we are ready to evaluate the PDF numerically.[4] The PDFs computed from (10.21) are plotted in Figure 10.4 for three values of the skewness parameter. The log-normal distribution predicted by the basic Rytov solution corresponds to $\gamma = 0$ and provides a useful reference. This figure demonstrates that the PDF is significantly modified for large positive and negative values of u even

---

[4] When one calculates $P(u, \gamma)$ for different values of γ, one finds that it goes negative for very large and very small values of irradiance relative to the mean irradiance. These negative values contradict our concept of a PDF. They represent a mathematical artifact that occurs whenever the characteristic function contains terms in the exponent stronger than q^2. The usual procedure is to ignore these regions. They do not cause a problem in our case because the parameter γ is very small and we are dealing with a limited range of positive and negative values of the logarithmic irradiance.

when the skewness parameter is very small. Experimental distributions of irradiance fluctuations should be skewed relative to the log-normal model in a measurable way.

It is significant that there are no adjustable constants in this description. The parameter γ is completely specified by the path length, frequency, C_n^2 and type of wave. The predicted distribution $P(u, \gamma)$ either agrees with experimental data or it does not.

10.2.2 Comparison with Numerical Simulations

We would like to test the predictions of Figure 10.4 against optical and microwave measurements. There are two problems in doing so. The first is that C_n^2 is not known well enough in most experimental situations to make the precise comparison that is required. Moreover, its value is seldom constant along the path. This is a particular problem for long microwave paths. For optical propagation, the Rytov approximation is valid only for short paths for which the Fresnel length $\sqrt{\lambda R}$ can be comparable to the inner scale length ℓ_0. One can easily correct the variance τ^2 for this effect, as we showed in Figure 3.5. On the other hand, the dissipation range of the turbulence spectrum was not included in evaluating the four-fold integrals that define K_3. Comparisons of our prediction with optical data therefore carry an unreasonable uncertainty.

We turn instead to numerical simulations that include all the relevant physics. These "numerical experiments" accurately describe the irradiance distribution for propagation through atmospheric turbulence [24][25]. It is usually assumed that the random medium is homogeneous and isotropic. The three-dimensional random medium is replaced by a series of independent screens that change the phases of passing signals but not their amplitudes. Propagation from one screen to the next is modeled by the parabolic-wave equation which represents small-angle scattering. A Monte Carlo approach is used that varies the properties of the individual screens in a random manner and then assembles the different outcomes. These simulations are ideal for our purpose because they hold C_n^2 and ℓ_0 fixed.

Several simulations for plane and spherical waves were compared with the predictions of Figure 10.4 as part of this development [22]. We illustrate that comparison here with a spherical-wave case. The logarithmic irradiance distributions generated by the simulations depend only on the Rytov variance and the inner scale length [26][27]. We select the simulation for $\beta_0^2 = 4\tau^2 = 0.06$ and $\ell_0 = 0$ which represents weak scattering. The simulated distribution is reproduced in Figure 10.5. The skewness parameter is estimated from (10.23) to be $\gamma = -0.01456$ and the corresponding PDF prediction is also plotted in Figure 10.5. It agrees with the simulation for positive values out to $u = 4$. It also gives good agreement down to $u = -3$,

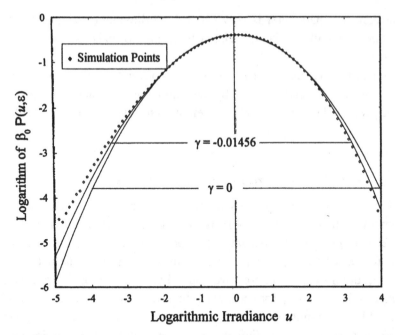

Figure 10.5: Comparison of the predicted distribution of logarithmic irradiance with a spherical-wave numerical simulation for $\beta_0^2 = 0.06$ and $\ell_0 = 0$. The log-normal distribution corresponds to $\gamma = 0$.

but the prediction and simulation separate rapidly below that point. Although the binned simulation points are sparse in the region of negative logarithmic irradiance, the difference is real. To describe it accurately, one must include cumulants beyond the third order in the characteristic function. Nonetheless, it is clear that the prediction provides a significant improvement on the traditional log-normal model. This is all the more remarkable because it contains no adjustable parameters.

10.3 Non-stationary Conditions

In all of our discussions so far we have assumed that C_n^2 is constant during the experiments which measure the amplitude or intensity distribution. On the other hand, we know from direct measurements that C_n^2 does change with time. This variability distorts the signal-level distribution and should influence how we interpret such data.

If a weather front or cloud moves through the propagation path, a sudden meteorological change can occur in C_n^2. This change is soon reflected in the logarithmic amplitude, as suggested by Figure 10.6. The amplitude variance τ^2 is a useful surrogate for C_n^2 because it contains also the relevant information on wavelength and transmission distance. Since τ sets the slope of the cumulative-probability curve for a log-normal distribution, one would expect to see the experimental points begin to

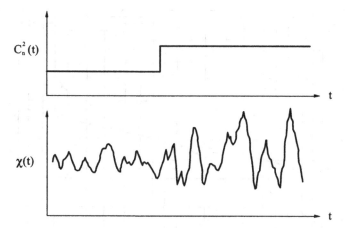

Figure 10.6: A sudden change in C_n^2 and the amplitude fluctuations induced by it when a weather front moves through the propagation path.

fall on a different straight line soon after the change in C_n^2. Interpretation of such data becomes quite confused if the shift is not noticed. The distribution of amplitude fluctuations depends on a time-varying parameter $\tau(t)$ in these situations and can be expressed as the *sample-averaged probability density function*:

$$P(A)\,dA = \frac{1}{\sqrt{2\pi}}\,\frac{dA}{A}\,\frac{1}{T}\int_0^T dt\,\frac{1}{\tau(t)}\exp\left\{-\frac{1}{2\tau^2(t)}\left[\log\left(\frac{A}{A_0}\right)\right]^2\right\} \qquad (10.24)$$

When a sudden change occurs midway through the measurement period, the sample average can be expressed as the sum of two terms:

$$P(A)\,dA = \frac{1}{\sqrt{2\pi}}\,\frac{dA}{A}\left\{\frac{1}{2\tau_1}\exp\left\{-\frac{1}{2\tau_1^2}\left[\log\left(\frac{A}{A_0}\right)\right]^2\right\}\right.$$
$$\left.+\frac{1}{2\tau_2}\exp\left\{-\frac{1}{2\tau_2^2}\left[\log\left(\frac{A}{A_0}\right)\right]^2\right\}\right\} \qquad (10.25)$$

The cumulative probability is represented by error functions:

$$\mathcal{P}(A<L) = \frac{1}{4}\left\{2+\text{erf}\left[\frac{1}{\tau_1\sqrt{2}}\log\left(\frac{L}{A_0}\right)\right]+\text{erf}\left[\frac{1}{\tau_2\sqrt{2}}\log\left(\frac{L}{A_0}\right)\right]\right\} \qquad (10.26)$$

This result is plotted in Figure 10.7 for two combinations of τ_1 and τ_2, which characterize the situations before and after. A discontinuity in C_n^2 thus distorts the straight-line relationship of Figure 10.1 rather significantly.

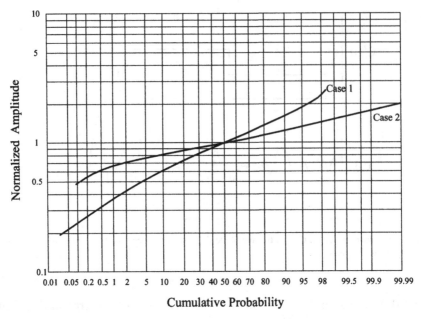

Figure 10.7: Cumulative probability distributions for a propagation path that undergoes a sudden change in C_n^2 midway through the data sample. In Case 1 we have taken $\tau_1 = 0.50$ and $\tau_2 = 0.25$. Case 2 corresponds to $\tau_1 = 0.1$ and $\tau_2 = 0.2$.

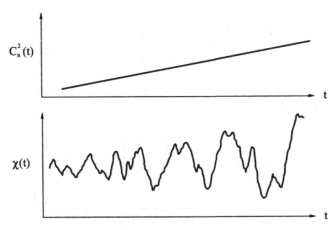

Figure 10.8: A gradual change in amplitude fluctuations caused by increasing levels of turbulent activity along the path.

Meteorological conditions along a propagation path can also change gradually as suggested in Figure 10.8 and this too distorts the measured probability distribution. The level of turbulent activity often increases or decreases linearly with time in these situations:

$$\tau(t) = \tau_0(1 + \eta t)$$

Using (10.24) to describe the sample-averaged PDF, we find that

$$P(A)\,dA = \frac{dA}{A\sqrt{2\pi}}\frac{1}{2\eta T}\left\{\mathrm{Ei}\left\{\frac{1}{2\tau_0^2}\left[\log\left(\frac{A}{A_0}\right)\right]^2\right\}\right.$$
$$\left.-\mathrm{Ei}\left\{\frac{(1+\eta T)^{-1}}{2\tau_0^2}\left[\log\left(\frac{A}{A_0}\right)\right]^2\right\}\right\} \tag{10.27}$$

where $\mathrm{Ei}(x)$ is the exponential integral function.

A rather different situation arises if one measures the distribution over an extended period of time. Communication-service providers are primarily interested in the performance of microwave and satellite links over periods of weeks, months or even years. These operations necessarily encounter a wide variety of turbulence conditions. Direct measurements show that C_n^2 varies in a slow but unpredictable manner. All possible values of τ^2 are encountered over an extended period. In these situations, it is necessary to consider a distribution of C_n^2 that describes the changing level of turbulent activity over a long period of time. We can again use τ as a convenient measure of this variability and describe its distribution by $Q(\tau)$. The log-normal expression (10.4) must now be regarded as a *conditional probability*:

$$P(A)\,dA = dA\int_0^\infty d\tau\, Q(\tau)P(A|\tau) \tag{10.28}$$

The probability function $P(A|\tau)$ means that *if* the amplitude variance has the value τ *then* the PDF for the amplitude of the signal is given by (10.4). The probability density that would be measured in a long survey is therefore represented by the weighted average of the log-normal distribution:

$$P(A)\,dA = \frac{dA}{A\sqrt{2\pi}}\int_0^\infty d\tau\, Q(\tau)\frac{1}{\tau}\exp\left\{-\frac{1}{2\tau^2}\left[\log\left(\frac{A}{A_0}\right)\right]^2\right\} \tag{10.29}$$

This convolution seems to describe long-term measurements made on terrestrial links if reasonable assumptions are made for $Q(\tau)$ [28]. It also characterizes the scintillation imposed on microwave signals transmitted by synchronous communication satellites [29]. If one measures the amplitude distribution of the signal, one should be able to invert this integral equation and establish the long-term distribution $Q(\tau)$ from communication-performance data.

10.4 Intermittency

The situation is complicated still further if one examines propagation data over very short time scales. We learn from such measurements that C_n^2 and other turbulence parameters are changing continuously. We introduced the concept of intermittency

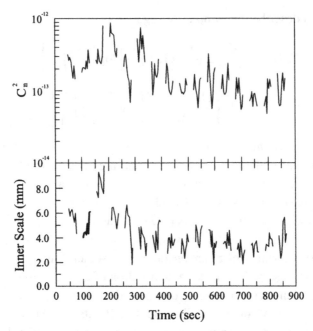

Figure 10.9: Measurements of C_n^2 and ℓ_0 made on a 50-m path by Frehlich [30] with two laser signals monitored by an array of photo diodes. These records were extracted from 4-s data samples and illustrate intermittent behavior near the surface.

in Volume 1 by noting that the defining properties of atmospheric turbulence appear to change rapidly. Electromagnetic and meteorological measurements show that this variation is random and can occur quite suddenly.

This phenomenon has been investigated experimentally with considerable precision [30]. Two laser signals and an array of photo diodes were used to measure the spatial covariance on a short path with short data samples. Nearly instantaneous values for C_n^2 and ℓ_0 could be extracted from this data. Typical time series for them are reproduced in Figure 10.9, which shows that random changes occur in both quantities. Pulsations of C_n^2 were also evident in high-definition airborne temperature measurements made at 400 m on a clear day, thereby demonstrating that intermittency is not limited to the surface [31]. These fluctuations are different than the diurnal patterns and the synoptic variations of C_n^2 and ℓ_0 discussed in Section 2.3.2 of Volume 1. The physics of these spontaneous events is not understood.

Transient structures can cause rapid changes in refractivity when they move through a propagation path. This implies that the values of C_n^2 and ℓ_0 can change substantially during the course of an experiment. A statistical distribution of the turbulence parameters is therefore required in order to characterize the short-term behavior of electromagnetic signals. Since we have already averaged over the ensemble of atmospheric configurations by expressing the spatial covariance of $\delta\varepsilon$ in terms of the spectrum of irregularities in (2.35) and (2.45), we need to understand how these fluctuations relate to our previous descriptions.

Random pulsations of temperature are manifestations of intermittency and are an integral property of the turbulent flow field. They can be regarded as events of enhanced turbulence. For the velocity field, they represent bursts of kinetic energy occurring at random times and locations throughout the fluid. This behavior is believed to be a natural consequence of the Navier–Stokes equation of hydrodynamics and is overlaid on the familiar process of subdivision of eddies. No one has yet predicted this behavior from first principles. The velocity bursts also cause pulsations and concentrations of passive scalars. As a result, the spectrum and other second-order statistics of the refractive index can change in a few seconds over spatial scales of approximately 50 m.

The basic concept is that turbulence itself gives rise to bursts of energy at both large and small scale sizes [32][33]. Intermittent structures can occur over a continuous range of scale sizes, but they are often divided into two general classes. *Microscale intermittency* refers to bursts that occur inside individual coherent eddies but propagation experiments usually average out their effects. *Global intermittency* is the term applied to pulsations whose scales are larger than those of the main coherent eddies. Its influence is important for short-path measurements that are often used to ensure that conditions of weak scattering pertain and for short data samples, which are chosen in order to avoid problems with nonstationarity. In the experiment reflected in Figure 10.9 Frehlich intentionally chose $R = 50$ m and $T = 4$ s in order to expose burst-like behavior. The influence of global intermittency is reduced roughly by the ratio ℓ/R and in most propagation experiments the influence of intermittency is masked because much longer path lengths and averaging times are used.

Thermal plumes can occur spontaneously anywhere in an otherwise uniform turbulent atmosphere. They are best represented by cloud-like structures whose horizontal dimensions are measured in tens or even hundreds of meters. These buoyant thermal plumes can extend to considerable heights and are often felt by low-flying aircraft. They can dominate a propagation experiment while they fill the path. Once they have been created, these cells can move across the path as suggested in Figure 10.10. In doing so, they can change C_n^2 and ℓ_0 in a matter of a few seconds.

The reality of intermittent structures causes two problems for the description of short-path propagation. The path length need not be large relative to the size of the influential eddies and one must question the convenient assumption $\kappa R \gg 1$ that is often used. When one or two large intermittent structures dominate the propagation, one cannot invoke the central-limit theorem to argue that the scattered field components should be GRVs. The challenge is to understand how these

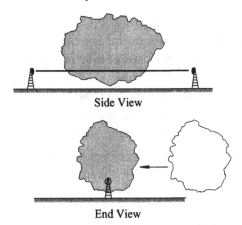

Side View

End View

Figure 10.10: An intermittent structure is shown moving through an electromagnetic path. These structures are often identified with thermal plumes or temperature ramps. They are associated with columns of water vapor at microwave frequencies.

changes are reflected in the distributions of signals that pass through the lower atmosphere.

The first studies of intermittency effects were focused on back-scattering experiments and tropospheric forward-scatter communication systems [31][34]. The scattered power is proportional to a weighted integral of C_n^2 taken over the scattering volume defined by the antenna patterns and range gate.[5] If one accepts the notion that C_n^2 varies between small and large values in a random manner, one would expect to measure large returns occasionally. In fact, bursts of enhanced signal strength are observed under clear atmospheric conditions and they are called *angels* or *point reflections*. It was initially hoped that one could invert the Rayleigh distribution of the scattered power to establish the short-term distribution of C_n^2 at various altitudes. This process is complicated by the rapid fading generated by interference among waves scattered from different parts of the scattering volume.

The next applications concentrated on image quality of optical transmissions near the surface [34][35]. The mutual coherence function was emphasized because it best describes this feature. The first studies focused on intermittent changes of C_n^2 and ignored variations in ℓ_0. We developed a general expression for the mutual coherence function in (9.54). For a path near the surface we can set $\vartheta = 0$ and $z = x$ in that expression to find

$$\Gamma(\rho) = I_0 \exp\left(-4\pi^2 k^2 \int_0^\infty dz\, C_n^2(z) \int_0^\infty d\kappa\, \kappa\, \Omega(\kappa)[1 - J_0(\kappa\rho)]\right) \quad (10.30)$$

---

[5] See the discussion on pages 64–70 of Volume 1.

The Kolmogorov model yields the familiar expression

$$\Gamma(\rho) = I_0 \exp\left(-2.914\rho^{\frac{5}{3}}k^2 \int_0^\infty dz\, C_n^2(z)\right) \qquad (10.31)$$

This expression must be averaged over the large scales of the intermittency if one acknowledges the random variability of C_n^2:

$$\langle\Gamma(\rho)\rangle_{1s} = I_0\left\langle\exp\left(-2.914\rho^{\frac{5}{3}}k^2 \int_0^\infty dz\, C_n^2(z)\right)\right\rangle_{1s} \qquad (10.32)$$

Jensen's inequality bounds the average of an exponential [36],

$$\langle\exp(-\eta x)\rangle > \exp(-\eta\langle x\rangle) \qquad (10.33)$$

and suggests that the image quality can occasionally be much better than it would be if C_n^2 were constant. This remarkable prediction is confirmed by the sudden enhancements of image quality that are observed on laser links from time to time. Occasional bursts of smaller-than-normal values for C_n^2 can increase the mutual coherence function very substantially because of the exponential dependence. Tatarskii developed this theme using a gamma distribution[6] for C_n^2 and elaborated on the predicted image enhancement [34].

The next approach to intermittency was empirical. Churnside and Frehlich measured irradiance distributions directly with a laser signal transmitted over a 50-m path using 60-s data samples [37]. The amplitude variance should be small on such a link and one would expect the log-normal distribution to describe such data. It was therefore surprising that the measured points fell on a curve that is both narrower and higher than the predicted distribution. It was noted that bursts of strong turbulence can produce instantaneous irradiance values with relatively large deviations from the mean. To investigate this possibility, the variance was estimated for each half second of data. The results showed that there was a wide distribution of instantaneous values. The PDF data was then broken into three groups corresponding to

$$\beta_0^2 = 4\tau^2 = 10^{-4}, 1.2 \times 10^{-3} \text{ and } 7.2 \times 10^{-3}$$

The resulting distributions are reproduced in Figure 10.11 and are in good agreement with individual log normal models that correspond to these values.

---

[6] If one defines

$$\xi = \int_0^\infty dz\, C_n^2(z)$$

this model is expressed as

$$P(\xi) = \frac{1}{\Gamma(\alpha)}\beta^\alpha \xi^{\alpha-1} \exp(-\beta\xi)$$

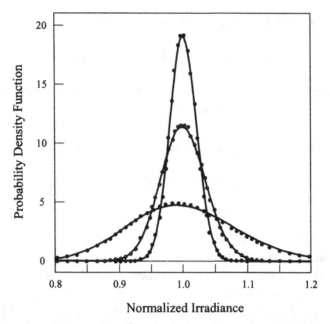

Figure 10.11: Comparison of log-normal irradiance distributions with experimental PDF data taken on a 50-m laser link by Churnside and Frehlich [37]. The data samples were grouped into three sets that had the same instantaneous values of rms irradiance. The solid theoretical curves were plotted using these short-term values and the agreement with the experimental data is striking.

The PDF for short-term variations of τ was measured in this way and found to be log-normal:

$$S(\tau)\, d\tau = \frac{d\tau}{\eta \sqrt{2\pi}} \exp\left\{ -\frac{1}{2\eta^2}\left[\log\left(\frac{\tau}{\tau_0}\right)\right]^2 \right\} \qquad (10.34)$$

The irradiance distribution should be a convolution of this result and the original distribution (10.5), which we now regard as a conditional probability:

$$P(I)\, dI = \frac{dI}{2I\sqrt{2\pi}} \int_0^\infty d\tau \, S(\tau) \frac{1}{\tau} \exp\left\{ -\frac{1}{8\tau^2}\left[\log\left(\frac{I}{I_0}\right)\right]^2 \right\} \qquad (10.35)$$

This integration was done numerically and Figure 10.12 compares the result with data measured in a 30-min data set. The agreement is satisfactory. A log-normal distribution of short-term values for C_n^2 therefore provides a good description of intermittent phenomenon.

Recent studies by Frehlich have been based primarily on such experiments [30][38]. He included the intermittent behavior of ℓ_0 as well as C_n^2 since both are important parameters in establishing τ^2 for short links. He found that C_n^2 and ℓ_0 are each distributed as log-normal variables. He assumed the validity of a joint

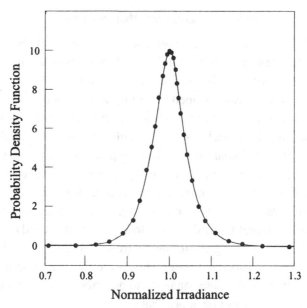

Figure 10.12: The distribution of irradiance variations measured with a laser signal on a 50-m path by Churnside and Frehlich [37]. The data is compared with a log-normal distribution computed from short-term measurements of the irradiance intensity.

log-normal distribution for these parameters and found reasonable agreement with measurements if their cross correlation is approximately 0.6. More recently, he has used numerical simulations to predict the mutual coherence functions and the irradiance power spectrum for this model. He concludes that one can ignore intermittency if the path length is more than ten times the correlation scale or roughly 200 m.

Intermittent structures therefore can have a profound impact on the interpretation of intensity-distribution data. Such measurements clearly depend on the path length and on the way in which the data is processed. When the path is relatively long, the influence of these structures is averaged with all the other irregularities and the distribution is log-normal. If the path is relatively short and the averaging time small, a large cell can act like a refractive lens and produce spikes in the signal that dominate the distribution. In a very real sense, what one measures depends on how one averages the data.

It is becoming clear that a wide variety of turbulence phenomena are at work in the lower atmosphere. The Kolmogorov model with a constant value for C_n^2 characterizes only a portion of the relevant physics. The practical problem is to distinguish modifications of the PDF caused by intermittent structures from the nonstationary effects discussed previously. The interpretation of such experiments places one at the awkward boundary between large-scale meteorology and small-scale turbulence. This is a regime for which our present understanding is not adequate.

10.5 Astronomical Observations

Intensity scintillations are readily observed during astronomical observations made with ground-based telescopes. A typical time history of these fluctuations is reproduced in Figure 3.19 for five different angles of elevation. The short-term distribution of intensity values was assumed to be log-normal on the basis of results of short-path optical experiments performed close to the surface. It was reasoned that inclined propagation paths to stellar sources would rapidly pass out of the region of strong turbulence in the boundary layer. The long portions of these paths lie primarily in the upper troposphere where C_n^2 is greatly reduced relative to surface values.[7] The importance of these higher regions is magnified by the altitude-weighting term $z^{\frac{5}{6}}$ in the amplitude-variance expression (3.119). Of course, sources near the horizon involve paths that travel a long distance close to the surface and these signals are usually characterized by strong scattering. Astronomers try to avoid these source positions for the same reason that observatories are invariably built on mountain tops. The result is that weak scattering tends to describe most observations and the log-normal distribution should be appropriate.

This hypothesis proved difficult to verify with astronomical telescopes for several reasons. It was recognized that such experiments should be done with small openings and a narrow-band optical filter to avoid suppressing large signals. These conditions exceeded the dynamic range and sensitivity capabilities of film and early photomultiplier tubes for the faint optical signals that would be available. This situation changed dramatically with the development of photon counting and correlation techniques based on the use of sensitive photomultiplier tubes and digital signal processing. They provided a sensitivity near the theoretical limit and made it possible to measure the distribution of astronomical scintillation very accurately.

This technique was first exploited using Sirius and Vega as sources [39]. It provided 2.5×10^4 photons per second with an aperture diameter of only 0.84 cm. The measured distributions were decidedly non-Gaussian for samples of 1–2 min. This feature had not been observed previously, probably because of slow response times and aperture averaging.

The same technology was next used to measure the first five moments of the intensity distribution [40]. A diaphragm diameter of only 9 mm and narrow-band optical filters were employed. The measured moments were plotted versus the second moment $\langle I^2 \rangle$ and compared both with the log-normal model and with the K distribution. The data agreed with the log-normal moment predictions

$$\langle I^n \rangle = \frac{\langle I^2 \rangle^{\frac{1}{2}n(n-1)}}{\langle I \rangle^{2-n}} \tag{10.36}$$

---

[7] See the profile measurments reproduced in Figure 2.18 of Volume 1.

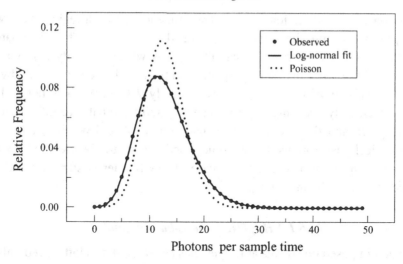

Figure 10.13: The distribution of scintillation levels measured by Dravins, Lindegren, Mezey and Young with Vega at $\vartheta = 31°$ using a 5-cm aperture [41]. The solid curve is a log-normal distribution. The data circles represent weak scattering and differ from the log-normal curve by less than 2%. ©1997, Astronomical Society of the Pacific; reproduced with permission.

when

$$\langle I^2 \rangle < 1.65 \langle I \rangle^2$$

This corresponds to weak scattering, for which the Rytov approximation should be valid. It is significant that the log-normal distribution is not uniquely determined by its moments, as noted in Appendix E.

This problem was addressed recently using the modern facilities at the La Palma observatory [41]. The PDF itself was measured, which is more definitive than the moments of the distribution. The opening was 5 cm and 100-s samples were used. A typical measurement is reproduced in Figure 10.13, where the data points are compared with a log-normal and a Poisson distribution. The latter corresponds to pure photon noise. The dimensionless intensity variance was

$$\sigma_I^2 = \frac{\langle I^2 \rangle - \langle I \rangle^2}{\langle I \rangle^2} = 0.058$$

and therefore within the range of validity for the Rytov approximation for this run. The agreement between the data points and the log-normal model is quite good, with differences of less than 2% over the entire range. The reader who wishes to examine this problem in greater depth should consult [41], which contains an excellent discussion and references to recent research.

10.6 Satellite Signals

The situation is considerably different for microwave signals that go to and from earth-orbiting satellites. These signals can be affected by irregularities both in the troposphere and in the ionosphere. The influences of both regions vary with wavelength, but in opposite ways as we learned in Chapter 3. There is a natural frequency division between those satellite signals whose intensity distributions are influenced primarily by the troposphere and those influenced primarily by the ionosphere. This division occurs at approximately 2 GHz but it is not a sharp boundary. Generally speaking, high-frequency satellite signals are influenced by the troposphere and their short-sample distributions should be similar to those of astronomical observations. On the other hand, we should be prepared for new physics and a different outcome in the case of low-frequency signals, which are influenced primarily by the ionosphere.

10.6.1 VHF, UHF and L-band Signals

Scintillation imposed on signals at frequencies below 2 GHz is influenced only by fluctuations in electron density in the ionosphere. Scintillation can enhance satellite signals by 6–8 dB, and can also bring them down to the level of the noise background for the sky. Amplitude variations created in the troposphere are inconsequential for several reasons. Molecular absorption by gaseous oxygen and water vapor is not important in this range, as the laboratory measurements reproduced in Figure 8.2 demonstrate. Scattering by rainfall can also be ignored because these wavelengths are much greater than the range of drop sizes. Scintillation imposed by refractive-index irregularities is the only source of significant modulation. The absolute value of tropospheric scintillation caused by refractive-index irregularities was measured at 1046 MHz on a 16-km path near the surface [42] and found to be quite modest:

$$0.01 < \langle \chi^2 \rangle < 0.10$$

Scintillation can be completely ignored at lower frequencies because the amplitude variance scales as $f^{\frac{7}{6}}$. This means that we can focus on the ionosphere as the source of amplitude distributions measured at these frequencies.

10.6.1.1 Transionospheric Signals

We return to the generic vector voltage diagram in Figure 9.1 which describes the complex signal received after propagation through a random medium. The mean field denoted by $\langle E_0 \rangle$ in that figure is the coherent component of the original field strength that reaches the ground station after passing through the ionosphere. Its polarization is changed by Faraday rotation as it passes through the F region, but that is not our concern because most receivers sense both polarization components. The original signal's amplitude is also decreased by deviative absorption in the E region but this is a trivial effect at the microwave frequencies we are examining. Our interest therefore centers on attenuation of the original signal caused by out-scattering from electron-density fluctuations. This is described quite generally by (9.7), which we repeat here:

$$\langle E \rangle = E_0 \exp\left(-\tfrac{1}{2}\langle \varphi_0^2 \rangle\right) \qquad (10.37)$$

If we knew the phase variance for transionospheric propagation, we could estimate the coherent component of the field strength shown in Figure 9.1. We discussed this question in Section 4.2.2 of Volume 1 and established an analytical expression for $\langle \varphi_0^2 \rangle$. If the irregularities are confined within an elevated layer of thickness ΔH, we found[8]

Ionosphere: $\qquad \langle \varphi_0^2 \rangle = \dfrac{\pi}{4} \lambda^2 r_{\mathrm{e}}^2 \langle \delta N^2 \rangle \, \Delta H \, \sec \vartheta \, L_0 \, \mathrm{cosec}\, \gamma \qquad (10.38)$

This shows that the mean field should become stronger as the frequency increases. Unfortunately, our knowledge of the profiles for L_0 and $\langle \delta N^2 \rangle$ is not accurate enough for us to estimate the phase variance with confidence. We therefore rely on experimental data from the DNA Wideband Satellite program to estimate the phase variance [43]. Simultaneous distributions of phase fluctuations were measured at 138, 380 and 1239 MHz. Those measurements gave rms values of approximately 180° at 138 MHz, 90° at 380 MHz, and 20° at 1239 MHz. These values scale linearly with wavelength and indicate that a strong coherent signal reaches the ground stations according to (10.37). The original signal is altered very little at L band and an appreciable mean field strength should be available even for VHF frequencies.[9] The situation for transionospheric propagation is therefore similar to the microwave example illustrated in the first panel of Figure 9.1.

The coherent field-strength component $\langle E \rangle$ provides a sturdy phase and amplitude reference since the original signal is not heavily attenuated by out-scattering during its passage through the ionosphere. If $\mathcal{G}$ is the receiver gain the mean field induces a constant voltage A_0 in the ground receiver described by

$$A_0 = E_0 \sqrt{\mathcal{G}} \exp\left(-\tfrac{1}{2}\langle \varphi_0^2 \rangle\right) \qquad (10.39)$$

This is used as the foundation for the vector voltage diagram of Figure 10.14 which is specific to transionospheric transmission. To complete that description we must next address the scattered field in E_s indicated in Figure 9.1. It is composed of all fields that are scattered to the receiver by electron-density fluctuations that are illuminated by the arriving signal. This incoherent field induces stochastic voltage components u and v in the receiver that are in phase and out of phase with respect

---

[8] This expression is based on geometrical optics with a representative spectral index of $\nu = 4$. Here L_0 is the effective outer scale length, r_{e} is the classical electron radius, ϑ is the zenith angle and γ is the angle between the magnetic field and line-of-sight vectors.

[9] This situation is quite different than that for optical signals traveling near the surface. The reference signal is completely destroyed by out-scattering in that case, as we learned in Section 9.1.2.

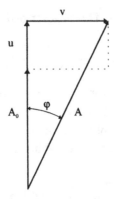

Figure 10.14: A vector voltage diagram showing a reference voltage generated by the coherent mean field and the orthogonal signal components u and v induced by the incoherent scattered field.

to the reference voltage. The measured amplitude and phase are related to the three voltage components in Figure 10.14 by

$$A \cos \varphi = u + A_0 \quad \text{and} \quad A \sin \varphi = v \qquad (10.40)$$

We should be able to predict the distributions for the composite amplitude A and phase φ if we know the distributions of u and v.

In developing the Rytov approximation to various orders we found that the surrogate function Ψ in the exponential field-strength expression (2.2) can be related to terms in the Born series by the algebraic expressions (2.83). Because the measured phase shift for transionospheric propagation is modest relative to those for terrestrial links, we can use the Born series to represent the field strength:

$$E = E_0(1 + B_1 + B_2 + B_3 + B_4 + \cdots) \qquad (10.41)$$

On substituting from (2.18) for B_1 and from (2.60) for B_2 we can express the orthogonal voltage components as follows:

$$u + iv = \sqrt{\mathcal{G}} \left(-k^2 \int d^3r \, G(\mathbf{R}, \mathbf{r}) \, \delta\varepsilon(\mathbf{r}) \, E_0(\mathbf{r}) + k^4 \int d^3r_1 \, G(\mathbf{R}, \mathbf{r}_1) \, \delta\varepsilon(\mathbf{r}_1) \right.$$
$$\left. \times \int d^3r_2 \, G(\mathbf{r}_1, \mathbf{r}_2) \, \delta\varepsilon(\mathbf{r}_2) \, E_0(\mathbf{r}_2) \right) \qquad (10.42)$$

We discussed the first term in Section 10.1 and concluded that it is likely to be a GRV even if $\delta\varepsilon$ itself is not Gaussian. The second term describes double scattering and is determined by the product of two $\delta\varepsilon$ measured at different locations. We learned earlier that this product is not a GRV even if the $\delta\varepsilon$ are.[10] Fortunately the

---

[10] See Problem 3 in Chapter 2 of Volume 1.

volume integrations operate twice on this product and the central-limit theorem again shows that this contribution to u and v is nearly normal.

This line of reasoning suggests that u and v can be approximated as GRVs. This is our first approximation and involves some uncertainty, but we will follow it through to examine the consequences. The average values of u and v vanish because $\delta\varepsilon$ is chosen to have zero mean and their joint PDF can be written as

$$W(u,v)\,du\,dv = \frac{du\,dv}{2\pi\sigma_u\sigma_v\sqrt{(1-v^2)}}\exp\left[-\frac{1}{2(1-v^2)}\left(\frac{u^2}{\sigma_u^2}+\frac{v^2}{\sigma_v^2}-2v\frac{uv}{\sigma_u\sigma_v}\right)\right]$$

(10.43)

where the cross correlation of the signal components is defined by

$$v = \frac{\langle uv\rangle}{\sigma_u\sigma_v}$$

(10.44)

Notice that these voltage components are proportional to the dimensionless phase and logarithmic amplitude fluctuations:

$$\sigma_u^2 = A_0^2\langle\chi^2\rangle \quad \text{and} \quad \sigma_v^2 = A_0^2\langle\varphi^2\rangle$$

(10.45)

One is not free to assume values for these parameters. Rather, they must be estimated for the propagation path and turbulent conditions appropriate to each physical situation [44].

The orthogonal signal components u and v are seldom measured directly. Experiments usually focus on the amplitude or intensity of the signal. One can transform the signal components to the phase and amplitude using the relationships (10.40) and the following joint PDF emerges:

$$P(A,\varphi)\,dA\,d\varphi = \frac{A\,dA\,d\varphi}{2\pi\sigma_u\sigma_v\sqrt{1-v^2}}\exp\left(-\frac{Q(A,\varphi)}{2(1-v^2)}\right)$$

(10.46)

where

$$Q(A,\varphi) = A^2\left(\frac{\cos^2\varphi}{\sigma_u^2}+\frac{\sin^2\varphi}{\sigma_v^2}-2v\frac{\cos\varphi\sin\varphi}{\sigma_u\sigma_v}\right)$$

$$-2AA_0\left(\frac{\cos\varphi}{\sigma_u^2}-v\frac{\sin\varphi}{\sigma_u\sigma_v}\right)+\frac{A_0^2}{\sigma_u^2}$$

(10.47)

One can calculate the distribution of amplitude fluctuations by integrating this result over all possible values of the phase angle. Sadly, that integration cannot be done analytically and one is compelled to (a) study special cases, (b) use numerical methods, or (c) depend on experimental determinations.

10.6.1.2 The Rice Distribution

The phase integration can be done analytically for one important class of propagation problems. They correspond to Fraunhofer scattering for which the Fresnel length is greater than the influential irregularities. This means that the Fresnel length is comparable to the outer scale length or much greater. The variances of the orthogonal signal components are equal in this regime and the cross correlation vanishes:

$$\text{Fraunhofer:} \qquad \sigma_u = \sigma_v = \sigma \qquad \text{and} \qquad \nu = 0 \qquad (10.48)$$

With these relationships the joint phase and amplitude distribution simplifies rather considerably:

$$P(A, \varphi) \, dA \, d\varphi = \frac{A \, dA \, d\varphi}{2\pi \sigma^2} \exp\left(-\frac{1}{2\sigma^2}(A^2 - 2AA_0 \cos\varphi + A_0^2)\right) \qquad (10.49)$$

One can now integrate over all possible values of the phase and express the amplitude distribution in terms of a modified Bessel function:

$$P(A) \, dA = \frac{A \, dA}{\sigma^2} I_0\left(\frac{AA_0}{\sigma^2}\right) \exp\left(-\frac{A^2 + A_0^2}{2\sigma^2}\right) \qquad (10.50)$$

This is the *Rice distribution*, which was first used to describe noisy signals in electric circuits [45].

The measured quantity is usually the cumulative probability distribution which gives the probability that the amplitude of the signal is less than or equal to a prescribed level L:

$$P(A \leq L) = \frac{1}{\sigma^2} \int_0^L dA \, A I_0\left(\frac{AA_0}{\sigma^2}\right) \exp\left(-\frac{A^2 + A_0^2}{2\sigma^2}\right) \qquad (10.51)$$

With a change of variable this can be written as

$$P(A \leq L) = \frac{1}{\tau^2} \int_0^{L/A_0} du \, u I_0\left(\frac{u}{\tau^2}\right) \exp\left(-\frac{1 + u^2}{2\tau^2}\right) \qquad (10.52)$$

where the parameter

$$\tau = \sigma/A_0 = \chi_{\text{rms}} \qquad (10.53)$$

describes the relative strength of the scattered and mean field strengths. The integral (10.52) cannot be done analytically but is easily evaluated numerically [46]. The result is plotted in Figure 10.15 for several values of τ. These curves are quite different than the straight-line probabilities plotted in Figure 10.1 and it should be easy to identify that difference in experimental data.

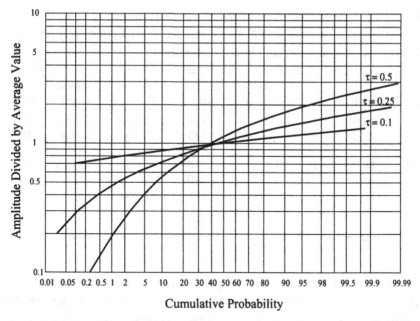

Figure 10.15: The cumulative probability for the Rice distribution. The parameter τ describes the relative strength of the scattered wave compared with the mean field.

The Rice distribution can be further simplified when the scattering is weak. If the signal-to-noise ratio is large, the asymptotic expansion for the modified Bessel function,

$$\lim_{z \to \infty} [I_0(z)] \simeq \frac{1}{\sqrt{2\pi z}} \exp z$$

can be used to show that the amplitude distribution is

Weak scattering: $\quad P(A)\, dA \simeq \dfrac{dA}{\sigma \sqrt{2\pi}} \sqrt{\dfrac{A}{A_0}} \exp\left(-\dfrac{1}{2\sigma^2}(A - A_0)^2\right) \quad (10.54)$

This indicates that the amplitude must be quite close to the reference value when A_0 is considerably larger than σ. Moreover, the ratio A/A_0 is nearly unity in this case and the distribution is essentially Gaussian.

10.6.1.3 Experimental Indications

In an early radio-astronomy experiment 53- and 108-MHz signals from Cygnus-A were used to measure the distribution of intensity fluctuations [47]. The Fresnel lengths for these signals are greater than 1 km for an overhead source and increase

Table 10.3: *Satellite measurements of the distribution of scintillation levels imposed on VHF, UHF and L-band signals by the ionosphere; N is the number of observing ground stations*

Year	Satellite	N	Elevation (degrees)	Frequency (MHz)	Duration	Sample length	Ref.
1971	ATS-3	1	37	138	2 years	15 min	48
1976	ATS-5	1	31	138	3 days	10 min	49
				412			
1977	NNSS	1	0–90	150	2 years	1 min	50
				400			
1980	DNA	4	0–90	138	1 year	20 s	43
				400			
				1239			

with zenith angle. These values are comparable to the outer scale length for the F region[11] and the assumed relationships (10.48) are reasonable. The measured scintillations were consistent with the Rice distribution. That gives one some confidence in the foregoing model for low-VHF signals.

The ability to place artificial satellites in earth orbit after 1957 provided a strong motivation to understand the distribution of scintillation levels. More importantly, spacecraft provided the practical means to measure it accurately. The relatively strong, well-defined signals transmitted by spacecraft beacons and transponders provide an ideal source for scintillation experiments. Operational signals from navigation satellites presented an unexpected opportunity to measure the intensity distribution. In addition, special-purpose satellites were placed in synchronous and other high earth orbits to facilitate accurate, steady measurements. A summary of early satellite experiments that measured the distribution of scintillation levels is provided in Table 10.3.

The first survey covered a two-year period and observed the 138-MHz signal from the ATS-3 synchronous satellite signal at a facility near Boston [48]. The corresponding Fresnel lengths were not much different than those in the radio-astronomy experiment mentioned above. Intensity distributions constructed from 15-min data samples were similar to the curves plotted in Figure 10.15 and were clearly *not log-normal*. When the ATS-5 synchronous satellite became available, its 138- and 412-MHz signals were also observed near Boston during daylight hours [49]. Each 10-min data sample was chosen to ensure that conditions of weak scattering and long periods of stationarity pertained. A very large number of data

---

[11] See Section 2.4.3 of Volume 1.

samples was used to construct intensity PDFs for both signals and the results favored a Gaussian model over the log-normal distribution – but by a small margin.

The availability of phase-coherent signals at 150 and 400 MHz radiated by the U. S. Navy Navigational Systems satellites in 1100-km orbits provided another important opportunity to measure the amplitude distribution. The 400-MHz signal was monitored for two years by an 84-ft dish at the Millstone Hill radar facility, while the 150-MHz signal was observed with an eleven-element Yagi antenna [50]. Observations were made for almost 2400 satellite passes, ensuring that a wide variety of zenith angles and ionospheric conditions was encountered. The Nakagami-*m* distribution provided a reasonable – but not perfect – description for the data taken at 400 MHz. We have not yet discussed this model but will soon do so.

The fourth experiment was the most extensive and revealing. Because the influence of ionospheric irregularities is important for the operation of military communication satellites, a special-purpose satellite was developed to investigate the scintillations they generate. The DNA Wideband Satellite was launched into a high-inclination 1000-km circular orbit, and provided periodic access to ground stations at virtually all latitudes.[12] The spacecraft radiated ten mutually coherent signals at frequencies ranging from 138 to 2891 MHz. The S-band signal provided an undisturbed phase reference for the other signals. The two quadrature components of the received signals could be extracted in this way. The signals were sampled a hundred times each second and used to construct PDFs for phase and intensity at each frequency and ground station. The intensity measurements were compared with four models: log-normal, Nakagami-*m*, a generalized Gaussian and a two-component model. The chi-squared test was used in 83 cases to rank the models. The authors noted "the consistency with which the Nakagami-*m* distribution provides a good fit to intensity data" and it emerged as the preferred description of ionospheric scintillation [43].

10.6.1.4 The Nakagami-m Distribution

The *Nakagami-m distribution* was discovered by analyzing the fading of short-wave radio signals reflected in the ionosphere [51]. We now know that it also describes a wider class of electromagnetic signals that are modulated randomly by transmission or reflection. In terms of signal intensity it can be written as

$$\mathsf{P}(I)\,dI = dI\,\frac{m^m I^{m-1}}{\Gamma(m)\langle I\rangle^m}\exp\!\left(-m\,\frac{I}{\langle I\rangle}\right) \qquad (10.55)$$

---

[12] The initial ground stations were located at Poker Flats in Alaska, Ancon in Peru, Stanford in California and Kwajalein in the Marshall Islands. The program was later extended with additonal equatorial receivers at Ascension Island and Manila in the Phillipines.

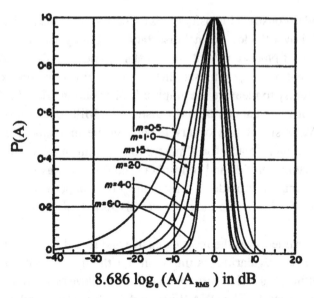

Figure 10.16: The PDF of the Nakagami-*m* distribution for various values of the parameter *m*. The curves are each normalized so that the maximum likelihood corresponds to unit probability [51].

where $\langle I \rangle$ is the average intensity or mean irradiance. The only adjustable parameter is *m* and it can assume noninteger values so long as $m > 0.5$. This parameter is related to the first and second moments of intensity by

$$m^{-1} = \frac{\langle (I - \langle I \rangle)^2 \rangle}{\langle I \rangle^2} \tag{10.56}$$

and is recognized as the reciprocal of the scintillation index introduced in (9.78) to measure the scintillation level:

$$m = 1/S_4 \tag{10.57}$$

The lone parameter is therefore determined by the level of fluctuations. The Nakagami model can also be expressed in terms of the amplitude of the signal.[13]

$$P(A)\,dA = dA \, \frac{2m^m A^{2m-1}}{\Gamma(m)\langle A^2 \rangle^m} \exp\left(-m \frac{A^2}{\langle A^2 \rangle}\right) \tag{10.58}$$

This version of the PDF is plotted in Figure 10.16 for various values of *m*. We see there that the range of likely amplitude values narrows as this parameter increases.

It is interesting that the Nakagami-*m* distribution reduces to a Gaussian form when the scattering is weak. The scintillation index is small in this case and the

---

[13] Notice that $m = 1$ corresponds to a Rayleigh distribution and $m = \frac{1}{2}$ is the one-sided Gaussian distribution.

value of m is large. We rearrange the terms in (10.58) as

$$P(A)\,dA = \frac{2\,dA}{A}\left(\frac{m^m}{\Gamma(m)}\right)\exp\left\{-m\left[\frac{A^2}{\langle A^2\rangle} - \log\left(\frac{A^2}{\langle A^2\rangle}\right)\right]\right\}$$

From Figure 10.16 we see that A^2 must be quite similar to $\langle A^2\rangle$ when m is large. If we write A in terms of the logarithmic amplitude,

$$A = A_0(1 + \chi)$$

we have to first order

$$A^2 = A_0(1 + 2\chi) \quad \text{and} \quad \langle A^2\rangle = A_0^2$$

This allows one to rewrite the exponent in terms of the difference in amplitude:

$$P(A)\,dA = \frac{2\,dA}{A}\left(\frac{m^m}{\Gamma(m)}\right)\exp\left(-\frac{m}{2}\frac{(A - A_0)^2}{A_0^2}\right)$$

The asymptotic expansion for the gamma function

$$\lim_{m\to\infty}\Gamma(m) = (m - 1)^m\sqrt{\frac{2\pi}{m - 1}}\,\exp(-m + 1)$$

gives

$$m \gg 1 \qquad P(A)\,dA = \frac{2\,dA}{A}\sqrt{\frac{m}{2\pi}}\,\exp\left(-\frac{m}{2}\frac{(A - A_0)^2}{A_0^2}\right)$$

The parameter m can be expressed in terms of A_0 and σ as follows:

$$m = A_0^2/(4\sigma^2) \tag{10.59}$$

In this limit the Nakagami-m distribution becomes

Weak scattering: $\quad P(A)\,dA = \dfrac{dA}{\sigma\sqrt{2\pi}}\left(\dfrac{A_0}{A}\right)\exp\left(-\dfrac{1}{2\sigma^2}(A - A_0)^2\right)$ (10.60)

The amplitude must be very similar to the reference value since $A_0 \gg \sigma$. This means that the ratio A/A_0 is almost unity and the distribution is essentially Gaussian. The result is virtually the same as the weak-scattering limit of the Rice distribution.

It is important to note that the Nakagami-m distribution is a purely phenomenological description. It was developed to fit experimental data. It has no apparent connection to the Rytov approximation and one must look elsewhere to explain this feature of ionospheric scintillation. No one has yet succeeded in deriving it from first principles, in the way that we were led to the Rice distribution. This does not detract from its utility or validity. It is evidently a good description of scintillation levels in the bands we are examining and it should be used to describe the random variations in signal level that are encountered.

10.6.2 Communication Satellites

Communication satellites were introduced in 1962 and rapidly deployed in geosynchronous orbit. Generous allocations of frequency bandwidth were made to civilian and government users at the following pairs of frequencies: (a) 4 and 6 GHz, (b) 7 and 8 GHz, (c) 11 and 14 GHz, and (d) 20 and 30 GHz. The higher frequency in each pair is used for uplink transmissions. The downlink signal is necessarily weak because the power of spacecraft is limited and the transmission distance is great. Signal-strength variations were therefore of concern and experimental programs were undertaken to establish the probability distribution of amplitude fluctuations for each band.

Scintillation imposed by the ionosphere is not very important above 2 GHz and its residual influence rapidly disappears as the frequency increases. The influence of intermittent structures in the troposphere is suppressed by path averaging along the long route between satellite and ground station. The two causes of variation in signal level are attenuation by rain and scattering by clear-air turbulence. Both occur in the troposphere. Early work showed that the 4- and 6-GHz bands are not sensitive to attenuation by rain but exhibit bothersome amplitude fluctuations caused by clear-air turbulence. As the frequency increases, satellite downlink signals are increasingly affected both by rain and by turbulence.

Signal attenuation by rainfall is especially important for the valuable frequencies above 10 GHz and special-purpose satellites were launched to measure this effect. Atmospheric turbulence generates rapid scintillations but rain-induced fluctuations change quite slowly. This difference is visually apparent in amplitude records and in power spectra [52]. Experiments showed that the two components are statistically independent [53]. The rain component of scintillation can be eliminated by filtering out low-frequency components and the residual data stream can then be evaluated like clear-air scintillation.[14] This was confirmed by measurements made in West Germany with the 11.575-GHz signal from the OTS satellite [52] and later using the 12.5-, 20- and 30-GHz signals from Olympus [54]. Similar conclusions were reached in Japan using 11.452- and 14.226-GHz signals from Telsat V [55]. This separability means that scintillations induced by rain need not influence probability-distribution measurements if the data is filtered – as it usually is in such experiments and in operational service.

---

[14] Note that the absolute signal level *is* strongly influenced by absorption by rain, especially at 11 and 20 GHz. This effect cannot be filtered out. It must be compensated by increasing the power of satellite transmitters, narrowing the information bandwidth or going to ground-station diversity.

Scattering of satellite signals by refractive-index irregularities in the troposphere is qualitatively different than it is for terrestrial links. The line of sight is close to the surface for the latter and strong turbulence is experienced along the entire route. In the satellite case, the wave passes quickly through regions of strong turbulence near the surface and the effective path length is quite small. Peak-to-peak amplitude fluctuations seldom exceed 2 dB for elevations above 20° and these levels have a negligible effect on satellite communications. There is little difference among the Rice, log-normal and Nakagami-*m* distributions at these low levels, as we have noted. However, the motivation to extend satellite coverage as far as possible drives one to low angles of elevation. Satellite signals often experience strong scintillation below 10° elevation and this can impair their quality. Its impact depends on the type of modulation employed and is most important for amplitude-modulated voice and television signals. The impact of scintillation on satellite communication systems is explained more fully in [56].

10.6.2.1 Short-sample Distributions

The intensity distribution should be log-normal if the measurements are made while the level of turbulent activity is stationary. This prediction was tested by measuring scintillation imposed on the 20-GHz beacon signal from the Olympus satellite [57]. The elevation of the source was 29.2° from the monitoring station in the UK and weak scattering was the predominant mode. A very large amount of data was gathered and processed for one year: 389 383 samples 1 min long, 12 913 samples 30 min long, and 6438 samples 60 min long. The log-normal distribution was consistently produced by the 1-min data samples under weak-scattering conditions. Distributions estimated from the 30-min samples were somewhat less likely to be log-normal, and the departure was still more pronounced for the 60-min samples. From these observations it was clear that the amplitude distribution is fundamentally different for short and long data samples. The transition is set by the time it takes for significant meteorological changes to occur along the path from the satellite. As a practical matter, this is less than 30 min.

Strong fluctuations were also encountered during this campaign and it is not surprising that intense fluctuations did not fit the log-normal model [57]. The distributions became visibly skewed when the scintillation variance exceeded 0.5 dB. Negative skewness in the PDF was confirmed by 11-GHz beacon observations made in the UK at elevation angles of 8.9° and 7.1° [56].

10.6.2.2 Synoptic Distributions

The level of scintillation experienced by satellite signals changes seasonally and these variations are important for planning communication-link margins. The strongest scattering occurs during summer months – primarily near the receiving ground station. Signals gathered over many months or years are governed primarily by changing meteorological conditions. The availability of more than a hundred communication satellites made it possible to gather large data sets and to establish synoptic distributions of scintillation level. These measurements have been made for receiving stations around the world, often for time spans of several years [29][58][59][60][61].

The observed distributions for long sample lengths can be explained if one notes that the signal variance τ^2 in the short-sample distribution itself changes appreciably with time. This suggests that we should regard (10.4) as a conditional probability for a prescribed instantaneous value of τ, as we did in treating nonstationary changes in C_n^2 for terrestrial links. The long-term distribution of signal amplitudes should then be described by (10.28), which is a convolution of (10.4) and the probability distribution of τ. Log-normal and gamma distributions for τ have been used to predict the long-term distribution of signal fading [29][56][62]. This modeling has reached a point where the agreement with satellite data is satisfactory [53][54][63][64].

10.7 Problem

Problem 1

Compute the Rice and Nakagami-m PDFs given by (10.50) and (10.55). Compare them with the log-normal version

$$P(A)\,dA = \frac{dA}{\tau A\sqrt{2\pi}} \exp\left\{-\frac{1}{2\tau^2}\left[\ln\left(\frac{A}{A_0}\right)\right]^2\right\}$$

How different are these distributions under weak-scattering conditions? What does this tell one about the possibility of using distribution data to identify the appropriate signal model?

References

[1] V. Tatarskii, private communication on 23 May 2001.
[2] M. Abramowitz and I. A. Stegun, *Handbook of Mathematical Functions* (Dover Publications, New York, 1972), 295–326.
[3] G. M. Babler, "Scintillation Effects at 4 and 6 GHz on a Line-of-Sight Microwave Link," *IEEE Transactions on Antennas and Propagation*, **AP-19**, No. 4, 574–575 (July 1971).
[4] H. B. Janes and M. C. Thompson, "Fading at 9.6 GHz on an Experimentally Simulated Aircraft-to-Ground Path," *IEEE Transactions on Antennas and Propagation*, **AP-26**, No. 5, 715–719 (September 1978).
[5] M. C. Thompson, L. E. Wood, H. B. Janes and D. Smith, "Phase and Amplitude Scintillations in the 10 to 40 GHz Band," *IEEE Transactions on Antennas and Propagation*, **AP-23**, No. 6, 792–797 (November 1975).

[6] K. L. Ho, N. D. Mavrokoukoulakis and R. S. Cole, "Propagation Studies on a Line-of-Sight Microwave Link at 36 GHz and 110 GHz," *Microwaves, Optics and Acoustics*, **3**, No. 3, 93–98 (May 1979).

[7] M. E. Gracheva and A. S. Gurvich, "Strong Fluctuations in the Intensity of Light Propagated Through the Atmosphere Close to the Earth," *Izvestiya Vysshikh Uchebnykh Zavedenii Radiofizika (Soviet Radiophysics)*, **8**, No. 4, 511–515 (July–August 1965).

[8] M. E. Gracheva, "Investigation of the Statistical Properties of Strong Fluctuations in the Intensity of Light Propagated Through the Atmosphere Near the Earth," *Izvestiya Vysshikh Uchebnykh Zavedenii Radiofizika (Radiophysics and Quantum Electronics)*, **10**, No. 6, 424–433 (June 1967).

[9] D. H. Höhn, "Effects of Atmospheric Turbulence on the Transmission of a Laser Beam at 6328 Å . 1 – Distribution of Intensity," *Applied Optics*, **5**, No. 9, 1427–1431 (September 1966).

[10] A. S. Gurvich, M. A. Kallistratova and N. S. Time, "Fluctuations in the Parameters of a Light Wave from a Laser During Propagation in the Atmosphere," *Izvestiya Vysshikh Uchebnykh Zavedenii Radiofizika (Radiophysics and Quantum Electronics)*, **11**, No. 9, 771–776 (September 1968).

[11] M. W. Fitzmaurice, J. L. Bufton and P. O. Minott, "Wavelength Dependence of Laser-Beam Scintillation," *Journal of the Optical Society of America*, **59**, No. 1, 7–10 (January 1969).

[12] M. W. Fitzmaurice and J. L. Bufton, "Measurement of Log-Amplitude Variance," *Journal of the Optical Society of America*, **59**, No. 4, 462–463 (April 1969).

[13] G. R. Ochs and R. S. Lawrence, "Saturation of Laser-Beam Scintillation under Conditions of Strong Atmospheric Turbulence," *Journal of the Optical Society of America*, **59**, No. 2, 226–227 (February 1969).

[14] G. R. Ochs, R. R. Bergman and J. R. Snyder, "Laser-Beam Scintillation over Horizontal Paths from 5.5 to 145 Kilometers," *Journal of the Optical Society of America*, **59**, No. 2, 231–234 (February 1969).

[15] M. E. Gracheva, A. S. Gurvich, S. O. Lomadze, V. V. Pokasov and A. S. Khrupin, "Probability Distribution of 'Strong' Fluctuations of Light Intensity in the Atmosphere," *Izvestiya Vysshikh Uchebnykh Zavedenii Radiofizika (Radiophysics and Quantum Electronics)*, **17**, No. 14, 83–87 (January 1974).

[16] J. H. Churnside, "Joint Probability-Density Function of Irradiance Scintillations in the Turbulent Atmosphere," *Journal of the Optical Society of America A*, **6**, No. 12, 1931–1940 (December 1989).

[17] D. L. Fried, G. E. Mevers and M. P. Keister, "Measurements of Laser-Beam Scintillation in the Atmosphere," *Journal of the Optical Society of America*, **57**, No. 6, 787–797 (June 1967).

[18] A. Consortini and G. Conforti, "Detector Saturation Effect on Higher-Order Moments of Intensity Fluctuations in Atmospheric Laser Propagation Measurement," *Journal of the Optical Society of America A*, **1**, No. 11, 1075–1077 (November 1984).

[19] A. Consortini, E. Briccolani and G. Conforti, "Strong-Scintillation-Statistics Deterioration due to Detector Saturation," *Journal of the Optical Society of America A*, **3**, No. 1, 101–107 (January 1986).

[20] A. Consortini and R. J. Hill, "Reduction of the Moments of Intensity Fluctuations Caused by Amplifier Saturation for Both the K and Log-Normally Modulated Exponential Probability Densities," *Optics Letters*, **12**, No. 5, 304–306 (May 1987).

[21] E. Goldner and N. Ben-Yosef, "Sample Size Influence on Optical Scintillation Analysis. 2: Simulation Approach," *Applied Optics*, **27**, No. 11, 2172–2177 (1 June 1988).

[22] A. D. Wheelon, "Skewed Distribution of Irradiance Predicted by the Second-Order Rytov Approximation," *Journal of the Optical Society of America A*, **18**, No. 11, 2789–2798 (November 2001).

[23] V. I. Tatarskii, private communication, June 1995.

[24] J. M. Martin and S. M. Flatté, "Intensity Images and Statistics from Numerical Simulation of Wave Propagation in 3-D Random Media," *Applied Optics*, **27**, No. 11, 2111–2126 (1 June 1988).

[25] S. M. Flatté C. Bracher and G. Y. Wang, "Probability-Density Functions of Irradiance for Waves in Atmospheric Turbulence Calculated by Numerical Simulation," *Journal of the Optical Society of America A*, **11**, No. 7, 2080–2092 (July 1994).

[26] R. J. Hill and R. G. Frehlich, "Probability Distribution of Irradiance for the Onset of Strong Scintillation," *Journal of the Optical Society of America A*, **14**, No. 7, 1530–1540 (July 1997).

[27] R. J. Hill, R. G. Frehlich and W. D. Otto, "The Probability Distribution of Irradiance Scintillation," NOAA Technical Memorandum ERL ETL-274, NOAA Environmental Technology Laboratories, Boulder, Colorado (September 1996). (Available from the National Technical Information Service, 5285 Port Royal Road, Springfield, VA 22161).

[28] P. Beckmann, *Probability in Communication Engineering* (Harcourt, Brace and World, New York, 1967), 124 *et seq.* and 154.

[29] T. J. Moulsley and E. Vilar, "Experimental and Theoretical Statistics of Microwave Amplitude Scintillations on Satellite Down-Links," *IEEE Transactions on Antennas and Propagation*, **AP-30**, No. 6, 1099–1106 (November 1982).

[30] R. Frehlich, "Laser Scintillation Measurements of the Temperature Spectrum in the Atmospheric Surface Layer," *Journal of Atmospheric Sciences*, **49**, No. 16, 1494–1509 (15 August 1992).

[31] A. S. Gurvich and V. P. Kukharets, "The Influence of Intermittence of Atmospheric Turbulence on the Scattering of Radio Waves," *Radiotekhnika i Electronika* (*Radio Engineering and Electronic Physics*), **30**, No. 8, 52–58 (1985).

[32] M. Nelkin, "Universality and Scaling in Fully Developed Turbulence," *Advances in Physics*, **43**, No. 2, 143–181 (1994). A less challenging description is given in the following article by the same author: "In What Sense Is Turbulence an Unsolved Problem?" *Science*, **255**, 566–570 (31 January 1992).

[33] U. Frisch, P. L. Sulem and M. Nelkin, "A Simple Dynamical Model of Intermittent Fully Developed Turbulence," *Journal of Fluid Mechanics*, **87**, part 4, 719–736 (29 August 1978).

[34] V. I. Tatarskii, "Some New Aspects in the Problem of Waves and Turbulence," *Radio Science*, **22**, No. 6, 859–865 (November 1987).

[35] V. I. Tatarskii and V. U. Zavorotnyi, "Wave Propagation in Random Media with Fluctuating Turbulent Parameters," *Journal of the Optical Society of America A*, **2**, No. 12, 2069–2076 (December 1985).

[36] A. Stuart and J. K. Ord, *Kendall's Advanced Theory of Statistics* (Oxford University Press, New York, 1994), 45 and 67.

[37] J. H. Churnside and R. G. Frehlich, "Experimental Evaluation of Log-Normally Modulated Rician and *I K* Models of Optical Scintillation in the Atmosphere," *Journal of the Optical Society of America A*, **6**, No. 11, 1760–1766 (November 1989).

[38] R. Frehlich, "Effects of Global Intermittency on Laser Propagation in the Atmosphere," *Applied Optics*, **33**, No. 24, 5764–5769 (20 August 1994).

[39] E. Jakeman, E. R. Pike and P. N. Pusey, "Photon Correlation Study of Stellar Scintillation," *Nature*, **263**, 215–216 (16 September 1976).

[40] G. Parry and J. G. Walker, "Statistics of Stellar Scintillation," *Journal of the Optical Society of America*, **70**, No. 9, 1157–1766 (September 1980).

[41] D. Dravins, L. Lindegren, E. Mezey and A. T. Young, "Atmospheric Intensity Scintillation of Stars. I. Statistical Distributions and Temporal Properties," *Publications of the Astronomical Society of the Pacific*, **109**, No. 732, 173–207 (February 1997).

[42] J. W. Herbstreit and M. C. Thompson, "Measurements of the Phase of Radio Waves Received over Transmission Paths with Electrical Lengths Varying as a Result of Atmospheric Turbulence," *Proceedings of the IRE*, **43**, No. 10, 1391–1401 (October 1955).

[43] E. J. Fremouw, R. C. Livingston and D. A. Miller, "On the Statistics of Scintillating Signals," *Journal of Atmospheric and Terrestrial Physics*, **42**, No. 8, 717–731 (August 1980).

[44] A. D. Wheelon, "Radiowave Scattering by Tropospheric Irregularities," *Journal of Research of the NBS – D. Radio Propagation*, **63D**, No. 2, 205–233 (September–October 1959).

[45] S. O. Rice, "Mathematical Analysis of Random Noise," *Bell System Technical Journal*, **23** and **24**, 1–162 (1945).

[46] K. A. Norton, L. E. Vogler, W. V. Mansfield and P. J. Short, "The Probability Distribution of the Amplitude of a Constant Vector Plus a Rayleigh-Distributed Vector," *Proceedings of the IRE*, **43**, No. 10, 1354–1361 (October 1955).

[47] R. S. Lawrence, J. L. Jesperson and R. C. Lamb, "Amplitude and Angular Scintillations of the Radio Source Cygnus-A Observed at Boulder, Colorado," *Journal of Research of the NBS – D. Radio Propagation*, **65D**, No. 4, 333–350 (July–August 1961).

[48] J. Aarons, H. E. Whitney and R. S. Allen, "Global Morphology of Ionospheric Scintillations," *Proceedings of the IEEE*, **59**, No. 2, 159–172 (February 1971).

[49] C. L. Rino, R. C. Livingston and H. E. Whitney, "Some New Results on the Statistics of Radio Wave Scintillation: 1. Empirical Evidence for Gaussian Statistics," *Journal of Geophysical Research*, **81**, No. 13, 2051–2057 (1 May 1976).

[50] R. K. Crane, "Ionospheric Scintillation," *Proceedings of the IEEE*, **65**, No. 2, 180–199 (February 1977).

[51] M. Nakagami, "The *m*-Distribution – A General Formula of Intensity Distribution of Rapid Fading," in *Statistical Methods in Radio Wave Propagation*, edited by W. C. Hoffman (Pergamon Press, London, 1960), 3–36.

[52] G. Ortgies, "Amplitude Scintillations Occurring Simultaneously with Rain Attenuation on Satellite Links in the 11 GHz Band," *IEE Proceedings 4th International Conference on Antennas and Propagation*, Coventry, UK (1985), 72–75.

[53] E. T. Salonen, J. K. Tervonen and W. J. Vogel, "Scintillation Effects on Total Fade Distributions for Earth–Satellite Links," *IEEE Transactions on Antennas and Propagation*, **44**, No. 1, 23–27 (January 1996).

[54] F. Dintelmann, G. Ortgies, F. Rücker and R. Jakoby, "Results from 12- to 30-GHz German Propagation Experiments Carried Out with Radiometers and the OLYMPUS Satellite," *Proceedings of the IEEE*, **81**, No. 6, 876–884 (June 1993).

[55] Y. Karasawa and T. Matsudo, "Characteristics of Fading on Low-Elevation Angle Earth–Space Paths with Concurrent Rain Attenuation and Scintillation," *IEEE Transactions on Antennas and Propagation*, **39**, No. 5, 657–661 (May 1991).

[56] O. P. Banjo and E. Vilar, "Measurement and Modeling of Amplitude Scintillations on Low-Elevation Earth–Space Paths and Impact on Communication Systems," *IEEE Transactions on Communications*, **COM-34**, No. 8, 774–780 (August 1986).

[57] I. E. Otung and B. G. Evans, "Short Term Distribution of Amplitude Scintillation on a Satellite Link," *Electronic Letters*, **31**, No. 16, 1328–1329 (3 August 1995).

[58] R. R. Taur, "Ionospheric Scintillation at 4 and 6 GHz," *Comsat Technical Review*, **3**, No. 1, 145–163 (Spring 1973).

[59] C. N. Wang, F. S. Chen, C. H. Liu and D. J. Fang, "Tropospheric Amplitude Scintillations at C-Band Along Satellite Up-Link," *Electronic Letters*, **20**, No. 2, 90–91 (19 January 1984).

[60] P. S. Lo, O. P. Banjo and E. Vilar, "Observations of Amplitude Scintillations on a Low-Elevation Earth–Space Path," *Electronic Letters*, **20**, No. 7, 307–308 (29 March 1984).

[61] D. C. Cox, H. W. Arnold and H. H. Hoffman, "Observations of Cloud-Produced Amplitude Scintillation on 19- and 28-GHz Earth–Space Paths," *Radio Science*, **16**, No. 5, 885–907 (September–October 1981).

[62] E. Vilar, J. Haddon, P. Lo and T. J. Moulsley, "Measurement and Modelling of Amplitude and Phase Scintillations in an Earth–Space Path," *Journal of the Institution of Electronic and Radio Engineers*, **55**, No. 3, 87–96 (March 1985).

[63] Y. Karasawa, M. Yamada and J. E. Allnutt, "A New Prediction Method for Tropospheric Scintillation on Earth–Space Paths," *IEEE Transactions on Antennas and Propagation*, **36**, No. 11, 1608–1614 (November 1988).

[64] I. E. Otung, "Prediction of Tropospheric Amplitude Scintillation on a Satellite Link," *IEEE Transactions on Antennas and Propagation*, **44**, No. 12, 1600–1608 (December 1996).

11

Changes in Polarization

The electromagnetic field is characterized by the electric and magnetic fields which are vector quantities. The direction taken by the electric-field vector at each point along the path defines the polarization of the field. Faraday rotation of the electric field occurs when a signal propagates through the ionosphere. This rotation is readily observed at microwave frequencies and provides a useful way to measure the integrated electron density along the path.

We are concerned here with the more subtle changes in polarization that are caused by scattering in the lower atmosphere. Significant changes in polarization that are caused by wide-angle scattering in the troposphere are observed for scatter propagation beyond the horizon. By contrast, line-of-sight propagation is dominated by very-small-angle forward scattering. We have assumed so far that the change in polarization for this type of propagation is negligible. That assumption permits one to describe the propagation of light and microwaves in terms of a single scalar quantity. We now need to test this assumption by calculating the depolarization of the incident field.

If the transmitted wave is linearly polarized, we want to estimate how much the electric field rotates as the signal travels through the random medium. Two rather different descriptions of depolarization have been developed and apparently describe different aspects of the same phenomenon. The first is based on diffraction theory. The second description was developed from geometrical optics. It is significant that both predictions give estimates that are well below the measurement threshold for optical and microwave systems. This conclusion gives one confidence that line-of-sight propagation can be described in terms of a single scalar quantity determined by the random-wave equation.

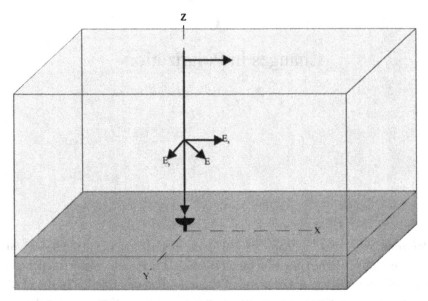

Figure 11.1: Rectangular coordinates used to describe the depolarization of a downcoming plane wave by scattering in the troposphere.

11.1 The Diffraction-theory Description

When we combine Maxwell's four equations for the electromagnetic fields,[1] the following wave equation results when the transmitter current is modulated harmonically:

$$\nabla^2 \mathbf{E} + k^2(1 + \delta\varepsilon)\mathbf{E} = -4\pi i k \mathbf{j}(\mathbf{r}) - \nabla[\mathbf{E} \cdot \nabla(\delta\varepsilon)] \tag{11.1}$$

The last term describes *depolarization* and our challenge is to estimate its magnitude. To do so we consider a linearly polarized plane wave falling on a region of homogeneous irregularities as illustrated in Figure 11.1. This is a crude approximation to radio and optical-astronomy observations of an overhead source. We assume that the source establishes the polarization of the transmitted wave in the *x* direction. The two components of the electric field propagating in the random medium are then described by coupled equations:

$$\nabla^2 E_x + k^2(1 + \delta\varepsilon)E_x = -4\pi i j_x(\mathbf{r}) - \frac{\partial}{\partial x}[\mathbf{E} \cdot \nabla(\delta\varepsilon)] \tag{11.2}$$

$$\nabla^2 E_y + k^2(1 + \delta\varepsilon)E_y = -\frac{\partial}{\partial y}[\mathbf{E} \cdot \nabla(\delta\varepsilon)] \tag{11.3}$$

The depolarization terms are first order in $\delta\varepsilon$ and one can replace the total field by the unperturbed field on the right-hand sides of both equations.

---

[1] As explained in Section 2.1 of Volume 1.

We concentrate on the component of the field E_y which is not excited by the source. This orthogonal component is generated by scattering in the random medium and is driven by the gradient of dielectric variations:

$$\nabla^2 E_y^1 + k^2 E_y^1 = -E_x^0(z) \frac{\partial^2}{\partial x\,\partial y} \delta\varepsilon(\mathbf{r})$$

since one can ignore $\delta\varepsilon$ on the left-hand side of (11.3). This equation can be solved with Green's function:

$$E_y^1(\mathbf{R}) = -\int d^3r\, G(\mathbf{R},\mathbf{r})E_x^0(z) \frac{\partial^2}{\partial x\,\partial y} \delta\varepsilon(\mathbf{r}) \tag{11.4}$$

It is customary to measure depolarization as the average intensity of the orthogonal field component divided by that of the incident plane wave:

$$\delta Pol_1 = \frac{\left\langle \left|E_y^1(\mathbf{R})\right|^2 \right\rangle}{\left|E_x^0(\mathbf{R})\right|^2} \tag{11.5}$$

On introducing expression (11.4) for the cross-polarized component we find that the depolarization is described by the following combination of terms:

$$\delta Pol_1 = \int d^3r \left(G(\mathbf{R},\mathbf{r})\frac{E_x^0(z)}{E_0(R)} \right) \int d^3r' \left(G(\mathbf{R},\mathbf{r}')\frac{E_x^0(z')}{E_x^0(R)} \right)^* \left\langle \frac{\partial^2\delta\varepsilon(\mathbf{r})}{\partial x\,\partial y} \frac{\partial^2\delta\varepsilon(\mathbf{r}')}{\partial x'\,\partial y'} \right\rangle$$

The ensemble average of the double-gradient product can be expressed in terms of the wavenumber spectrum. We make the replacement $\delta\varepsilon = 2\,\delta n$ and use the spectrum of refractive-index fluctuations because we are primarily interested in the tropospheric influence:

$$\left\langle \frac{\partial^2\delta\varepsilon}{\partial x\,\partial y} \frac{\partial^2\delta\varepsilon}{\partial x'\,\partial y'} \right\rangle = 4 \frac{\partial^2}{\partial x\,\partial y} \frac{\partial^2}{\partial x'\,\partial y'} \int d^3\kappa\, \Phi_n(\kappa)$$

$$\times \exp\{i[\kappa_x(x-x') + \kappa_y(y-y') + \kappa_z(z-z')]\}$$

$$= 4 \int d^3\kappa\, \Phi_n(\kappa)\kappa_x^2\kappa_y^2 \exp[i\kappa\cdot(\mathbf{r}-\mathbf{r}')]$$

This allows one to write the depolarization in terms of the complex weighting function introduced in (2.42):

$$\delta Pol_1 = 4\int d^3\kappa\, \Phi_n(\kappa)\kappa_x^2\kappa_y^2|\Lambda(\kappa)|^2 \tag{11.6}$$

The small-scattering-angle approximation (3.10) for $\Lambda(\kappa)$ is sufficiently accurate for our purpose. It is reasonable to assume that the irregularities are isotropic and we use spherical wavenumber coordinates:

$$\delta Pol_1 = \frac{\pi^2 R}{2k^2} \int_0^\infty d\kappa \, \kappa^5 \Phi_n(\kappa) \tag{11.7}$$

The spectrum's fifth moment depends on the dissipation region and on aperture averaging. The Kolmogorov model with an inner-scale wavenumber cutoff gives the following result for a point receiver [1][2]:

$$\delta Pol_1 = 0.070 R C_n^2 (\kappa_s)^{\frac{7}{3}} k^{-2} \qquad \text{for} \qquad a_r = 0 \tag{11.8}$$

To include the effect of aperture averaging one replaces the inner scale length by the radius of the receiver a_r. Receiver smoothing therefore reduces the effect still further and one is more likely to be able to measure depolarization with a small detector.

One can use the expression (11.8) to make numerical estimates for various propagation situations. From the scintillometer measurements reproduced in Figure 2.14 in Volume 1, we observe that $C_n^2 = 10^{-13}$ m$^{-\frac{2}{3}}$ is a reasonable estimate for the optical structure constant near the surface. For a 1-km path and an inner scale length of 1 cm,

$$\delta Pol_1 = 2 \times 10^{-19} \qquad \text{for} \qquad \lambda = 0.6 \, \mu m \tag{11.9}$$

which is well below the threshold of changes in polarization that can be measured with optical systems. The effect is ten times greater for a collimated beam but is still below the detection threshold [3]. The depolarization expression developed from diffraction theory is proportional to λ^2 and one would expect a substantially larger effect on microwave links. Near the surface a typical microwave value is $C_n^2 = 0.5 \times 10^{-14}$ m$^{-\frac{2}{3}}$ according to Figure 2.13 in Volume 1. For a 16-km path the estimate is

$$\delta Pol_1 = 4.3 \times 10^{-8} \qquad \text{for} \qquad \lambda = 30 \, cm \tag{11.10}$$

This is also well below the detection threshold of microwave receivers. Lee and Harp made comparable estimates for millimeter-wave links using a somewhat different but more intuitive description of the scattering [4]. These examples explain why the depolarization term is consistently omitted in analyzing line-of-sight propagation through turbulent media.

One should also examine the influence of random irregularities on the field component E_x which carries the initial polarization. To estimate this effect we replace the total field by $\mathbf{E}_0$ in (11.2) when it is multiplied by $\delta\varepsilon$:

$$\nabla^2 E_x^1 + k^2 E_x^1 = -4\pi i j(r) - E_x^0(z)\left(k^2 + \frac{\partial^2}{\partial x^2}\right)\delta\varepsilon \tag{11.11}$$

The term in large parentheses can be approximated by

$$\left(\frac{2\pi}{\lambda}\right)^2 + \left(\frac{2\pi}{\ell}\right)^2$$

where ℓ is the eddy size. The wavelength is smaller than the turbulent eddies in the optical range and the first term dominates. In the microwave range λ is larger than the smallest eddy but less than the outer scale length. In this case the depolarization term depends on the relative influence of turbulent eddies – which are both larger and smaller than the radiation wavelength [5][6].

11.2 Experimental Tests

Most researchers accepted the theoretical estimates and considered it unlikely that the depolarization effect could be detected. Experimental tests were also discouraged because scattering by aerosols, dust and insects should produce effects that are larger than those expected from refractive-index irregularities. Nonetheless, a few experimenters have tried to measure the depolarization caused by the turbulent atmosphere.

11.2.1 Horizontal Propagation

Saleh transmitted a laser beam with $\lambda = 0.6\ \mu m$ on a 2.6-km path but could detect no polarization change down to -42 to -45 dB [7]. A chance polarization experiment was performed during a program to measure phase fluctuations at X band on Maui [8]. The narrow-band signal could not be detected by the very sensitive receivers when the experiment began. It was soon discovered that the dipole feeds of the transmitter and receiver antennas had been set to opposite linear polarizations. This observation is entirely consistent with the theoretical model.

11.2.2 Satellite Signals

A different type of test was conducted using microwave signals radiated by a synchronous communication satellite [9]. Linearly polarized signals at 19 GHz were transmitted simultaneously on two orthogonal polarizations. The ionospheric influences are negligible at these frequencies. Ground measurements with a 7-m dish showed that the cross-polarized signal was greater than 40 dB below the co-polarized component. To interpret this experiment we must generalize the diffraction expression (11.7) to describe propagation along an inclined path through an inhomogeneous atmosphere.[2] It is also important to include aperture averaging.

---

[2] In analyzing satellite experiments, one must also consider depolarization by rain and clouds on the long slanted paths through the troposphere.

We showed how to include both effects in Chapter 3 in describing astronomical scintillation:

$$\delta Pol_1 = \frac{\pi^2 \sec \vartheta}{2k^2} \int_0^\infty dz \int_0^\infty d\kappa \, \kappa^5 \Phi_n(\kappa, z) \left(\frac{2J_1(\kappa a_r)}{\kappa a_r} \right)^2 \qquad (11.12)$$

Aperture smoothing limits the wavenumber integration and the Kolmogorov spectrum can be used to describe the local spectrum along the path:

$$\delta Pol_1 = 0.430 \frac{\sec \vartheta}{k^2} (a_r)^{-\frac{7}{3}} \int_0^\infty dz \, C_n^2(z) \qquad (11.13)$$

To evaluate this expression we use the values for integrated turbulence profiles which were reviewed in Section 7.4 of Volume 1. The largest value discussed there,

$$\int_0^\infty dz \, C_n^2(z) = 10^{-8} \text{ m}^{\frac{2}{3}}$$

suggests that

$$\delta Pol_1 \simeq 10^{-15} \qquad (11.14)$$

The depolarization should not be measurable in this situation and that is what the observations showed.

11.3 The Geometrical-optics Description

A second description of depolarization is based on geometrical optics [10] and the techniques of differential geometry [7]. This approach describes the random twisting of the electric-field vector as the signal travels along its ray path. In these formulations, the orthogonal polarization component appears first as a term that is second order in $\delta\varepsilon$. Its mean-square value is a fourth-order quantity and proportional to the square of the angle-of-arrival variance:

$$\delta Pol_2 = 2 \left(R\pi^2 \int_0^\infty d\kappa \, \kappa^3 \Phi_n(\kappa) \right)^2 \qquad (11.15)$$

This outcome is fundamentally different than the diffraction expression (11.7) with respect to frequency and distance scaling. For a Kolmogorov spectrum

$$\delta Pol_2 = 1.91 R^2 C_n^4 (\kappa_s)^{\frac{2}{3}} \qquad (11.16)$$

This result is independent of wavelength, varies quadratically with distance and depends on the square of the structure constant. It was subsequently found that the same result emerges from diffraction theory as a second-order term in the $\delta\varepsilon$ expansion of the orthogonal field component [11].

The predicted effect is exceedingly small. For the optical example cited above

$$\delta Pol_2 = 1.4 \times 10^{-18} \qquad \text{for} \qquad \lambda = 0.6 \ \mu m$$

which is greater than the diffraction estimate. The corresponding value for the microwave example discussed previously becomes

$$\delta Pol_2 = 10^{-20} \qquad \text{for} \qquad \lambda = 30 \ cm$$

and this is twelve orders of magnitude smaller than the diffraction result.

11.4 Problems

Problem 1

Use geometrical optics to estimate the depolarization of a plane wave. In this approach one defines the polarization as the unit vector:

$$\mathbf{p} = \mathbf{E}/E_0$$

Kravtsov showed that $\mathbf{p}$ is defined at each point along the ray path as the solution of the following differential equation [10]:

$$\frac{d\mathbf{p}}{ds} = -\mathbf{t}\left(\mathbf{p} \cdot \frac{d\mathbf{t}}{ds}\right)$$

The randomly curved ray path is described by its tangent vector $\mathbf{t}$, which is related to the local dielectric constant by the basic equation of geometrical optics:

$$\frac{d\mathbf{t}}{ds} = \frac{1}{2\epsilon}[\nabla\epsilon - \mathbf{t}(\nabla\varepsilon \cdot \mathbf{t})]$$

Combine these vector differential equations to confirm the result (11.15).

Problem 2

Calculate the second-order correction to the field component which is not excited by the transmitter. Use the defining equation for E_y given by (11.3), being careful to keep all terms that are quadratic in $\delta\varepsilon$. Show that the depolarization is the same as that predicted by geometrical optics.

Problem 3

The second-order polarization predicted by diffraction theory and geometrical optics is sometimes greater than the first-order result, as noted in the examples given in the text. Calculate the distance at which the crossover from one description to the other occurs as a function of frequency. Does the second-order term improve the chance that changes in polarization can be detected?

References

[1] V. I. Tatarskii, "Depolarization of Light by Turbulent Atmospheric Inhomogeneities," *Izvestiya Vysshikh Uchebnykh Zavedenii, Radiofizika (Radiophysics and Quantum Electronics)*, **10**, No. 12, 987–988 (December 1967).

[2] J. W. Strohbehn and S. F. Clifford, "Polarization and Angle-of-Arrival Fluctuations for a Plane Wave Propagated through a Turbulent Medium," *IEEE Transactions on Antennas and Propagation*, **AP-15**, No. 3, 416–421 (May 1967).

[3] E. Collett and R. Alferness, "Depolarization of a Laser Beam in a Turbulent Medium," *Journal of the Optical Society of America*, **62**, No. 4, 529–533 (April 1972).

[4] R. W. Lee and J. C. Harp, "Weak Scattering in Random Media, with Applications to Remote Probing," *Proceedings of the IEEE*, **57**, No. 4, 375–406 (April 1969).

[5] S. F. Clifford, "Wave Propagation in a Turbulent Medium," Ph.D. Dissertation, Thayer School of Engineering, Dartmouth College, Hanover, NH (June 1969).

[6] J. W. Strohbehn, "Line-of-Sight Wave Propagation Through the Turbulent Atmosphere," *Proceedings of the IEEE*, **56**, No. 8, 1301–1318 (August 1968).

[7] A. A. M. Saleh, "An Investigation of Laser Wave Depolarization due to Atmospheric Transmission," *IEEE Journal of Quantum Electronics*, **QE-3**, No. 11, 540–543 (November 1967).

[8] K. A. Norton, J. W. Herbstreit, H. B. Janes, K. O. Hornberg, C. F. Peterson, A. F. Barghausen, W. E. Johnson, P. I. Wells, M. C. Thompson, M. J. Vetter and A. W. Kirkpatrick, *An Experimental Study of Phase Variations in Line-of-Sight Microwave Transmissions* (National Bureau of Standards Monograph 33, U. S. Government Printing Office, 1 November 1961).

[9] D. C. Cox, H. W. Arnold and H. H. Hoffman, "Observations of Cloud-produced Amplitude Scintillation on 19- and 28-GHz Earth–Space Paths," *Radio Science*, **16**, No. 5, 885–907 (September–October 1981).

[10] Yu. A. Kravtsov, "Geometric Depolarization of Light in a Turbulent Atmosphere," *Izvestiya Vysshikh Uchebnykh Zavedenii, Radiofizika (Radiophysics and Quantum Electronics)*, **13**, No. 2, 217–220 (February 1970).

[11] B. Crosignani, P. DiPorto and S. F. Clifford, "Coupled-mode Theory Approach to Depolarization Associated with Propagation in Turbulent Media," *Applied Optics*, **27**, No. 11, 2183–2186 (1 June 1988).

12

The Validity of the Rytov Approximation

It is important to establish the conditions for which the Rytov approximation can provide a valid description of propagation through random media. Pioneering measurements by Gracheva, Gurvich and Kallistratova [1] showed that this method can describe the physics only when the logarithmic amplitude variance satisfies the following condition:

$$\langle \chi^2 \rangle < 1 \tag{12.1}$$

Data from their experiments is reproduced in Figure 9.11 and leaves little doubt that the Rytov prediction does not match the data beyond this limit. Tatarskii confirmed this result analytically using the second-order Rytov solution [2].

The condition (12.1) means that Rytov's approximation can be used to describe optical propagation only for short distances near the surface, as noted in Chapter 3. On the other hand, this condition is satisfied for almost all microwave and millimeter-wave links. It is also satisfied for most astronomical observations. Indeed, we have successfully interpreted a wide range of experiments with the Rytov approximation in this volume.

The picture presented by (12.1) is incomplete because it does not indicate the range of phase fluctuations for which the Rytov approximation is valid. That is an important question because the phase variances measured at optical wavelengths greatly exceed one radian – even on links that satisfy the amplitude condition [3]. To complete the picture, we must address the limits placed both on the amplitude variance and on the phase variance.

Pisareva was the first to do so [4]. She used qualitative estimates for the terms that appear in the transformed wave equation (2.7) that is used to generate the Rytov series. She established bounds both on the phase and on the amplitude in this way. The limits in the near field and far field are quite different. She found that the amplitude variance must be less than unity for Fresnel scattering. By contrast, the phase variance is essentially unbounded:

$$\text{Fresnel:} \qquad \langle \chi^2 \rangle < 1 \qquad \text{and} \qquad \langle \varphi^2 \rangle \quad \text{unbounded} \qquad (12.2)$$

This regime characterizes optical propagation on short paths. The Rytov method represents a significant improvement on the Born approximation since the latter is valid only when $\langle \varphi^2 \rangle < 1$. In the Fraunhofer regime she found that both the phase variance and the amplitude variance must be less than unity:

$$\text{Fraunhofer:} \qquad \langle \chi^2 \rangle < 1 \qquad \text{and} \qquad \langle \varphi^2 \rangle < 1 \qquad (12.3)$$

This is the same condition as that imposed by Born theory. Pisareva's insightful paper seems not to have been noticed in the west and a confusing debate surrounded this problem for almost twenty years [5][6][7][8][9][10].

The question of the validity of the Rytov approximation has attracted little attention in recent years, as attention shifted to theories of strong scattering. The important role played by the second-order Rytov solution in describing weak-scattering measurements has been demonstrated repeatedly in this volume. It is important to know when that correction is needed and when it is adequate.

12.1 A Qualitative Approach

We begin our study by presenting Pisareva's derivation of the limits placed on the Rytov approximation [4]. She begins with the transformed wave equation (2.7), which we repeat here:

$$\nabla^2 \Psi + (\nabla \Psi)^2 + 2 \nabla \psi_0 \cdot \nabla \Psi + k^2 \, \delta\varepsilon = 0 \qquad (12.4)$$

The solution of this equation is usually sought by neglecting the nonlinear term $(\nabla \Psi)^2$ and this step is called *closure*. When one does so, the resulting equation for Ψ is identical to the first field equation in (2.14) that generates the first-order Rytov solution:

$$\nabla^2 \Psi + 2 \nabla \psi_0 \cdot \nabla \Psi + k^2 \, \delta\varepsilon = 0 \qquad (12.5)$$

Our challenge is to establish when it is legitimate to omit the nonlinear term. One cannot do so casually. In the study of nonlinear oscillations one learns that even a tiny nonlinear term can have a profound influence on the system's behavior over long time spans. Closer to our interest in turbulent media, we recall that the nonlinear term in the Navier–Stokes equations,

$$\frac{\partial}{\partial t} \mathbf{v}(\mathbf{r}, t) + [\mathbf{v}(\mathbf{r}, t) \cdot \nabla] \mathbf{v}(\mathbf{r}, t) = \nu \, \nabla^2 \mathbf{v}(\mathbf{r}, t) \qquad (12.6)$$

plays the decisive role in generating turbulence from steady flow fields.

If one can neglect the term $(\nabla\Psi)^2$ in (12.4), the solution is identical to the first-order Rytov approximation derived from (12.5). This means that we should be able to deduce the conditions of validity for the Rytov approximation by finding out when the nonlinear term can be ignored. In doing so, we are working back from the approximate equation to the complete equation. We will then use the steps required in that process to identify the conditions for which the solution of (12.5) is valid.

This process focuses attention on the importance of the nonlinear term relative to the other terms in (12.4). The traditional argument for ignoring the nonlinear term emerges on comparing it with the second term:

$$(\nabla\Psi)^2 \ll 2\,\nabla\psi_0\cdot\nabla\Psi \tag{12.7}$$

This inequality can be expressed in terms of the magnitude of the surrogate's gradient since $\psi_0 = ikx$ for a plane wave:

$$\lambda|\nabla\Psi| \ll \pi$$

The surrogate function defines the signal's phase and amplitude through the following relationship:

$$\Psi(\mathbf{r}) = \log\left(\frac{A(\mathbf{r})}{A_0}\right) + i\phi(\mathbf{r}) \tag{12.8}$$

The change in phase and the *relative* change in amplitude should both be small over a distance of one wavelength. That condition is easily satisfied for actual experiments. Approximating the relative change in amplitude by its mean value divided by a scale length ℓ, we have

$$\lambda|\nabla\ln(A/A_0)| \simeq \lambda\left|\frac{1}{A}\left(\frac{A}{\ell}\right)\right| = \frac{\lambda}{\ell}$$

This ratio is small for optical and infrared signals since all atmospheric eddies are large relative to their wavelengths. At microwave frequencies this need not be true. However, we know that phase fluctuations are determined by the largest irregularities in concentration of water vapor and this ratio is small for them. Moreover, eddies that are important for generating amplitude fluctuations are comparable to the Fresnel length, so

$$\lambda|\nabla\ln(A/A_0)| \approx \sqrt{\frac{\lambda}{R}}$$

and this ratio is small for microwave experiments. Unfortunately, this line of reasoning places no useful bound on the phase and amplitude of the signals. To do so we must look at other terms in (12.4). The considerations are different in the Fresnel and Fraunhofer regimes.

12.1.1 Fraunhofer Scattering

From Figure 7.2 we observe that diffraction influences both the phase and the amplitude of the field strength in the Fraunhofer regime. This means that the Laplacian term in (12.4) plays a vital role in defining both quantities. Pisareva argued that the important validity condition arises from neglecting the squared gradient in comparison with the Laplacian term:

$$\text{Fraunhofer:} \qquad (\nabla\Psi)^2 \ll \nabla^2\Psi \qquad\qquad (12.9)$$

She first assumed that the surrogate function changes by a considerable amount

$$\delta\psi = \chi + i\varphi$$

over a scale length ℓ and approximated the gradient and Laplacian terms as

$$\nabla\Psi \approx |\delta\psi|/\ell \qquad \text{and} \qquad \nabla^2\Psi \approx |\delta\psi|/\ell^2$$

The inequality (12.9) is satisfied only if the change $\delta\psi$ itself is small:

$$|\delta\psi| < 1$$

This condition is independent of the assumed scale length[1] and leads to a stringent bound on the logarithmic amplitude and phase [6][7]:

$$\text{Fraunhofer:} \qquad \langle\chi^2\rangle + \langle\varphi^2\rangle < 1 \qquad\qquad (12.10)$$

This result is consistent with (12.3) since the phase variance and amplitude variance are the same in this region.

12.1.2 Fresnel Scattering

In the region of small scattering parameters, we showed in (3.32) that the amplitude variance approaches the geometrical-optics expression. This is also true for the phase variance since the spectral weighting function for the phase variance (7.10) approaches unity, giving

$$\text{Fresnel:} \qquad \langle\varphi^2\rangle = 4\pi^2 R k^2 \int_0^\infty d\kappa\,\kappa\,\Phi_n(\kappa) \qquad\qquad (12.11)$$

which is the geometrical-optics description. The Laplacian term in (12.4) influences amplitude fluctuations but has little effect on the phase in this regime.

---

[1] Pisareva assumed that the scale length ℓ falls in the inertial range and identified it with the Fresnel length:

$$\ell_0 \ll \ell \ll L_0$$

The solution of the linearized equation (12.5) agrees with the geometrical-optics description when the scattering parameter is small. According to (3.32) the amplitude variance for a plane wave in that limit can be expressed as

$$\text{Fresnel:} \qquad \langle \chi^2 \rangle = \frac{1}{3} \pi^2 R^3 \int_0^\infty d\kappa \, \kappa^5 \Phi_n(\kappa) \qquad (12.12)$$

The corresponding angle-of-arrival variance is given in (7.8) of Volume 1 by

$$\langle \delta\theta^2 \rangle = 2\pi^2 R \int_0^\infty d\kappa \, \kappa^3 \Phi_n(\kappa)$$

and the mean-square lateral displacement of the rays becomes

$$\text{Fresnel:} \qquad \langle \rho^2 \rangle = \frac{2}{3} \pi^2 R^3 \int_0^\infty d\kappa \, \kappa^3 \Phi_n(\kappa) \qquad (12.13)$$

Each power of κ in the wavenumber integration has approximately the same effect as multiplying the result by $1/\ell$, where ℓ is the scale length introduced previously. With this crude replacement the logarithmic amplitude variance becomes

$$\langle \chi^2 \rangle = 2\frac{\langle \rho^2 \rangle}{\ell^2}$$

The ray displacement should be less than the scale length of the irregularities if diffraction effects are to be ignored, so that

$$\text{Fresnel:} \qquad \langle \chi^2 \rangle < 1 \qquad (12.14)$$

which agrees with (12.2). Notice that this approach sets the same generous limit on the phase variance as that established for geometrical optics.

12.2 An Analytical Approach

One can also use an analytical approach to establish the conditions for which the Rytov approximation is valid. The question of its validity is directly related to the convergence of the series expansion for the surrogate function (2.13). We judge that the Rytov approximation can be used to describe the field strength if each term in the Rytov series is smaller than the preceding one – in some suitable statistical sense. As a practical matter, we can focus our attention on the first and second terms that are established in (2.64),

$$\Psi_2 = a + ib + \left(c - \tfrac{1}{2}a^2 + \tfrac{1}{2}b^2\right) + i(d - ab) \qquad (12.15)$$

and insist that

$$\left(c - \tfrac{1}{2}a^2 + \tfrac{1}{2}b^2\right) + i(d - ab) < a + ib \qquad (12.16)$$

Because the terms in this inequality are stochastic functions of time, we must express the conditions in terms of statistical moments of the two sides. In doing so, it is helpful to identify the amplitude and phase using the connection (12.8):

$$\log\left(\frac{A_2}{A_0}\right) = a + c - \frac{1}{2}a^2 + \frac{1}{2}b^2 \qquad (12.17)$$

and

$$\phi_2 = b + d - ab \qquad (12.18)$$

12.2.1 The Amplitude Limit

The first attempt to gauge the convergence of the Rytov series was based on the *average value* of the second-order surrogate function [2]:

$$\langle \Psi_2 \rangle = \left\langle c - \tfrac{1}{2}a^2 + \tfrac{1}{2}b^2 \right\rangle + i\langle d - ab \rangle \qquad (12.19)$$

For both plane and spherical waves we learned in (8.53) that $\langle d \rangle$ vanishes and $\langle c \rangle$ can be expressed in terms of the first-order variances. The result is that

$$\langle \Psi_2 \rangle = -\langle \chi^2 \rangle - i\langle \varphi \chi \rangle \qquad (12.20)$$

and the average logarithmic amplitude to second order becomes

$$\left\langle \log\left(\frac{A_2}{A_0}\right) \right\rangle = -\langle \chi^2 \rangle \qquad (12.21)$$

Tatarskii argues that the quantity on the left-hand side should be less than unity for weak scattering [2]. That leads to

$$\langle \chi^2 \rangle < 1$$

which he identifies with the condition for validity of the Rytov approximation.

One can extend this approach by using a fourth-order expression for the average value of the surrogate function. Tatarskii gave the following description for plane-wave propagation [2]:

$$\langle \Psi_4 \rangle = -\langle \chi^2 \rangle - i\langle \varphi \chi \rangle - 2.9\left(1 + i\sqrt{3}\right)\langle \chi^2 \rangle^2 \qquad (12.22)$$

The correlation of fluctuations in phase and amplitude is proportional to $\langle \chi^2 \rangle$ in the Fresnel regime, as noted in (7.76):

$$\langle \Psi_4 \rangle = -(1 + i3.73)\langle \chi^2 \rangle - 2.9\left(1 + i\sqrt{3}\right)\langle \chi^2 \rangle^2 \qquad (12.23)$$

This suggests that higher-order corrections are proportional to successively higher powers of the amplitude variance. Each term in this series must be smaller than the previous term for this expansion to converge. Comparing the real and imaginary parts of the second- and fourth-order terms yields

$$\langle \chi^2 \rangle < 0.34 \quad \text{and} \quad \langle \chi^2 \rangle < 0.74 \tag{12.24}$$

The second condition is consistent with the experimental result (12.1). The first condition is more restrictive and does not agree with the data. Tatarskii has recently indicated that the coefficients of the fourth-order term may be wrong by a factor of two [11], which may solve the problem.

12.2.2 The Phase Limit

We need to extend this calculation to find an approximate limit on the phase. We found in Section 7.1 that the phase variance computed with the Rytov approximation is little different than the geometrical-optics result in the Fresnel regime. However, the situation is basically different when Fraunhofer scattering describes the transmission.

To bound the phase variance we exploit the sum rule (7.11) that is anchored by the phase variance calculated with geometrical optics:

$$\langle \varphi^2 \rangle = \langle \varphi_0^2 \rangle - \langle \chi^2 \rangle$$

In the Fresnel regime one can use the explicit description (7.13) for $\langle \chi^2 \rangle$ to write

$$\text{Fresnel:} \quad \langle \varphi^2 \rangle = \langle \varphi_0^2 \rangle \left[1 - 0.393 \left(\frac{R\kappa_0^2}{k} \right)^{\frac{5}{6}} \right]$$

Both expressions tell us that

$$\text{Fresnel:} \quad \langle \varphi^2 \rangle < \langle \varphi_0^2 \rangle \tag{12.25}$$

The phase variance $\langle \varphi_0^2 \rangle$ is quite large for optical signals and thus equivalent to the "unbounded" condition presented in (12.2). Notice that the phase variance is not much smaller than $\langle \varphi_0^2 \rangle$ because $\langle \chi^2 \rangle$ is less than unity. This analysis provides a more specific estimate for the phase-variance limit.

In the Fraunhofer regime we can establish the limit on the phase variance quite easily. We learned in (7.14) that the variances of phase and amplitude are equal when $R\kappa_0^2 > k$. This means that φ has the same bound as χ as suggested in (12.3):

$$\text{Fraunhofer:} \quad \langle \varphi^2 \rangle < 1 \tag{12.26}$$

References

[1] M. E. Gracheva, A. S. Gurvich and M. A. Kallistratova, "Dispersion of 'Strong' Atmospheric Fluctuations in the Intensity of Laser Radiation," *Izvestiya Vysshikh Uchebnykh Zavedenii Radiofizika (Radiophysics and Quantum Electronics)*, **13**, No. 1, 40–42 (January 1970).

[2] V. I. Tatarskii, *The Effects of the Turbulent Atmosphere on Wave Propagation* (translated from the Russian and issued by the National Technical Information Office, U. S. Department of Commerce, Springfield, VA 22161, 1971), 253–258.

[3] V. P. Lukin, V. V. Pokasov and S. S. Khmelevtsov, "Investigation of the Time Characteristics of Fluctuations of the Phases of Optical Waves Propagating in the Bottom Layer of the Atmosphere," *Izvestiya Vysshikh Uchebnykh Zavedenii, Radiofizika (Radiophysics and Quantum Electronics)*, **15**, No. 12, 1426–1430 (December 1972).

[4] V. V. Pisareva, "Limits of Applicability of the Method of 'Smooth' Perturbations in the Problem of Radiation Propagation through a Medium Containing Inhomogeneities," *Akusticheskii Zhurnal (Soviet Physics – Acoustics)*, **6**, No. 1, 81–86 (July–September 1960).

[5] D. A. deWolf, "Wave Propagation Through Quasi-Optical Irregularities," *Journal of the Optical Society of America*, **55**, No. 7, 812–817 (July 1965).

[6] W. P. Brown, "Validity of the Rytov Approximation in Optical Propagation Calculations," *Journal of the Optical Society of America*, **56**, No. 8, 1045–1052 (August 1966).

[7] W. P. Brown, "Validity of the Rytov Approximation," *Journal of the Optical Society of America*, **57**, No. 12, 1539–1543 (December 1967).

[8] L. S. Taylor, "On Rytov's Method," *Radio Science*, **2** (New Series), No. 4, 437–441 (April 1967)

[9] G. R. Heidbreder, "Multiple Scattering and the Method of Rytov," *Journal of the Optical Society of America*, **57**, No. 12, 1477–1479 (December 1967).

[10] J. B. Keller, "Accuracy and Validity of the Born and Rytov Approximations," *Journal of the Optical Society of America*, **59**, No. 8, 1003–1004 (August 1969, Part I).

[11] V. I. Tatarskii, private communication on 25 May 2001.

Appendix A

Glossary of Symbols

The equation numbers below refer to Volume 2.

a	Real part of B_1 defined by (2.27)
	Anisotropic scaling parameter in the x direction
a_r	Radius of circular receiver
$\mathbf{a}$	Axial ratio parameter defined by (4.71)
$A(\mathbf{r})$	Amplitude of electric field strength
$A_n(\mathbf{r})$	Terms in expansion of amplitude in inverse powers of k
$A(\mathbf{R}, \mathbf{r})$	Real part of the scattering kernel (2.30)
A	Area of circular receiving aperture
	Cross-sectional area of ray bundle
$\mathcal{A}$	Axial ratio of field-aligned plasma irregularities
$\mathbb{A}(\tau)$	Allan variance
b	Imaginary part of B_1 defined by (2.27)
	Anisotropic scaling parameter in the y direction
$B(\mathbf{R}, \mathbf{r})$	Imaginary part of the scattering kernel (2.30)
$\mathbf{B}(\mathbf{r})$	Induction field vector at $\mathbf{r}$
B_n	Normalized Born approximation of order n: E_n/E_0
$\mathsf{B}_\varepsilon(\mathbf{r}, \mathbf{r}')$	Spatial covariance of dielectric fluctuations
$\mathsf{B}_\varepsilon(\tau)$	Temporal covariance of dielectric fluctuations
$\mathcal{B}$	Medium bandwidth
B_ν	Combination of ionospheric parameters (3.154)
$\mathfrak{B}$	Combination of ionospheric parameters (4.69)
c	Speed of light
c	Real part of B_2 defined by (2.63)
	Anisotropic scaling parameter in the z direction
$C(x)$	Fresnel integral defined in (3.52)

$C_\chi(\rho)$	Spatial correlation of amplitude fluctuations
$C_\chi(\rho, a_r)$	Spatial correlation including aperture averaging
$C_\chi(\rho_1, \rho_2)$	Spatial correlation for beam waves (4.33)
$C(\tau)$	Autocorrelation of dielectric variations
$C(\rho)$	Spatial correlation of dielectric variations
$C_{\Delta\varphi}(\rho)$	Spatial correlation function of phase difference
$C_n^2(z)$	Strength of turbulent irregularities at height z
$\mathrm{Coh}(k_1, k_2, \omega)$	Amplitude coherence for different wavelengths (6.15)
d	Imaginary part of B_2 defined by (2.63)
	Differential of following quantity
$\mathbf{D(r)}$	Displacement field at $\mathbf{r}$
d	Astronomical angle-of-arrival variance
D_r	Diameter of circular receiver
D	Diffusion constant
D_{12}	Separation between two points on adjacent rays
$D(\kappa)$	Amplitude weighting function defined in (3.26)
$\mathcal{D}(\rho)$	Wave structure function
$\mathcal{D}_\varphi(\rho)$	Phase structure function
$\mathcal{D}_\chi(\rho)$	Amplitude structure function
$\mathcal{D}_n(\rho)$	Refractive-index structure function
$\mathcal{D}(\rho_1, \rho_2)$	Wave structure function for beam waves
$\mathcal{D}_\varphi(\rho_1, \rho_2)$	Phase structure function for beam waves
$\mathcal{D}_\chi(\rho_1, \rho_2)$	Amplitude structure function for beam waves
e	Electric charge
	Real part of B_3 defined by (2.74)
$E(\kappa)$	Phase weighting function defined in (3.29)
$\mathbf{E(r)}$	Electric field at $\mathbf{r}$
$E(\mathbf{r})$	Principal scalar component of electric-field vector
$E_0(\mathbf{r})$	Unperturbed scalar field strength
$E_n(\mathbf{r})$	Born-series term representing n scatterings
$\mathrm{erf}(x)$	Error function defined in Appendix B
$\mathcal{E}_0$	Reference level of field strength
f	Electromagnetic frequency in cycles per second
	Imaginary part of B_3 defined by (2.74)
δf	Frequency fluctuation caused by random medium
$F(\kappa R)$	Finite-path wavenumber weighting function for phase
$F_\chi(\kappa)$	Amplitude-variance wavenumber weighting function
$F_\varphi(\kappa)$	Phase-variance wavenumber weighting function
$F_{\varphi\chi}(\kappa)$	Cross-correlation wavenumber weighting function
$_2F_1(a, b, c; z)$	Hypergeometric function defined in Appendix H

$\mathcal{F}(\kappa\ell_0)$	Spectrum factor that describes the dissipation region
g	Real part of B_4 defined by (2.79)
$G(\mathbf{R}, \mathbf{r})$	Green's function connecting $\mathbf{r}$ and $\mathbf{R}$: Appendix L
$G(\eta)$	Receiver gain factor (3.66) and (3.72)
$\mathbb{G}(\eta)$	Thin-screen receiver gain factor (3.129)
$\mathcal{G}$	Voltage gain of receiver
g	Variance of frequency flucutations
$\mathcal{G}(\kappa, \rho, v\tau)$	Phase-difference weighting function
h	Imaginary part of B_4 defined by (2.79)
	Height of receiver
h	Line-of-sight height above ground
H	Effective height of troposphere
	Layer height for ionospheric irregularities
ΔH	Thickness of turbulent layer
$\mathbf{H}(\mathbf{r})$	Magnetic field at $\mathbf{r}$
$\mathbb{H}$	Logarithmic intensity (9.67)
$\mathcal{H}(\kappa, \vartheta)$	Thick-layer weighting function for amplitude variance (3.126)
$\mathcal{H}(\eta)$	Phase-variance weighting function for satellite paths
i	$\sqrt{-1}$
I	Intensity of electric field: $I = |E|^2$
I_0	Intensity of unperturbed field: $I_0 = |E_0|^2$
$I_\nu(x)$	Modified Bessel function of the first kind
$\Im(z)$	Imaginary part of z
$\mathcal{I}(x)$	Inner scale factor for amplitude variance (3.39) and (3.54)
$J_\nu(x)$	Ordinary Bessel function of the first kind
$\mathbf{j}(\mathbf{r})$	Current-density distribution at transmitter
$\mathcal{J}$	Symbol used for integrals to be evaluated
$\mathcal{J}(x, \kappa)$	Amplitude-covariance weighting for beam waves (4.30)
k	Electromagnetic wavenumber $k = 2\pi/\lambda$
K_n	Cumulant of order n
$K_\nu(x)$	Modified Bessel function of the second kind
$K(z, \Delta\lambda, \vartheta)$	Wavelength-correlation weighting for $C_n^2(z)$ in (6.28)
$K(\rho, z)$	Kernel for Peskoff inversion solution (4.56)
$K(\eta)$	Weighting function in $\mathcal{D}_\varphi(\rho)$ for satellite paths
$\mathbb{K}(x)$	Cross-correlation outer scale factor
$\mathcal{K}(\omega, \omega')$	Sample-size factor for phase power spectrum
ℓ	Eddy size
ℓ_0	Inner scale length

L_0	Outer scale length
$L(\rho_e, z)$	Weighting of $C_n^2(z)$ to yield spatial covariance (4.50)
L	Reference level for signal amplitude (10.6)
$\mathfrak{L}(\zeta_0, \varepsilon, v)$	Integrated thick-layer weighting function (3.151)
$\mathbb{L}_v(x)$	Struve function
$\mathcal{L}(\kappa_0 R/a)$	Finite-path-distance factor for phase variance
m	Single parameter in Nakagami distribution (10.55)
	Parameter combination (5.39)
$M(a, b; x)$	First solution of Kummer's differential equation
$\mathcal{M}(\kappa)$	Aperture-averaging factor for satellite paths
$\mathfrak{M}(\kappa, x)$	First term in beam-wave amplitude weighting function (3.92)
M^2	Sample-averaged mean value of phase difference
n	Refractive index
δn	Variation in refractive index
n	Vector normal to ray path
$N(\mathbf{r})$	Electron density in ionosphere or interstellar plasma
$N_0(z)$	Ambient electron density
δN	Fluctuation in electron density
$N(z, \mu)$	Altitude weighting of $C_n^2(z)$ to yield $S(\mu)$
$\mathfrak{N}(\kappa, x)$	Second term in beam-wave amplitude weighting function (3.92)
$N(\lambda, w_0, \mathcal{R}_0)$	Combination of beam-wave parameters (9.39)
$\mathcal{N}$	Normalization for correlation functions
$\mathbb{O}(x)$	Outer scale factor for amplitude variance (3.41)
p	Displacement of centroid of image
	Effective baseline divided by the Fresnel length (4.73)
p	Polarization unit vector used in geometrical otpics
P	Beam-wave parameter
p	Anisotropy term for satellite paths
$P(x)$	Probability density function of x
$P(x\|y)$	Conditional probability for x given that y is fixed
$\mathcal{P}(x)$	Cumulative probability of x
$\mathbb{P}(x)$	Plane-wave power-spectrum factor (5.13)
$\wp(z)$	Normalized turbulence profile (2.48)
q	Inter-receiver spacing divided by the Fresnel length
	Argument of characteristic function
q	Stretched coordinates used to describe anisotropic media
q(x)	Interferometer-baseline and sample-length function
Q_v	Normalization for ionospheric spectrum models
Q	Inter-receiver-separation parameter for beam waves (4.32)

$\mathcal{Q}(t, t', \rho)$	Ensemble average of time-shifted phase differences
$Q(A, \varphi)$	Combination of phase and amplitude terms (10.47)
$Q(\kappa)$	Aperture-averaging wavenumber-weighting function (4.22)
$Q(\kappa, x)$	Wavenumber- and altitude-weighting function (6.13)
$Q(\tau)$	Long term distribution of τ values (10.28)
$\mathbb{Q}(x)$	Altitude-weighting factor for power spectrum (5.35)
Q	Constant in interferometer scaling law
r	Radial distance to point in random medium
	Frequency ratio defined by (6.19)
r_E	Radius of earth: $r_E = 6378$ km
r_i	Radial coordinate defining position on receiver surface
r_e	Electron radius: $r_e = 2.8 \times 10^{-13}$ cm
r_0	Fried's coherence radius: $r_0 = 2.099\rho_0$
$\mathbf{r}$	Position vector defining point in random medium
$\mathbf{r}(x)$	Interferometer-baseline and sample-length function
R	Length of propagation path
ΔR	Entry plane-to-focus distance for parabolic receiver
$\mathbf{R}$	Vector position of receiver
$\Re(z)$	Real part of z
$\mathcal{R}_0$	Initial radius of curvature for beam wave (3.79)
$\mathcal{R}(k_1, k_2)$	Frequency or wavelength correlation of amplitude (6.1)
$\mathcal{R}(\Delta\lambda, \vartheta)$	Astronomical wavelength-difference correlation (6.26)
$\mathcal{R}_1(a, b, c, \vartheta)$	Anisotropy factor for in-plane angular error
$\mathcal{R}_2(a, b, c, \vartheta)$	Anisotropy factor for out-of-plane angular error
R	Constant in scaling law for an interferometer
s	Arc length along line of sight to source
$\Delta\mathbf{s}$	Vector separation between stellar lines of sight (4.35)
S_4	Scintillation index defined by (3.155)
$S(x)$	Fresnel integral defined by (3.52)
S	Total signal phase for geometrical optics
$\mathbf{S}$	Poynting vector (9.20)
$S(\mu)$	Spatial spectrum of arriving wave-front (4.57)
$S(\kappa, \tau)$	Self-motion term in turbulence spectrum
$S(\tau)$	Short-term distribution of τ values (10.34)
$\mathbb{S}(x)$	Spherical-wave power-spectrum factor (5.23)
$S(\kappa vT)$	Sample-length wavenumber weighting function
$S(\kappa_0 vT)$	Sample-length factor for phase variance
t	Time in seconds or minutes
$\mathbf{t}$	Tangent vector to ray path

$\mathbf{T}$	Vector defining transmitter position
T	Sample length in seconds or minutes
$T(\mathbf{r}, t)$	Temperature measured at point $\mathbf{r}$ and time t
δT	Temperature fluctuation about mean value
$\mathbb{T}(x)$	Spatial correlation for thin screen of irregularities (4.43)
T	Decorrelation time for turbulent velocity components
u	Difference coordinate
$\mathbf{U}$	Gradient of surrogate function (2.8)
$U(a, b; z)$	Second solution of Kummer's differential equation
$\mathcal{U}(u, \rho, \tau, \vartheta, \mathbf{v})$	Term in time-shifted structure function
v	Surface wind speed normal to the propagation path
	Sum coordinate
$v(z)$	Vertical profile of horizontal wind speed
$\mathbf{v}$	Wind vector
$\mathbf{v}_0$	Average wind vector
$\delta \mathbf{v}$	Wind-velocity fluctuation
$w(x)$	Width of beam wave as a function of distance
w_0	Initial width of beam wave
$W_\varphi(\omega)$	Power spectrum of single-path phase fluctuations
$W_{\Delta\varphi}(\omega)$	Power spectrum of phase-difference fluctuations
$W_\chi(\omega)$	Power spectrum of amplitude fluctuations (5.2)
$W_\chi(k_1, k_2, \omega)$	Cross-spectral density for different frequencies (6.16)
$W(\kappa, k, R, \alpha)$	Weighting function for Σ_v including attentuation (8.32)
$\mathcal{W}_\nu(a)$	Ionospheric aperture factor for angle of arrival
x	Horizontal coordinate along propagation path
$\mathcal{X}(\xi, p)$	Spectrum factor describing variable wind speed
y	Horizontal coordinate normal to propagation path
$Y_\nu(x)$	Second solution of Bessel's equation
$\mathcal{Y}(\zeta, \rho, \tau, \vartheta, \mathbf{v})$	Weighting function for phase-difference covariance
z	Vertical coordinate
$\mathcal{Z}(\zeta, \omega)$	Weighting function for astronomical phase power spectrum
∇	Gradient operator
$\nabla_\perp$	Gradient operator in direction normal to line of sight
∇^2	Laplacian operator
$\nabla_\perp^2$	Laplacian operator taken normal to path
α	Complex parameter describing beam waves (3.81)
	Dimensionless attenuation parameter defined by (8.21)
	Local inclination of ray path

α_0	Critical angular separation for source averaging (3.135)
	Inclination of transmitted signal
	Angle of elevation of received signal
β	Azimuth orientation of receiver baseline or velocity
	Half-angular width of source
β_0^2	Rytov variance: $\beta_0^2 = 4\tau^2$
γ	Attenuation measured in dB km^{-1} (8.21)
	Skewness parameter defined by (10.22)
	Angle between magnetic field and line-of-sight vectors
$\gamma(x)$	Complex function defining evolution of beam wave
δ	Variation of following quantity
δ	Angle between two nearby stars
$\delta(x)$	Dirac delta function defined in Appendix F
ϵ	Small parameter – usually taken to be zero
	Normalized wavelength difference (6.25)
ε	Dielectric constant
	Layer-thickness-to-height ratio
ε_0	Initial or average value of dielectric constant
$\delta\varepsilon$	Variation in the dielectric constant
ε_n	One or two in series of Bessel functions
ζ	Scattering parameter defined in (3.31)
ζ_0	Outer-scale scattering parameter (3.33)
ζ_m	Inner-scale scattering parameter (3.39)
η	Aperture radius divided by Fresnel length (3.132)
	Slope of phase or amplitude trend
	Parameter used in distance scaling: $\eta = \kappa_0 R/a$
	Kolmogorov microscale
θ	Polar-angle defining point in medium
	Angular width of scattered energy pattern (3.7)
	Angle between rotated ray paths
θ_i	Polar-angle defining points on extended source
$\delta\boldsymbol{\theta}$	Vector defining angular error components
$\delta\theta_n$	Angular error in the ray plane and normal to the ray
$\delta\theta_\perp$	Angular error out of the ray plane
ϑ	Zenith angle
ι	Inner scale length divided by the Fresnel length in Figure 4.8
κ	Turbulence wavenumber: $\kappa = 2\pi/\ell$
$\boldsymbol{\kappa}$	Turbulence wavenumber vector
κ_0	Outer-scale wavenumber cutoff: $\kappa_0 = 2\pi/L_0$
κ_s	Inner-scale wavenumber cutoff: $\kappa_s = 2\pi/\ell_0$

κ_m	$\kappa_m = 0.942\kappa_s$
κ_B	Bragg wavenumber for scattering defined in (9.8)
λ	Wavelength of electromagnetic radiation
μ	Exponent in structure-function separation scaling
$\mu_\chi(\tau)$	Autocorrelation of amplitude fluctuations (5.5)
$\mu(\tau)$	Autocorrelation of phase
$\mu(\rho)$	Spatial correlation of phase
$\mu(\tau, T)$	Autocorrelation of phase for sample length T
μ_m	Magnetic permeability connecting $\mathbf{E}$ and $\mathbf{H}$
ν	Exponent in ionospheric spectrum model
	Cross correlation of phase and amplitude (7.70)
	Index of Bessel functions
π	Numerical constant: $\pi = 3.141\,59$
ρ	Vector connecting adjacent receivers
ρ	Scalar separation between adjacent receivers
ρ_e	Charge density in Maxwell's equations
	Effective baseline for oblique propagation (4.39)
ρ_0	Coherence radius: $\rho_0 = r_0/2.099$
$\rho(\lambda_1, \lambda_2, \vartheta)$	Ray-path separation due to tropospheric refraction (6.21)
σ^2	Phase variance
σ_v^2	Variance of Gaussian wind-speed distribution
$d^2\sigma$	Surface element on receiver aperture
τ	Time delay used in autocorrelation functions
τ^2	Logarithmic amplitude variance
υ	Exponent in zenith-angle scaling law
$\upsilon(\rho, R)$	Normalized amplitude variance for a beam wave (3.95)
ϕ	Azimuth angle defining a point in a random medium
ϕ_i	Polar coordinate of points on receiving aperture
	Polar angle defining point on extended source
$\varphi(\mathbf{r})$	Phase fluctuation of electric-field strength
$\overline{\varphi}$	Average phase shift estimated from sample length T
$\Delta\varphi$	Phase difference measured between adjacent receivers
$\langle\varphi_0^2\rangle$	Phase variance computed with geometrical optics (7.4)
χ	Logarithmic amplitude fluctuation of signal
ψ	Polar angle defining wavenumber vector
$\psi_0(\mathbf{r})$	Rytov surrogate for unperturbed field (2.5)
$\psi_n(\mathbf{r})$	Succesive terms in Rytov series (2.13)
$\delta\psi$	Substantial change in surrogate function over distance ℓ

ω	Frequency in radians per second: $\omega = 2\pi f$
	Azimuthal angle defining wavenumber vector
ω_F	Fresnel frequency (5.11)
ω_0	Threshold frequency in single-path phase spectrum
ω_c	Corner frequency in phase-difference spectrum
ϱ	Displacement for locally frozen random medium
$\Gamma(x)$	Gamma function
$\Gamma(\mathbf{R}, \rho_1, \rho_2)$	Mutual coherence function (9.46)
$\Delta(\mathbf{R}, \rho_1, \rho_2)$	Beam-wave mutual-coherence-function term (9.64)
Θ	Angle between magnetic field and wavenumber vector
$\Theta(\kappa, \mathbf{R}, r)$	Double-scattering weighting: terrestrial links (8.6)
$\Lambda(\kappa)$	Complex weighting function (2.42)
Π	Normalized difference of intensity fluctuations (4.46)
$\Pi(\kappa, \mathbf{R}, r)$	Double-scattering weighting: astronomical (8.10)
Σ	Summation of following terms
Σ_v	Sum of phase and amplitude variances (8.19)
$\Upsilon(\rho)$	Difference of vertical and horizontal angular errors
$\Phi_\varepsilon(\kappa)$	Wavenumber spectrum of dielectric variations
$\Psi(\mathbf{r})$	Surrogate function in Rytov approximation (2.2)
	Eikonal function in geometrical optics
$\Phi_n(\kappa)$	Wavenumber spectrum of refractive-index variations
$\Psi_N(\kappa)$	Wavenumber spectrum of electron-density variations
$\Omega(\kappa)$	Normalized wavenumber spectrum: $\Phi_n(\kappa) = C_n^2 \, \Omega(\kappa)$
$F(I_1, I_2, C_\chi(\rho))$	Exponential in log-normal bivariate distribution (10.8)

$$\Phi_\varepsilon(\kappa) = \begin{cases} 4\Phi_n(\kappa) & \text{in the troposphere} \\ \\ r_0^2 \lambda^4 \Psi_N(\kappa) & \text{in the ionosphere} \end{cases}$$

Appendix B

Integrals of Elementary Functions

Numerical values for the following integrals with specific exponents are needed in our work.

1. $\displaystyle\int_0^\infty \frac{\sin x}{x^{\frac{2}{3}}}\, dx = 1.339\,47$

2. $\displaystyle\int_0^\infty \frac{\sin x}{x^{\frac{11}{6}}}\, dx = 6.451\,98$

3. $\displaystyle\int_0^\infty \frac{\sin^2 x}{x^{\frac{11}{6}}}\, dx = 1.540\,19$

4. $\displaystyle\int_0^\infty \frac{\sin^2 x}{x^{\frac{7}{3}}}\, dx = 4.836\,44$

5. $\displaystyle\int_0^\infty \frac{\sin^2 x}{x^{\frac{17}{6}}}\, dx = 6.270\,61$

6. $\displaystyle\int_0^\infty \frac{1}{x^{\frac{5}{3}}}\left(1 - \frac{\sin x}{x}\right) dx = 1.205\,52$

7. $\displaystyle\int_0^\infty \frac{1}{x^{\frac{11}{6}}}\left(1 - \frac{\sin x}{x}\right) dx = 0.942\,98$

8. $\displaystyle\int_0^\infty \frac{1}{x^{\frac{7}{3}}}\left(1 - \frac{\sin x}{x}\right) dx = 0.652\,87$

9. $\displaystyle\int_0^\infty \frac{1 - \cos x}{x^{\frac{5}{3}}}\, dx = 2.009\,20$

10. $\displaystyle\int_0^\infty \frac{1 - \cos x}{x^2}\, dx = \frac{\pi}{2} = 1.570\,80$

11. $\int_0^\infty \dfrac{1-\cos x}{x^{\frac{11}{6}}}\,dx = 1.728\,80$

12. $\int_0^1 x^{\frac{5}{6}}(1-x)^{\frac{5}{6}}\,dx = 0.220\,54$

13. $\int_0^1 x^{\frac{8}{3}}(1-x)^{\frac{8}{3}}\,dx = 0.011\,88$

14. $\int_1^\infty \dfrac{1}{x^{\frac{8}{3}}\sqrt{x^2-1}}\,dx = 0.841\,31$

15. $\int_1^\infty \dfrac{1}{x^{\frac{11}{3}}\sqrt{x^2-1}}\,dx = 0.739\,18$

16. $\int_0^\infty \dfrac{1}{\left(x^2+a^2\right)^{\frac{4}{3}}}\,dx = 1.120\,25a^{-\frac{5}{3}}$

17. $\int_0^\infty \dfrac{1}{\left(x^2+a^2\right)^{\frac{11}{6}}}\,dx = 0.841\,31a^{-\frac{8}{3}}$

18. $\int_0^\infty \dfrac{x^2}{\left(x^2+a^2\right)^{\frac{11}{6}}}\,dx = 1.261\,96a^{-\frac{2}{3}}$

19. $\int_0^\infty \dfrac{dx}{x^{\frac{2}{3}}(a^2+x^2)^{\frac{1}{3}}} = 5.782\,87a^{-\frac{1}{3}}$

20. $\int_0^\infty \left[(x^2+a^2)^{\frac{1}{3}} - \left(x^2\right)^{\frac{1}{3}}\right] dx = 1.457\,19a^{\frac{5}{3}}$

21. $\int_z^\infty \dfrac{x^{\mu-1}}{(1+\beta x)^\nu}\,dx = \dfrac{z^{\mu-\nu}}{\beta^\nu(\nu-\mu)}{}_2F_1\!\left(\nu,\nu-\mu;\nu-\mu+1;-\dfrac{1}{\beta z}\right) \quad \mu > \nu$

22. $\int_0^\infty \dfrac{x^{\mu-1}}{(1+\beta x)^\nu}\,dx - \dfrac{B(\mu,\nu-\mu)}{\beta^\mu} \quad\quad \nu > \mu$

In some applications it is important to have general expressions for these integrals that are valid for arbitrary exponents.

1. $\int_0^\infty x^{\mu-1}\sin(ax)\,dx = a^{-\mu}\Gamma(\mu)\sin\!\left(\dfrac{\pi}{2}\mu\right) \quad\quad 0 < \mu < 1, \quad a > 0$

2. $\int_0^\infty x^{\mu-1}\cos(ax)\,dx = a^{-\mu}\Gamma(\mu)\cos\!\left(\dfrac{\pi}{2}\mu\right) \quad\quad 0 < \mu < 1, \quad a > 0$

3. $\int_0^\infty x^{\mu-1} \sin^2(ax)\,dx = -a^{-\mu} \dfrac{\Gamma(\mu)}{2^{1+\mu}} \cos\left(\dfrac{\pi}{2}\mu\right)$

$$-2 < \mu < 0, \quad a > 0$$

4. $\int_0^\infty \dfrac{1}{x^\mu}\left(1 - \dfrac{\sin(ax)}{ax}\right) dx = a^{\mu-1} \dfrac{-\pi}{2\Gamma(1+\mu)\cos(\frac{1}{2}\pi\mu)}$

$$1 < \mu < 3, \quad a > 0$$

5. $\int_0^\infty \dfrac{1 - \cos(ax)}{x^\mu}\,dx = a^{\mu-1} \dfrac{-\pi}{2\Gamma(\mu)\cos(\frac{1}{2}\pi\mu)}$

$$1 < \mu < 3, \quad a > 0$$

6. $\int_0^\infty \dfrac{dx}{(x^2 + a^2)^\mu} = a^{1-2\mu} \dfrac{\sqrt{\pi}\,\Gamma(\mu - \frac{1}{2})}{2\Gamma(\mu)} \qquad |\arg a| < \dfrac{\pi}{2}, \quad \mu > \dfrac{1}{2}$

7. $\int_0^1 [x(1-x)]^\mu\,dx = \dfrac{[\Gamma(\mu+1)]^2}{\Gamma(2\mu+2)}$

8. $\int_0^\infty \dfrac{\sin(bx)}{(x^2 + a^2)^\mu}\,dx = (2a)^{\frac{1}{2}-\mu}\,(b)^{\mu-\frac{1}{2}} \dfrac{\sqrt{\pi}\,\Gamma(1-\mu)}{2}$

$$\times \left[I_{\mu-\frac{1}{2}}(ab) - \mathbf{L}_{\mu-\frac{1}{2}}(ab) \right]$$

$$b > 0, \quad \mu > 0, \quad a > 0$$

9. $\int_0^\infty \dfrac{\cos(bx)}{(x^2 + a^2)^\mu}\,dx = \left(\dfrac{2a}{b}\right)^{\frac{1}{2}-\mu} \dfrac{\sqrt{\pi}}{\Gamma(\mu)} K_{\frac{1}{2}-\mu}(ab)$

$$b > 0, \quad \mu > 0, \quad a > 0$$

10. $\int_0^{\frac{\pi}{2}} (\sin x)^{2\mu-1}(\cos x)^{2\nu-1}\,dx = \dfrac{1}{2}B(\mu, \nu) = \dfrac{1}{2}\dfrac{\Gamma(\mu)\Gamma(\nu)}{\Gamma(\mu+\nu)}$

$$\mu > 0, \quad \nu > 0$$

11. $\int_0^1 x^{a-1}(1-x)^{b-1}\,dx = B(a, b)$ $\qquad a > 0, \quad b > 0$

12. $\int_0^{\frac{\pi}{2}} \dfrac{1}{a^2 \cos^2 \omega + b^2 \sin^2 \omega}\,d\omega = \dfrac{\pi}{2ab}$

13. $\int_0^{\frac{\pi}{2}} \dfrac{\sin^2 \omega}{(a^2 \cos^2 \omega + b^2 \sin^2 \omega)^2}\,d\omega = \dfrac{\pi}{4ab^3}$

14. $\int_0^{\frac{\pi}{2}} \dfrac{\cos^2 \omega}{(a^2 \cos^2 \omega + b^2 \sin^2 \omega)^2}\,d\omega = \dfrac{\pi}{4a^3 b}$

15. $$\int_0^{\frac{\pi}{2}} \frac{1}{(a^2 \cos^2 \omega + b^2 \sin^2 \omega)^2} \, d\omega = \frac{\pi}{4ab}\left(\frac{1}{a^2} + \frac{1}{b^2}\right)$$

16. $$\int_0^{\frac{\pi}{2}} \frac{1}{c^2 + a^2 \cos^2 \omega + b^2 \sin^2 \omega} \, d\omega = \frac{\pi}{2\sqrt{(c^2 + a^2)(c^2 + b^2)}}$$

17. $$\int_0^{\infty} \cos(\omega x) \frac{\sin\left(b\sqrt{x^2 + a^2}\right)}{\sqrt{x^2 + a^2}} \, dx = \begin{cases} \dfrac{\pi}{2} J_0\left(a\sqrt{b^2 - \omega^2}\right), \\ \qquad \text{for } 0 < \omega < b \\ 0, \qquad \text{for } \omega > b \end{cases}$$

The following parametric representations are useful in evaluating the small difference of large quantities.

1. $$\left[u^2 + a^2\right]^{\frac{1}{3}} - u^{\frac{2}{3}} = \frac{1}{3}\int_0^{a^2} \frac{dx}{\left(u^2 + x\right)^{\frac{2}{3}}}$$

2. $$\int_0^z \left[(u^2 + a^2)^{\frac{1}{3}} - u^{\frac{2}{3}}\right] du = \frac{3}{5}z\left[(z^2 + \rho^2)^{\frac{1}{3}} - z^{\frac{2}{3}}\right]$$
$$+ \frac{2}{5}a^2 \int_0^z (x^2 + a^2)^{-\frac{2}{3}} \, dx$$

Appendix C

Integrals of Gaussian Functions

The *error function* is the simplest example of an integral involving the Gaussian function:

$$\text{erf}(x) = \frac{2}{\sqrt{\pi}} \int_0^x dt \, \exp(-t^2)$$

It can be evaluated using the following series expansion, although tables of numerical values are provided in many texts and references:[1]

$$\text{erf}(x) = \frac{2x}{\sqrt{\pi}} \left(1 - \frac{x^2}{1!\,3} + \frac{x^4}{2!\,5} - \frac{x^6}{3!\,7} + \cdots \right)$$

The following definite integrals are encountered in our work. They depend primarily on the error function and the Kummer function $M(a, b, z)$ which is discussed in Appendix G.

1. $\displaystyle \int_0^\infty x^\mu e^{-ax^2} dx = \frac{1}{2} \Gamma\left(\frac{1+\mu}{2}\right) \frac{1}{a^{\frac{\mu+1}{2}}}$ $\qquad \mu > -1, \quad a > 0$

2. $\displaystyle \int_0^\infty \cos(bx)\, e^{-ax^2} dx = \frac{1}{2}\sqrt{\frac{\pi}{a}} \exp\left(-\frac{b^2}{4a}\right) \qquad a > 0$

3. $\displaystyle \int_0^\infty x \sin(bx)\, e^{-ax^2} dx = \frac{b}{4a}\sqrt{\frac{\pi}{a}} \exp\left(-\frac{b^2}{4a}\right) \qquad a > 0$

4. $\displaystyle \int_0^\infty \frac{\sin(bx)}{x} e^{-ax^2} dx = \frac{\pi}{2} \text{erf}\left(\frac{b}{2\sqrt{a}}\right) \qquad a > 0$

[1] For example, see M. Abramowitz and I. A. Stegun, *Handbook of Mathematical Functions* (Dover, New York, 1964), 295–329.

5. $$\int_0^\infty \frac{1-\cos(bx)}{x^2} e^{-ax^2}\, dx = \frac{\pi}{2}b\ \text{erf}\left(\frac{b}{2\sqrt{a}}\right) - \sqrt{\pi a}\left[1 - \exp\left(-\frac{b^2}{4a}\right)\right]$$

$$a > 0$$

6. $$\int_0^\infty x^{\mu-1}\sin(bx)\, e^{-ax^2}\, dx = \frac{b}{2a^{\frac{\mu+1}{2}}}\Gamma\left(\frac{1+\mu}{2}\right) M\left(\frac{1+\mu}{2}, \frac{3}{2}, -\frac{b^2}{4a}\right)$$

$$a > 0, \quad \mu > -1$$

7. $$\int_0^\infty x^{\mu-1}\cos(bx)\, e^{-ax^2}\, dx = \frac{1}{2a^{\frac{\mu}{2}}}\Gamma\left(\frac{\mu}{2}\right) M\left(\frac{\mu}{2}, \frac{1}{2}, -\frac{b^2}{4a}\right)$$

$$a > 0, \quad b > 0, \quad \mu > 0$$

8. $$\int_0^\infty \frac{1}{x^{\frac{5}{3}}}\left(1 - \frac{\sin(bx)}{bx}\right) e^{-ax^2}\, dx = \frac{3}{2}\Gamma\left(\frac{2}{3}\right) a^{\frac{1}{3}}\left[M\left(-\frac{1}{3}, \frac{3}{2}, -\frac{b^2}{4a}\right) - 1\right]$$

9. $$\int_{-\infty}^\infty e^{-ax^2+bx}\, dx = \sqrt{\frac{\pi}{a}}\ \exp\left(\frac{b^2}{4a}\right)$$

10. $$\int_{-\infty}^\infty xe^{-ax^2+bx}\, dx = \sqrt{\frac{\pi}{a}}\left(\frac{b}{2a}\right)\exp\left(\frac{b^2}{4a}\right)$$

11. $$\int_0^\infty \frac{1}{x^{\frac{8}{3}}}\left(1 - e^{-ax^2}\right)\, dx = \frac{3}{5}\Gamma\left(\frac{1}{6}\right) a^{\frac{5}{6}}$$

12. $$\int_0^\infty \frac{1}{x^{\frac{8}{3}}}[1 - J_0(ibx)]e^{-ax^2}\, dx = \frac{3}{5}\Gamma\left(\frac{1}{6}\right) a^{\frac{5}{6}}\left[M\left(-\frac{5}{6}, 1; \frac{b^2}{4a}\right) - 1\right]$$

Appendix D

Bessel Functions

Bessel functions play an important role in descriptions of propagation through random media. This occurs because we often need to solve the wave equation in cylindrical coordinates. There are two families of Bessel functions. They differ by having real or imaginary arguments and we discuss them separately.

D.1 Ordinary Bessel Functions

The ordinary Bessel functions satisfy the following second-order differential equation:

$$\frac{d^2w}{dz^2} + \frac{1}{z}\frac{dw}{dz} + \left(1 - \frac{\nu^2}{z^2}\right)w = 0$$

which has two independent solutions:

$$w(z) = A J_\nu(z) + B Y_\nu(z)$$

The first solution is regular at the origin and is described by a power-series expansion:

$$J_\nu(z) = \left(\frac{z}{2}\right)^\nu \sum_{n=0}^\infty \frac{(-z^2/4)^n}{n!\,\Gamma(\nu+n+1)} \qquad |\arg z| < \pi$$

Tables of numerical values for integral and half-integral values of the index ν are available in standard references.[1] The Bessel functions corresponding to $\nu = 0$ and $\nu = 1$ occur frequently in our work:

$$J_0(x) = 1 - \frac{1}{(1!)^2}\left(\frac{x}{2}\right)^2 + \frac{1}{(2!)^2}\left(\frac{x}{2}\right)^4 - \frac{1}{(3!)^2}\left(\frac{x}{2}\right)^6 \cdots$$

$$J_1(x) = \frac{x}{2} - \frac{1}{2}\left(\frac{x}{2}\right)^3 + \frac{1}{12}\left(\frac{x}{2}\right)^5 \cdots$$

---

[1] M. Abramowitz and I. A. Stegun, *Handbook of Mathematical Functions* (Dover, New York, 1964), 355–433 and 435–478.

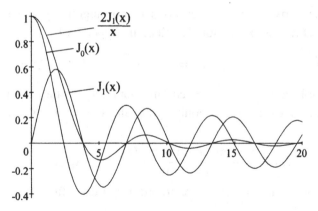

Figure D.1: A graph comparing the Bessel functions $J_0(x)$ and $J_1(x)$, and a common combination that occurs in descriptions of aperture averaging.

The following combination occurs in descriptions of aperture averaging:

$$\frac{2J_1(x)}{x} = 1 - \frac{1}{8}x^2 + \frac{1}{192}x^4 \cdots$$

These three functions are plotted together in Figure D.1.
For integral values of the index

$$J_n(-x) = (-1)^n J_n(x)$$

One can establish the following expressions for the derivatives of the first two Bessel functions from the power series:

$$\frac{d}{dx} J_0(x) = -J_1(x)$$

$$\frac{d}{dx} J_1(x) = J_0(x) - \frac{J_1(x)}{x}$$

The zeroth-order Bessel function has the following important property, which is called the *addition theorem*:

$$J_0\left(\kappa\sqrt{r^2 + \rho^2 - 2r\rho\cos(\alpha - \beta)}\right) = \sum_{0}^{\infty} \varepsilon_n J_n(\kappa r) J_n(\kappa \rho) \cos[n(\alpha - \beta)]$$

where

$$\varepsilon_n = \begin{cases} 1 & n = 0 \\ 2 & n \geq 1 \end{cases}$$

An important property of Bessel functions is their orthogonality, which is expressed in terms of the Dirac delta function described in Appendix F:

$$\int_0^\infty J_\nu(xa)J_\nu(xb)x\,dx = \frac{1}{a}\delta(a-b) \qquad \text{for} \quad \nu \geq -\frac{1}{2}$$

The second solution of Bessel's equation is not regular at the origin and we shall seldom need it in our studies. For nonintegral values of the index ν it is defined by the relationship[2]

$$Y_\nu(z) = \frac{\cos(\nu\pi)\,J_\nu(z) - J_{-\nu}(z)}{\sin(\nu\pi)} \qquad |\arg z| < \pi$$

The following definite integrals lead to ordinary Bessel functions.

1. $\displaystyle\int_0^{2\pi} \exp[i(a\sin x + b\cos x)]\,dx = 2\pi J_0(\sqrt{a^2+b^2})$

2. $\displaystyle\int_1^\infty \frac{\sin(ax)}{\sqrt{x^2-1}}\,dx = \frac{\pi}{2}J_0(a)$

3. $\displaystyle\int_0^1 \frac{\cos(ax)}{\sqrt{1-x^2}}\,dx = \frac{\pi}{2}J_0(a)$

4. $\displaystyle\int_{-\infty}^\infty \frac{\sin(\kappa\sqrt{x^2+a^2})}{\sqrt{x^2+a^2}}\,dx = \pi J_0(\kappa a)$

5. $\displaystyle\int_0^{\frac{\pi}{2}} \cos(a\cos\varphi)(\sin\varphi)^{2\nu}\,d\varphi = \frac{\Gamma(\frac{1}{2}+\nu)\Gamma(\frac{1}{2})}{2}\left(\frac{2}{a}\right)^\nu J_\nu(a) \quad \nu > -\frac{1}{2}$

We must carry out integrations that involve Bessel functions in many parts of our work. More than fifteen hundred integrals involving Bessel functions of all kinds have been solved analytically.[3] The following indefinite integral is valid for integral indices:

$$\int_0^z x^{n+1}J_n(x)\,dx = z^{n+1}J_{n+1}(x)$$

and the special case

$$\frac{1}{a^2}\int_0^a x J_0(x\kappa)\,dx = \frac{J_1(\kappa a)}{\kappa a}$$

is often used in describing aperture averaging by circular receivers.

---

[2] $Y_\nu(z)$ is denoted by $N_\nu(z)$ in some references.

[3] A. P. Prudnikov, Yu. A. Brychkov and O. I. Marichev, *Integrals and Series, Volume 2: Special Functions* (Gordon and Breach, New York and London, 1986), 168–420.
I. S. Gradshteyn and I. M. Ryzhik, *Table of Integrals, Series, and Products* (Academic Press, New York, 1980), 665–788.
A. D. Wheelon, *Tables of Summable Series and Integrals Involving Bessel Functions* (Holden-Day, San Francisco, 1968).

Definite integrals are needed frequently and the vast majority of the available analytical results falls in this category. In many cases the index v can assume complex values, but we need only real values in our applications. The other parameters that occur in these integrals can also be complex but are usually real in our work. We need only a small fraction of the available integrals and they are listed below.

1. $\displaystyle\int_0^\infty \frac{1}{x^{\frac{1}{3}}} J_1(ax)\,dx = 1.863\,88a^{\frac{2}{3}}$

2. $\displaystyle\int_0^\infty \frac{1}{x^{\frac{2}{3}}} J_1^2(ax)\,dx = 0.660\,28a^{-\frac{1}{3}}$

3. $\displaystyle\int_0^\infty \frac{1}{x^{\frac{8}{3}}} J_1^2(ax)\,dx = 0.864\,37a^{\frac{5}{3}}$

4. $\displaystyle\int_0^\infty \frac{1}{x^{\frac{8}{3}}}\left(1 - \frac{4J_1^2(ax)}{(ax)^2}\right) dx = 1.064\,98a^{\frac{5}{3}}$

5. $\displaystyle\int_0^\infty \frac{1}{x^{\frac{8}{3}}}[1 - J_0^2(ax) - J_1^2(ax)]\,dx = 1.037\,28a^{\frac{5}{3}}$

6. $\displaystyle\int_0^\infty \frac{1}{x^{\frac{8}{3}}}[1 - J_0(ax)]\,dx = 1.118\,33a^{\frac{5}{3}}$

7. $\displaystyle\int_0^\infty \frac{1}{x^{\frac{8}{3}}}\left(1 - \frac{2J_1(ax)}{ax}\right) dx = 0.609\,98a^{\frac{5}{3}}$

8. $\displaystyle\int_0^\infty \frac{1}{x^\mu} J_v(ax)\,dx = \frac{\Gamma\left(\dfrac{v - \mu + 1}{2}\right)}{2^\mu \Gamma\left(\dfrac{v + \mu + 1}{2}\right)} a^{\mu-1}$

$$-\tfrac{1}{2} < \mu < v + 1$$

9. $\displaystyle\int_0^\infty \frac{1}{x^\mu} J_v^2(ax)\,dx = \frac{\Gamma(\mu)\Gamma\left(\dfrac{2v + 1 - \mu}{2}\right)}{2^\mu\left[\Gamma\left(\dfrac{\mu + 1}{2}\right)\right]^2 \Gamma\left(\dfrac{2v + 1 + \mu}{2}\right)} a^{\mu-1}$

$$0 < \mu < 2v + 1, \quad a > 0$$

10. $\displaystyle\int_0^\infty \frac{1}{x^\mu}[1 - J_0(ax)]\,dx = \frac{\Gamma\left(\dfrac{3 - \mu}{2}\right)}{2^{\mu-1}(\mu - 1)\Gamma\left(\dfrac{\mu + 1}{2}\right)} a^{\mu-1}$

$$1 < \mu < 3$$

11. $\displaystyle\int_0^\infty J_\nu(ax)x^\mu\,dx = \frac{\Gamma\left(\dfrac{\nu+1+\mu}{2}\right)}{2\Gamma\left(\dfrac{\nu+1-\mu}{2}\right)}\left(\frac{2}{a}\right)^{\mu+1}$

$$\mu+\nu>-1, \quad a>0, \quad \mu<\tfrac{1}{2}$$

12. $\displaystyle\int_0^\infty [1-J_0(ax)]\frac{x}{(x^2+1)^{\frac{11}{6}}}\,dx = \frac{3}{5}\left(1 - \frac{2^{\frac{1}{6}}}{\Gamma\left(\frac{5}{6}\right)}a^{\frac{5}{6}}K_{\frac{5}{6}}(a)\right)$

13. $\displaystyle\int_0^\infty J_0(ax)\sin(bx^2)\,x\,dx = \frac{1}{2b}\cos\left(\frac{a^2}{4b}\right)$

14. $\displaystyle\int_0^\infty J_0(ax)\cos(bx^2)\,x\,dx = \frac{1}{2b}\sin\left(\frac{a^2}{4b}\right)$

15. $\displaystyle\int_0^\infty \frac{x^{\nu+1}J_\nu(ax)}{(x^2+b^2)^{\mu+1}}\,dx = \frac{a^\mu b^{\nu-\mu}}{2^\mu\Gamma(\mu+1)}K_{\mu-\nu}(ab)$

$$a,b>0, \quad -1<\nu<2\mu+\tfrac{3}{2}$$

16. $\displaystyle\int_0^\infty J_\nu(ax)\sin(bx)\,dx = \frac{\sin[\nu\sin^{-1}(b/a)]}{\sqrt{a^2-b^2}}$

$$0<b<a, \quad \nu>-2$$

$$= \frac{a^\nu\cos(\nu\pi/2)\left[b+\sqrt{b^2-a^2}\right]^{-\nu}}{\sqrt{b^2-a^2}}$$

$$0<a<b, \quad \nu>-2$$

17. $\displaystyle\int_0^\infty J_\nu(ax)\cos(bx)\,dx = \frac{\cos[\nu\sin^{-1}(b/a)]}{\sqrt{a^2-b^2}}, \quad \text{for} \quad 0<b<a$

$$= \frac{-a^\nu\sin(\nu\pi/2)\left[b+\sqrt{b^2-a^2}\right]^{-\nu}}{\sqrt{b^2-a^2}}$$

$$0<a<b, \quad \nu>-1$$

18. $\displaystyle\int_0^{\frac{\pi}{2}} J_{\mu+\nu}(2a\cos\varphi)\cos[(\mu-\nu)\varphi]\,d\varphi = \frac{\pi}{2}J_\mu(a)J_\nu(a)$

$$\mu+\nu>-1$$

19. $\displaystyle\int_0^{\frac{\pi}{2}} J_{m-n}(2a\cos\varphi)\cos[(m+n)\varphi]\,d\varphi = \frac{\pi}{2}(-1)^n J_m(a)J_n(a)$

$$n \text{ and } m \text{ integers}, \quad m-n>-1$$

20. $\displaystyle\int_0^\pi J_0(2a\cos\varphi)\cos(2n\varphi)\,d\varphi = (-1)^n\frac{\pi}{2}J_n^2(a)$ n an integer

21. $\displaystyle\int_0^{\frac{\pi}{2}} J_0(2a\cos\varphi)\,d\varphi = \frac{\pi}{2}J_0^2(a)$

22. $\displaystyle\int_0^{\frac{\pi}{2}} J_0(2a\cos\varphi)\cos(2\varphi)\,d\varphi = -\frac{\pi}{2}J_1^2(a)$

23. $\displaystyle\int_0^\pi J_{2n}(2a\sin\varphi)\,d\varphi = \pi J_n^2(a)$

24. $\displaystyle\int_0^{\frac{\pi}{2}} J_\mu(a\sin\varphi)(\sin\varphi)^{\mu+1}(\cos\varphi)^{2\nu+1}\,d\varphi = \frac{2^\nu\Gamma(\nu+1)}{a^{\nu+1}}J_{\mu+\nu+1}(a)$
$$\mu > -1,\quad \nu > -1$$

25. $\displaystyle\int_0^\infty J_0\left(a\sqrt{x^2+y^2}\right)\cos(bx)\,dx = \frac{\cos\left(y\sqrt{a^2-b^2}\right)}{\sqrt{a^2-b^2}}\quad a > b$
$$= 0 \qquad a \le b,\quad y > 0$$

26. $\displaystyle\int_0^\pi J_0(a\sin\varphi)\exp(ib\cos\varphi)\sin\varphi\,d\varphi = 2\frac{\sin\left(\sqrt{a^2+b^2}\right)}{\sqrt{a^2+b^2}}$

Weber's Integrals[4]

27. $\displaystyle\int_0^\infty J_0(ax)\exp(-p^2x^2)x\,dx = \frac{1}{2\,p^2}\exp\left(-\frac{a^2}{4p^2}\right)\quad \Re(p^2) > 0$

28. $\displaystyle\int_0^\infty J_\nu(ax)\exp(-p^2x^2)x^{\mu-1}\,dx$

$$= \frac{a^\nu\Gamma\left(\dfrac{\mu+\nu}{2}\right)}{2^{\nu+1}p^{\mu+\nu}\Gamma(\nu+1)}M\left(\frac{\mu+\nu}{2},\nu+1,-\frac{a^2}{4p^2}\right)$$

$$= \frac{a^\nu\Gamma\left(\dfrac{\mu+\nu}{2}\right)}{2^{\nu+1}p^{\mu+\nu}\Gamma(\nu+1)}\exp\left(-\frac{a^2}{4p^2}\right)M\left(\frac{\nu-\mu}{2}+1,\nu+1,\frac{a^2}{4p^2}\right)$$

$$\mu+\nu > 0,\quad \Re(p^2) > 0$$

[4] G. N. Watson, *A Treatise on the Theory of Bessel Functions* (Cambridge University Press, Cambridge, 1952), 393.

Lamb's Integral[5]

29. $\displaystyle\int_0^\infty J_0(ax)\frac{\exp\left(ik\sqrt{x^2+y^2}\right)}{\sqrt{x^2+y^2}}\,x\,dx$

$$= \frac{i}{\sqrt{k^2-a^2}}\exp\left(i|y|\sqrt{k^2-a^2}\right) \qquad a < k$$

$$= \frac{1}{\sqrt{a^2-k^2}}\exp\left(-|y|\sqrt{a^2-k^2}\right) \qquad a > k$$

Sommerfeld's Integral[6]

There is another way to describe the basic relationship suggested by the integral above and it is known as Sommerfeld's integral.

30. $\displaystyle\int_0^\infty \frac{J_0(xu)}{\sqrt{u^2-k^2}}\exp\left(-|y|\sqrt{x^2-k^2}\right)u\,du = \frac{\exp\left(ik\sqrt{x^2+y^2}\right)}{\sqrt{x^2+y^2}}$

$$0 \le \arg k < \pi, \quad -\pi/2 \le \arg\left(\sqrt{u^2-k^2}\right) < \pi/2, \quad x \text{ and } y \text{ real}$$

It can be established by taking the Fourier Bessel transform of Lamb's integral and using the orthogonality of the Bessel functions indicated earlier.

Series of Bessel Functions

Sometimes infinite series of Bessel functions are needed.

1. $\displaystyle\sum_0^\infty \varepsilon_{2n} J_{2n}(a)\cos(2n\theta) = \cos(a\sin\theta)$

2. $\displaystyle\sum_0^\infty \varepsilon_{2n}(-1)^n J_{2n}(a)\cos(2n\theta) = \cos(a\cos\theta)$

3. $\displaystyle\sum_0^\infty J_{2n+1}(a)\sin[(2n+1)\theta] = \sin(a\sin\theta)$

4. $\displaystyle\sum_0^\infty (-1)^n J_{2n+1}(a)\cos[(2n+1)\theta] = \sin(a\cos\theta)$

---

[5] H. Lamb, "On the Theory of Waves Propagated Vertically in the Atmosphere," *Proceedings of the London Mathematical Society* (2), **7**, 122–141 (10 December 1908). H. Bateman, *Partial Differential Equations of Mathematical Physics* (Dover, New York, 1944), 411.

[6] W. Magnus and F. Oberhettinger, *Formulas and Theorems for the Special Functions of Mathematical Physics* (Chelsea, New York, 1949), 34.

D.2 Modified Bessel Functions

These functions satisfy the modified Bessel differential equation:

$$\frac{d^2w}{dz^2} + \frac{1}{z}\frac{dw}{dz} - \left(1 + \frac{v^2}{z^2}\right)w = 0$$

and the two independent solutions are written as

$$w(z) = A I_v(z) + B K_v(z)$$

The first solution is regular at the origin and can be computed from its power-series expansion:

$$I_v(z) = \left(\frac{z}{2}\right)^v \sum_0^\infty \frac{1}{n!\,\Gamma(n+v+1)}\left(\frac{z^2}{4}\right)^n$$

From this it is apparent that

$$\lim_{z\to 0} [I_v(z)] = \frac{1}{\Gamma(v+1)}\left(\frac{z}{2}\right)^v$$

provided that v is not a negative integer. For real z this solution is related to the ordinary Bessel function by the relationship

$$I_v(z) = \exp\left(-i\frac{v\pi}{2}\right) J_v\left[z\exp\left(i\frac{v\pi}{2}\right)\right]$$

as one can verify by changing the independent variable in the power series or in the differential equation.

The second solution is known as the *MacDonald function* and diverges at the origin. It can be computed from the first solution using the following relationship, provided that v is not an integer:

$$K_v(z) = \frac{\pi}{2}\frac{I_{-v}(z) - I_v(z)}{\sin(v\pi)}$$

We are usually interested in fractional values of the index v so this form can be used in conjunction with the power series for $I_v(z)$ to calculate $K_v(z)$. When the argument is small

$$\lim_{z\to 0} [K_v(z)] = \frac{1}{2}\Gamma(v)\left(\frac{z}{2}\right)^{-v} \qquad v > 0$$

We often need the second term in the small-argument expansion and the following results are useful.

1. $\lim\limits_{x\to 0} [K_0(x)] = -\left[\ln\left(\dfrac{x}{2}\right) + \gamma\right]\left(1 + \dfrac{1}{4}x^2 + \cdots\right) + \dfrac{1}{4}x^2$

$$\text{Euler's constant } \gamma = 0.5772157$$

2. $\lim\limits_{x\to 0}\left[K_{\frac{1}{3}}(x)\right] = \dfrac{1}{2}\Gamma\left(\dfrac{1}{3}\right)\left(\dfrac{2}{x}\right)^{\frac{1}{3}}\left[1 - \left(\dfrac{x}{2}\right)^{\frac{2}{3}}\dfrac{\Gamma\left(\frac{2}{3}\right)}{\Gamma\left(\frac{4}{3}\right)}\cdots\right]$

3. $\lim\limits_{x\to 0}\left[K_{\frac{5}{6}}(x)\right] = \dfrac{1}{2}\Gamma\left(\dfrac{5}{6}\right)\left(\dfrac{2}{x}\right)^{\frac{5}{6}}\left(1 - x^{\frac{5}{3}}\dfrac{\Gamma\left(\frac{1}{6}\right)}{\Gamma\left(\frac{11}{6}\right)}\cdots\right)$

4. $\lim\limits_{x\to 0}\left[K_{\frac{4}{3}}(x)\right] = \dfrac{1}{2}\Gamma\left(\dfrac{4}{3}\right)\left(\dfrac{2}{x}\right)^{\frac{4}{3}}\left(1 - x^2\dfrac{\pi}{2\sqrt{3}\,\Gamma\left(\frac{2}{3}\right)\Gamma\left(\frac{4}{3}\right)}\cdots\right)$

5. $\lim\limits_{x\to 0}\left[K_{\frac{11}{6}}(x)\right] = \dfrac{1}{2}\Gamma\left(\dfrac{11}{6}\right)\left(\dfrac{2}{x}\right)^{\frac{11}{6}}\left(1 - x^2\dfrac{\pi}{2\Gamma\left(\frac{1}{6}\right)\Gamma\left(\frac{11}{6}\right)}\cdots\right)$

We encounter the modified Bessel functions primarily as definite integrals of elementary functions.

1. $\displaystyle\int_0^{\pi} \cos(n\varphi)\exp(b\cos\varphi)\,d\varphi = \pi I_n(b)$ n an integer

2. $\displaystyle\int_{-1}^{1} (1 - x^2)^{\nu - \frac{1}{2}}\exp(-bx)\,dx = \sqrt{\pi}\,\Gamma\left(\nu + \dfrac{1}{2}\right)\left(\dfrac{2}{b}\right)^{\nu} I_{\nu}(b)$

$$\nu > -\tfrac{1}{2}$$

3. $\displaystyle\int_0^{\infty} \exp\left(-ax - \dfrac{b}{x}\right) x^{\mu - 1}\,dx = 2\left(\dfrac{b}{a}\right)^{\frac{\mu}{2}} K_{\mu}\left(2\sqrt{ab}\right)$

4. $\displaystyle\int_1^{\infty} \dfrac{\exp(-bx)}{(x^2 - 1)^{\nu + \frac{1}{2}}}\,dx = \dfrac{1}{\sqrt{\pi}}\Gamma\left(\dfrac{1}{2} - \nu\right)\left(\dfrac{z}{2}\right)^{-\nu} K_{\nu}(b)$

$$\nu > -\tfrac{1}{2}, \quad |\arg b| < \pi/2$$

5. $\displaystyle\int_0^{\infty} \dfrac{\cos(bx)}{(x^2 + a^2)^{\nu + \frac{1}{2}}}\,dx = \dfrac{\sqrt{\pi}}{\Gamma\left(\nu + \frac{1}{2}\right)}\left(\dfrac{2a}{b}\right)^{-\nu} K_{\nu}(ab)$

$$\nu > -\tfrac{1}{2}, \quad b > 0, \quad |\arg a| < \pi/2$$

On rare occasions we need integrals involving the modified Bessel functions and the following are helpful.

1. $\displaystyle \int_0^z dx\, x^\nu K_{\nu-1}(x) = -z^\nu K_\nu(z) + 2^{\nu-1}\Gamma(\nu) \quad R(\nu) > 0$

2. $\displaystyle \int_0^\infty dx\, x^\nu K_\nu(x) = 2^{\nu-1}\sqrt{\pi}\,\Gamma\left(\nu + \frac{1}{2}\right)$

3. $\displaystyle \int_0^\infty dx\, \frac{x^{2\mu+1}}{\left(x^2 + z^2\right)^{\frac{\nu}{2}}} K_\nu\left(a\sqrt{x^2 + z^2}\right) = \frac{2^\mu \Gamma(\mu + 1)}{a^{\mu+1} z^{\nu-\mu-1}} K_{\nu-\mu-1}(az)$

$$a > 0, \quad R(\mu) > -1$$

4. $\displaystyle \int_{-\infty}^\infty dx\, |x|^\nu e^{-ax} K_\nu(b|x|) = \frac{\sqrt{\pi}\,\Gamma\left(\nu + \frac{1}{2}\right)(2b)^\nu}{\left(b^2 - a^2\right)^{\nu+\frac{1}{2}}} \quad a < b, \quad \nu < \frac{1}{2}$

Appendix E

Probability Distributions

E.1 The Gaussian Distribution

The Gaussian distribution is frequently used to describe random processes. The central-limit theorem tells us that a signal that is the additive result of a large number of independent contributions will be distributed as a Gaussian random variable. This is often the case in our work and the probability density function is written

$$P(x) = \frac{1}{\sigma\sqrt{2\pi}} \exp\left(-\frac{1}{2\sigma^2}(x - x_0)^2\right)$$

This is symmetrical around the most likely value x_0. The width and height of the distribution are determined by the parameter σ. The area under this function is unity for all values of x_0 and σ, as one can show by using the integrals in Appendix C:

$$\int_{-\infty}^{\infty} dx\, P(x) = 1$$

This model is widely used to describe thermal noise in electronic circuits. In our context, it provides a good description of phase fluctuations over a wide range of conditions. We will also use it to describe the distribution of wind speeds about a mean value. The average value of this distribution is independent of the parameter σ:

$$\langle x \rangle = \int_{-\infty}^{\infty} dx\, x P(x) = x_0$$

Notice that the average value coincides with the most likely value for a Gaussian distribution.

The mean-square value depends on both parameters:

$$\langle x^2 \rangle = \int_{-\infty}^{\infty} dx\, x^2 P(x) = x_0^2 + \sigma^2$$

This means that σ^2 is the variance of x and defines the spread of the distribution:

$$\left\langle (x - \langle x \rangle)^2 \right\rangle = \sigma^2$$

The Gaussian distribution becomes narrow and increases at the reference point $x = x_0$ as σ becomes small. In the limit as $\sigma \to 0$ this provides one definition for the delta function, as we shall see in Appendix F.

Using the integrals in Appendix C one can establish the following expressions for higher-order moments:

$$\left\langle x^3 \right\rangle = 3x_0\sigma^2 + x_0^3$$

$$\left\langle x^4 \right\rangle = 3\sigma^4 + 6\sigma^2 x_0^2 + x_0^4$$

$$\left\langle x^5 \right\rangle = 15x_0\sigma^4 + 10\sigma^2 x_0^3 + x_0^5$$

$$\left\langle x^6 \right\rangle = 15\sigma^6 + 45\sigma^4 x_0^2 + 15\sigma^2 x_0^4 + x_0^6$$

The characteristic function plays an important role in analyzing random variables and is defined by the ensemble average:

$$\langle \exp(iqx) \rangle = \int_{-\infty}^{\infty} dx \, \exp(iqx) \, P(x)$$

We can estimate this function by introducing the probability density function and carrying out the integrations:

$$\langle \exp(iqx) \rangle = \exp\left(iqx_0 - \tfrac{1}{2}q^2\sigma^2\right)$$

With this result we can express the moments of x as derivatives of the characteristic function:

$$\left\langle x^n \right\rangle = \int_{-\infty}^{\infty} dx \, P(x) \left(-i\frac{\partial}{\partial q}\right)^n \exp(iqx)\Bigg|_{q=0}$$

so that

$$\left\langle x^n \right\rangle = \left(-i\frac{\partial}{\partial q}\right)^n \exp\left(iqx_0 - \frac{1}{2}q^2\sigma^2\right)\Bigg|_{q=0}$$

Sometimes we need to average Gaussian random variables that occur in the exponent. We can do so by assigning special values to q in the characteristic function or differentiating with respect to q:

$$\langle \exp(\eta x) \rangle = \exp\left(\eta x_0 + \tfrac{1}{2}\eta^2\sigma^2\right)$$

$$\langle x \exp(\eta x) \rangle = \left(x_0 + \eta\sigma^2\right) \exp\left(\eta x_0 + \tfrac{1}{2}\eta^2\sigma^2\right)$$

The *probability distribution* is defined as the integral of the probability density function. It gives the probability that the random variable is somewhere in the range $-\infty < x < \eta$:

$$\mathcal{P}(x \leq \eta) = \int_{-\infty}^{\eta} dx\, \mathsf{P}(x)$$

This is also called the *cumulative probability* and is the quantity that is often measured. When the underlying variable x is Gaussian we can estimate it as follows:

$$\mathcal{P}(x \leq \eta) = \frac{1}{\sigma\sqrt{2\pi}} \int_{-\infty}^{\eta} dx\, \exp\left(-\frac{1}{2\sigma^2}(x - x_0)^2\right)$$

$$= \frac{1}{\sqrt{\pi}}\left(\int_{-\infty}^{0} du\, e^{-u^2} + \int_{0}^{\frac{\eta - x_0}{\sigma\sqrt{2}}} du\, e^{-u^2}\right)$$

The second integral is the error function defined in Appendix C, which is widely available in tabulated form:[1]

$$\mathcal{P}(x \leq \eta) = \frac{1}{2}\left[1 + \operatorname{erf}\left(\frac{\eta - x_0}{\sigma\sqrt{2}}\right)\right]$$

The complementary measure is the *exceedance probability* which describes the probability that the signal level exceeds a prescribed value. Since the two measures must add to give unity,

$$\mathcal{P}(x \geq \eta) + \mathcal{P}(x \leq \eta) = 1$$

and we can write

$$\mathcal{P}(x \geq \eta) = \frac{1}{2}\left[1 - \operatorname{erf}\left(\frac{\eta - x_0}{\sigma\sqrt{2}}\right)\right]$$

E.2 The Log-Normal Distribution

A log-normal distribution for signal amplitude is a natural consequence of the Rytov approximation, as noted in Section 10.1. That prediction is confirmed by optical and microwave measurements made in the weak-scattering regime and beyond. This description provides no information about the signal's phase. Its probability density function takes the following form:

$$\mathsf{P}(x) = \frac{1}{\sigma\sqrt{2\pi}}\frac{1}{x}\exp\left\{-\frac{1}{2\sigma^2}\left[\log\left(\frac{x}{x_0}\right)\right]^2\right\} \qquad \text{for} \quad 0 < x < \infty$$

---

[1] For example, see M. Abramowitz and I. A. Stegun, *Handbook of Mathematical Functions* (Dover, New York, 1964), 295–329.

where x_0 and σ are parameters to be determined. With the substitution $u = \log(x/x_0)$ one can show that this probability distribution function is normalized to unity for all values of the parameters.

$$\int_0^\infty dx\, P(x) = 1$$

The mean value of this distribution depends both on x_0 and on σ:

$$\langle x \rangle = x_0 \exp\left(\tfrac{1}{2}\sigma^2\right)$$

as do the other moments. The mean-square value is

$$\langle x^2 \rangle = x_0^2 \exp\left(2\sigma^2\right)$$

and the higher moments are given by the relationship

$$\langle x^n \rangle = x_0^n \exp\left(\tfrac{1}{2}n^2\sigma^2\right)$$

The cumulative probability for this distribution is described in terms of the error function:

$$\mathcal{P}(x \leq \mathsf{L}) = \frac{1}{2}\left\{1 + \mathrm{erf}\left[\frac{1}{\sigma\sqrt{2}}\log\left(\frac{\mathsf{L}}{x_0}\right)\right]\right\}$$

and plots as a straight line on probability coordinates as noted in Figure 10.1. The slope of this straight line is set exclusively by the dispersion parameter σ. The reference level x_0 defines the median amplitude, with half the values of x exceeding x_0 and half being less than x_0.

A bivariate distribution for the signal amplitudes measured at adjacent receivers was described in Section 10.3. The general moments of that distribution

$$\langle x^n y^m \rangle$$

can be described in closed form.[2]

E.3 The Nakagami-*m* Distribution

The Nakagami-*m* distribution was discovered by analyzing the fading of short-wave radio signals reflected in the ionosphere. We now know that it describes a wide class of electromagnetic signals that are modulated randomly by transmission or reflection. This model describes only positive values of the independent variable,

[2] R. Frehlich, "Laser Scintillation Measurements of the Temperature Spectrum in the Atmospheric Surface Layer," *Journal of Atmospheric Sciences*, **49**, No. 16, 1494–1509 (15 August 1992).

which are identified with the intensity or amplitude of a signal in our applications.[3]

$$P(x) = \frac{2m^m}{\sigma^{2m}\Gamma(m)} x^{2m-1} \exp\left(-m\frac{x^2}{\sigma^2}\right) \qquad \text{for} \quad 0 < x < \infty$$

A normalized version of this function is plotted in Figure 10.16. Using a result from Appendix C, one can demonstrate that this distribution is normalized to unity for all values of the parameters σ and m:

$$\int_0^\infty dx\, P(x) = 1$$

provided that

$$m \geq \tfrac{1}{2}$$

It is significant that this parameter need not be an integer. The Nakagami-m distribution reduces to familiar models for special values of m. When $m = \tfrac{1}{2}$ it yields the one-sided Gaussian model. The Rayleigh distribution emerges when one takes $m = 1$. We usually look to measurements to set the value of m and large values are often encountered.

The Nakagami-m distribution has no obvious reference value x_0, in contrast to the Gaussian model. The probability distribution function reaches its maximum value when

$$x_{\max} = \sigma\sqrt{1 - 1/(2m)}$$

but this depends on both parameters. The average value also has a joint dependence:

$$\langle x \rangle = \sigma\frac{\Gamma\left(m + \tfrac{1}{2}\right)}{\Gamma(m)\sqrt{m}}$$

On the other hand, the mean-square value depends only on σ:

$$\langle x^2 \rangle = \sigma^2$$

From these results, one can write the variance as follows:

$$\langle (x - \langle x \rangle)^2 \rangle = \sigma^2\left(1 - \frac{\Gamma^2\left(m + \tfrac{1}{2}\right)}{\Gamma^2(m)m}\right)$$

---

[3] M. Nakagami, "The *m*-Distribution – A General Formula of Intensity Distribution of Rapid Fading," in *Statistical Methods in Radio Wave Propagation*, edited by W. C. Hoffman (Pergamon Press, London, 1960), 3–36.

When m is very large one can exploit Sterling's asymptotic formula for the gamma function provided in Appendix K to express

$$\lim_{m\to\infty} \langle(x - \langle x\rangle)^2\rangle = \frac{\sigma^2}{4m}$$

and this is used to determine m from measurements of the scintillation index S_4 as explained in (10.57).

The higher-order moments can also be evaluated in closed form:

$$\langle x^k\rangle = \frac{\sigma^k \Gamma(m + k/2)}{m^{\frac{k}{2}}\Gamma(m)}$$

Nakagami also derived an integral expression for the probability distribution function, which is often helpful in evaluating ensemble averages:

$$P(x) = \frac{x^m}{2^{m-1}\Gamma(m)} \int_0^\infty d\zeta\, J_{m-1}(\zeta x)\zeta^m \exp\left(-m\frac{\zeta^2}{\sigma^2}\right)$$

This connection can be verified by using Weber's second integral, which is described in Appendix D. In his original paper, Nakagami also presented an expression for the bivariate distribution of two correlated random variables. This can be used to describe signals measured at adjacent receivers or time-delayed signals measured at a single site.

E.4 The Bivariate Gaussian Distribution

In our work we often need to describe the statistical properties of two quantities that are each distributed as Gaussian random variables. The bivariate probability density function is the key to these descriptions. To first order the Rytov approximation leads to this result, as we have found in Chapter 10. The probability density function for this situation is represented quite generally as the exponential of a quadratic function in the excursions of the two variables about their mean values:

$$P(x, y) = \frac{1}{N} \exp\left\{-\left[A(x - x_0)^2 + B(y - y_0)^2 + 2C(x - x_0)(y - y_0)\right]\right\}$$

This joint distribution must reduce to the simple Gaussian expression when it is integrated over all values of either x or y. This condition allows one to express the constants in terms of their variances and cross correlation:

$$\sigma_x^2 = \langle(x - x_0)^2\rangle$$
$$\sigma_y^2 = \langle(y - y_0)^2\rangle$$
$$\nu\sigma_x\sigma_y = \langle(x - x_0)(y - y_0)\rangle$$

The constants in the general expression for the bivariate probability distribution are defined in terms of these parameters by

$$N = 2\pi \sigma_x \sigma_y \sqrt{1 - v^2}$$

$$A = \left[2\sigma_x^2(1 - v^2) \right]^{-1}$$

$$B = \left[2\sigma_y^2(1 - v^2) \right]^{-1}$$

$$C = -v \left[\sigma_x \sigma_y (1 - v^2) \right]^{-1}$$

In our applications the average values of x and y vanish, so that

$$\langle x^2 \rangle = \sigma_x^2 \qquad \langle y^2 \rangle = \sigma_y^2 \qquad \text{and} \qquad \langle xy \rangle = v\sigma_x \sigma_y$$

and

$$P(x, y) = \frac{1}{2\pi \sigma_x \sigma_y \sqrt{1 - v^2}} \exp\left[-\frac{1}{2(1 - v^2)} \left(\frac{x^2}{\sigma_x^2} + \frac{y^2}{\sigma_y^2} - 2v \frac{xy}{\sigma_x \sigma_y} \right) \right]$$

Using integral results found in Appendix C, we can show that this is normalized to unity for all values of the parameters σ_x, σ_y and v :

$$\int_{-\infty}^{\infty} dx \int_{-\infty}^{\infty} dy \, P(x, y) = 1$$

The parameters in this distribution are the variances and cross correlation of the signal components. Higher-order moments can be computed by using results from Appendix C:

$$\langle x^3 y \rangle = 3v\sigma_x^3 \sigma_y$$

$$\langle xy^3 \rangle = 3v\sigma_x \sigma_y^3$$

$$\langle x^2 y^2 \rangle = \sigma_x^2 \sigma_y^2 (1 + 2v^2)$$

The joint characteristic function for the bivariate Gaussian distribution is calculated as the double Fourier integral of the bivariate probability density function:

$$\langle \exp[i(px + qy)] \rangle = \int_{-\infty}^{\infty} dx \int_{-\infty}^{\infty} dy \, P(x, y) \exp[i(px + qy)]$$

and is evaluated to give the following expression:

$$\langle \exp[i(px + qy)] \rangle = \exp\left[-\tfrac{1}{2} (p^2 \sigma_x^2 + 2vpq\sigma_x \sigma_y + q^2 \sigma_y^2) \right]$$

The moments of x and y can be expressed as partial derivatives of the characteristic function:

$$\langle x^n y^m \rangle = \frac{\partial^n}{\partial p^n} \frac{\partial^m}{\partial q^m} \langle \exp[i(px + qy)] \rangle \Big|_{p=q=0}$$

$$= \frac{\partial^n}{\partial p^n} \frac{\partial^m}{\partial q^m} \exp\left[-\tfrac{1}{2}(p^2\sigma_x^2 + 2vpq\sigma_x\sigma_y + q^2\sigma_y^2)\right] \Big|_{p=q=0}$$

and this often simplifies the required calculation.

In the first volume we identified the variables x and y with phase fluctuations measured at adjacent receivers:

$$x = \varphi(R) \quad \text{and} \quad y = \varphi(R + \rho)$$

We have also identified them with phase fluctuations measured at the same receiver but displaced in time:

$$x = \varphi(t) \quad \text{and} \quad y = \varphi(t + \tau)$$

One can assume that the variances are the same if the receivers are not too far apart or the time delay is not too great:

$$\sigma_x = \sigma_y = \sigma$$

The bivariate probability density function under these assumptions is

$$P(x, y) = \frac{1}{2\pi\sigma^2\sqrt{1 - v^2}} \exp\left(\frac{-1}{2\sigma^2(1 - v^2)}(x^2 + y^2 - 2vxy)\right)$$

In Section 8.2.1 of the first volume we showed that this leads to the following probability density function for their difference:

$$P(x - y) = \frac{1}{2\sigma\sqrt{\pi}\sqrt{1 - v}} \exp\left(-\frac{(x - y)^2}{4\sigma^2(1 - v)}\right)$$

The *conditional probability* density function for x given that its companion y has a fixed value is another useful measure:

$$P(x|y) = \frac{P(x, y)}{P(y)} = \frac{1}{\sigma\sqrt{2\pi}\sqrt{1 - v^2}} \exp\left(-\frac{(x - vy)^2}{2\sigma^2(1 - v^2)}\right)$$

In writing this we have used the simple Gaussian distribution for y given previously.

Appendix F

Delta Functions

The delta function was first widely used by Dirac in quantum mechanics[1] and now plays an important role in descriptions of propagation through random media. Of all the special functions used in mathematical physics the delta function is the most unusual. It is not a proper function in the mathematical sense since it is neither continuous nor differentiable. Our first approach to understanding it is to consider its influence when it is paired with a continuous function in the integrand of a definite integral:

$$\int_a^b \delta(x - x_0) f(x)\, dx = f(x_0) \qquad a < x_0 < b$$

We regard this expression as the basic definition of the delta function. Its unique property is that it allows one to select the value of a companion function at the reference point and completely disregard all other values in the range. For this reason it is sometimes called the *sifting integral*. The following normalization condition is an immediate consequence:

$$\int_{-\infty}^{\infty} \delta(x - x_0)\, dx = 1$$

This follows on setting $f(x) = 1$ and noting that the reference point must occur somewhere in the infinite range of integration.

   Another way to think about the delta function is to imagine that it is an infinitely high and infinitely narrow pulse. It would satisfy the basic definition if it had this characteristic. We approximate this unusual behavior by a limiting process applied

---

[1] P. A. M. Dirac, *The Principles of Quantum Mechanics* (Clarendon Press, Oxford, 1930). The delta function was known to Oliver Heaviside and others much earlier.

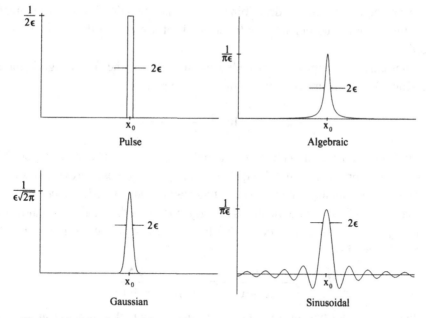

Figure F.1: Graphical descriptions of four models that approach the Dirac delta function in the limit $\epsilon \to 0$.

to the square wave illustrated in Figure F.1 above:

$$\delta(x - x_0) = \lim_{\epsilon \to 0} \left\{ \begin{array}{ll} \dfrac{1}{2\epsilon} & x_0 - \epsilon < x < x_0 + \epsilon \\ \\ 0 & \text{otherwise} \end{array} \right\}$$

On introducing this expression into the basic definition and expanding the companion function in a Taylor series about the reference point, we obtain

$$\int_a^b \delta(x - x_0) f(x) \, dx = \lim_{\epsilon \to 0} \left(\frac{1}{2\epsilon} \int_{x_0-\epsilon}^{x_0+\epsilon} dx \, [f(x_0) + (x - x_0) f'(x_0) \right.$$
$$\left. + \frac{1}{2}(x - x_0)^2 f''(x_0)] \right)$$

The second term is as often negative as it is positive in the integration interval and vanishes. The third term is of order ϵ^2 when it is integrated and disappears as $\epsilon \to 0$. The combined result is the original definition:

$$\int_a^b \delta(x - x_0) f(x) \, dx = f(x_0)$$

This description is useful in describing short-pulse signals. On the other hand, it is neither continuous nor differentiable, both of which are desirable physical properties.

An alternative description is provided by the limit of the following algebraic expression which is a smooth version of the sharp pulse:

$$\delta(x - x_0) = \frac{1}{\pi} \lim_{\epsilon \to 0} \left(\frac{\epsilon}{\epsilon^2 + (x - x_0)^2} \right)$$

This function is illustrated in the second panel of Figure F.1. It becomes higher and narrower as ϵ approaches zero. The quantity in large parentheses provides a simple description of frequency filters used to tune radio and television receivers to a particular station. In these applications $\epsilon = \Delta f$ corresponds to the filter bandwidth and $x_0 = f_0$ to its center frequency. The factor π ensures that the normalization condition is satisfied:

$$\frac{1}{\pi} \int_{-\infty}^{\infty} \frac{\epsilon \, dx}{\epsilon^2 + (x - x_0)^2} = \frac{2\epsilon}{\pi} \int_{0}^{\infty} \frac{du}{\epsilon^2 + u^2} = 1$$

The algebraic model has an important consequence, which follows from the following identity:

$$\frac{1}{z - i\epsilon} = \frac{z}{z^2 + \epsilon^2} + i \frac{\epsilon}{z^2 + \epsilon^2}$$

By taking the limit of both sides we can express the result in terms of the *principal value*[2] of z and the delta function:

$$\lim_{\epsilon \to 0} \left(\frac{1}{z - i\epsilon} \right) = \mathcal{P} \left(\frac{1}{z} \right) + i\pi \delta(z)$$

This means that we can also represent the delta function as

$$\delta(x - x_0) = \Im \left[\frac{1}{\pi} \lim_{\epsilon \to 0} \left(\frac{1}{x - x_0 - i\epsilon} \right) \right]$$

One encounters this combination in some physical problems in which ϵ represents the influence of a small attenuation term on the wavenumber and hence on Green's function.

Another representation of the delta function is to describe it as the limit of a sinusoidal function:

$$\delta(x - x_0) = \frac{1}{\pi} \lim_{\epsilon \to 0} \left(\frac{\sin[\epsilon^{-1}(x - x_0)]}{x - x_0} \right)$$

[2] E. T. Whittaker and G. N. Watson, *A Course of Modern Analysis* (Cambridge University Press, Cambridge, 1947), 75.

Like the previous models, it has a very large value at x_0 and falls rapidly to zero away from the reference point. This behavior is illustrated in the fourth panel of Figure F.1. The factor π^{-1} ensures that its integration over all values satisfies the normalization condition. This formulation leads to two important integral representations of the delta function. Noting that

$$\int_{-\epsilon^{-1}}^{\epsilon^{-1}} d\kappa \, \exp[i\kappa(x - x_0)] = 2\frac{\sin[\epsilon^{-1}(x - x_0)]}{x - x_0}$$

the delta function can be expressed as

$$\delta(x - x_0) = \lim_{\epsilon \to 0} \left(\frac{1}{2\pi} \int_{-\epsilon^{-1}}^{\epsilon^{-1}} d\kappa \, \exp[i\kappa(x - x_0)] \right)$$

or, finally,

$$\delta(x - x_0) = \frac{1}{2\pi} \int_{-\infty}^{\infty} d\kappa \, \exp[i\kappa(x - x_0)]$$

Since the imaginary term integrates to zero an equivalent version is

$$\delta(x - x_0) = \frac{1}{\pi} \int_{0}^{\infty} d\kappa \, \cos[\kappa(x - x_0)]$$

These integral definitions of the delta function provide a convenient way to represent the current distribution of a point source. In such cases we can write the current density as the product of three delta functions that localize the current density to the immediate position of the transmitter:

$$j(\mathbf{r}) = j_0 \delta(x - x_0)\delta(y - y_0)\delta(z - z_0)$$

The appropriate generalization of the infinite integral above allows us to express a point source in terms of a Fourier integral in three dimensions:

$$j(\mathbf{r}) = j_0 \frac{1}{8\pi^3} \int d^3\kappa \, \exp[i\kappa \cdot (\mathbf{r} - \mathbf{r}_0)]$$

This result is used to generate solutions for a point source (i.e. spherical waves) and to develop explicit expressions for Green's function.

There is a fourth way to describe the delta function. It follows by taking the limit of a Gaussian probability density function as the variance goes to zero:

$$\delta(x - x_0) = \lim_{\epsilon \to 0} \left[\frac{1}{\epsilon\sqrt{2\pi}} \exp\left(-\frac{1}{2\epsilon^2}(x - x_0)^2 \right) \right]$$

This is also illustrated in Figure F.1, where one sees that ϵ represents the width of the distribution. It is normalized to unity for all possible values of x_0 and ϵ.

This formulation occurs because the Gaussian distribution is quite common and the dispersion of the measured quantity is often small.

We have now found four analytical models that lead to the same conclusion in the appropriate limit. When we encounter any one of them we will replace it by $\delta(x - x_0)$. The resulting calculations are simplified by using the basic definition to collapse one of the remaining integrations.

Several important properties of the delta function can be established with the models advanced here. Sometimes the argument of the delta function is multiplied by a parameter. Appealing to the integral definition and rescaling the transform variable, we see that

$$\delta(\eta x) = \frac{1}{|\eta|}\delta(x)$$

and, in particular,

$$\delta(-x) = \delta(x)$$

By factoring the argument and using the rescaling relation we can also show that

$$\delta(x^2 - a^2) = \frac{1}{2|a|}[\delta(x - a) + \delta(x + a)]$$

In some applications we encounter delta functions whose arguments are more complicated functions of x:

$$\delta[g(x)]$$

If $g(x)$ vanishes at a point x_0 in the integration range, we can expand it in a Taylor series about that point:

$$g(x) = (x - x_0)g'(x_0) + \tfrac{1}{2}(x - x_0)^2 g''(x_0)$$

When the first derivative does not vanish at the reference point one can use the rescaling relation to write

$$\delta[g(x)] = \frac{1}{|g'(x_0)|}\delta(x - x_0)$$

The basic definition imagines that the reference point lies in the range of integration but does not coincide with the end points. If it does, only half of the area under the symmetrical delta function is included in the range of integration:

$$\int_a^b \delta(x - a)f(x)\,dx = \frac{1}{2}f(a)$$

$$\int_a^b \delta(x-b)f(x)\,dx = \frac{1}{2}f(b)$$

The derivative of a delta function occurs inside the integration in some applications:

$$\int_a^b f(x)\frac{d}{dx}\delta(x-x_0)\,dx \qquad a < x_0 < b$$

We integrate by parts to obtain

$$\int_a^b f(x)\frac{d}{dx}\delta(x-x_0)\,dx = f(x)\delta(x-x_0)\big|_a^b - \int_a^b \delta(x-x_0)\frac{df(x)}{dx}\,dx$$

The integrated term vanishes because x_0 falls in the range between a and b:

$$\int_a^b f(x)\frac{d}{dx}\delta(x-x_0)\,dx = -\frac{df(x)}{dx}\bigg|_{x=x_0}$$

The delta function occurs in expressing the inherent orthogonality of many special functions. In Appendix D we showed that two Bessel functions with different arguments, integrated over all positive values, give a delta function:

$$\int_0^\infty J_\nu(xa)J_\nu(xb)x\,dx = \frac{1}{a}\delta(a-b) \qquad \text{for} \quad \nu > -\frac{1}{2}$$

This is a special case of a general class of *eigenfunctions*, which have the following property:[3]

$$\int_0^\infty \psi(x,a)\psi(x,b)r(x)\,dx = \frac{1}{r(a)}\delta(a-b)$$

where $r(x)$ is a density function that is associated with the family of eigenfunctions $\psi(x,a)$.

---

[3] P. M. Morse and H. Feshbach, *Methods of Theoretical Physics: Part I* (McGraw-Hill, New York, 1953), 764.

Appendix G

Kummer Functions

Kummer functions play a small but important role in our studies of electromagnetic propagation through random media. The second-order differential equation

$$z \frac{d^2 w}{dz^2} + (b - z) \frac{dw}{dz} - aw = 0$$

has two independent solutions:

$$w(z) = AM(a, b, z) + BU(a, b, z)$$

The variable z can be real or complex and we will use both versions in our work. The first solution is well behaved at the origin but the second is not.

Our applications depend primarily on the first solution, which is called the *Kummer function*. It is also known as the *confluent hypergeometric function* and has two equivalent notations:

$$M(a, b, z) = {}_1F_1(a, b, z)$$

These functions reduce to elementary forms for various combinations of the parameters a and b.[1] This is seldom the case in our applications and hence we must use the power-series expansion to compute numerical values:

$$M(a, b, z) = 1 + \frac{a}{b} z + \frac{a(a+1)}{b(b+1)} \frac{z^2}{2!} + \cdots + \frac{(a)_n}{(b)_n} \frac{z^n}{n!} + \cdots$$

Here

$$a_0 = 1 \quad \text{and} \quad (a)_n = a(a+1)(a+2) \ldots (a+n-1)$$

with a similar expression for $(b)_n$. Several useful properties emerge from this series:

$$M(a, b, 0) = 1 \quad \text{for } b \neq -n$$

---

[1] M. Abramowitz and I. A. Stegun, *Handbook of Mathematical Functions* (Dover, New York, 1964), 509–510.

$$M(a, b, z) = e^z M(b - a, b, -z)$$

$$\frac{d}{dz} M(a, b, z) = \frac{a}{b} M(a + 1, b + 1, z)$$

$$\frac{d^n}{dz^n} M(a, b, z) = \frac{(a)_n}{(b)_n} M(a + n, b + n, z)$$

The second function is defined in terms of the Kummer function as follows:

$$U(a, b, z) = \frac{\pi}{\sin(b\pi)} \left(\frac{M(a, b, z)}{\Gamma(1 + a - b)\Gamma(b)} - z^{1-b} \frac{M(1 + a - b, 2 - b, z)}{\Gamma(a)\Gamma(2 - b)} \right)$$

The following asymptotic expansions of these functions are sometimes needed.

1. $\displaystyle \lim_{z \to \infty} M(a, b, z) = \frac{\Gamma(b)}{\Gamma(a)} e^z z^{a-b} [1 + O(|z|^{-1})] \quad \Re(z) > 0$

2. $\displaystyle \lim_{z \to \infty} M(a, b, z) = \frac{\Gamma(b)(-z)^{-a}}{\Gamma(b - a)} [1 + O(|z|^{-1})] \quad \Re(z) < 0$

3. $\displaystyle \lim_{z \to \infty} U(a, b, z) = z^{-a} [1 + O(|z|^{-1})]$

Kummer functions occur in our work primarily as definite integrals that express measured quantities. They also occur as weighting functions in integrals over the turbulence spectrum.

1. $\displaystyle \int_0^1 e^{zx} x^{a-1} (1 - x)^{b-a-1} \, dx = \frac{\Gamma(b - a)\Gamma(a)}{\Gamma(b)} M(a, b, z) \qquad \Re(b) > \Re(a) > 0$

2. $\displaystyle \int_{-1}^1 e^{-\frac{zx}{2}} (1 - x)^{a-1} (1 + x)^{b-a-1} \, dx = 2^{b-1} e^{-\frac{z}{2}} \frac{\Gamma(b - a)\Gamma(a)}{\Gamma(b)} M(a, b, z)$

$$\Re(b) > \Re(a) > 0$$

3. $\displaystyle \int_1^\infty e^{-zx} x^{b-a-1} (x - 1)^{a-1} \, dx = e^{-z} \Gamma(a) U(a, b, z) \qquad \Re(a) > 0, \quad \Re(z) > 0$

4. $\displaystyle \int_0^\infty e^{-zx} x^{a-1} (x + 1)^{b-a-1} \, dx = \Gamma(a) U(a, b, z) \qquad \Re(a) > 0, \quad \Re(z) > 0$

The following results are useful when the Kummer functions must be integrated over the independent variable.

1. $\displaystyle \int_0^\infty x^{b-1} M(a, c, -x) \, dx = \frac{\Gamma(b)\Gamma(c)\Gamma(a - b)}{\Gamma(a)\Gamma(c - b)} \qquad \Re(a) > \Re(b) > 0$

2. $\displaystyle \int_0^z x^{b-1} (z - x)^{c-b-1} M(a, b, x) \, dx = z^{c-1} \frac{\Gamma(b)\Gamma(c - b)}{\Gamma(c)} M(a, c, z)$

$$\Re(c) > \Re(b) > 0$$

Appendix H

Hypergeometric Functions

The *hypergeometric functions of Gauss* play a small role in descriptions of electromagnetic propagation. They are solutions of the second-order differential equation:

$$z(1 - z)\frac{d^2w}{dz^2} + [c - (a + b + 1)z]\frac{dw}{dz} - abw(z) = 0$$

The solution which is regular at the origin is denoted by

$$w(z) = {}_2F_1(a, b, c; z)$$

When the parameters a, b and c take on integral or half-integral values it reduces to elementary functions or one of the special functions of mathematical physics. These cases are seldom encountered in our applications and hence we must use the following series expansion to compute numerical values:

$$_2F_1(a, b, c; z) = 1 + \frac{a\,b}{c}\frac{z}{1!} + \frac{a(a + 1)b(b + 1)}{c(c + 1)}\frac{z^2}{2!} + \cdots$$

or, more compactly,

$$_2F_1(a, b, c; z) = \frac{\Gamma(c)}{\Gamma(a)\Gamma(b)}\sum_0^\infty \frac{\Gamma(a + n)\Gamma(b + n)}{\Gamma(c + n)}\frac{z^n}{n!} \qquad \text{for} \qquad |z| < 1$$

Values of z outside the unit circle can be established by analytical continuation. From the series expression one can establish the following special values:

$$_2F_1(a, b, c; 0) = 1$$

$$_2F_1(a, b, c; 1) = \frac{\Gamma(c)\Gamma(c - a - b)}{\Gamma(c - a)\Gamma(c - b)}$$

$$\Re(c - a - b) > 0, \quad c \neq 0 \text{ or } -n$$

The derivative relationship also follows from the power series:

$$\frac{d}{dz}\,_2F_1(a, b, c; z) = \frac{ab}{c}\,_2F_1(a + 1, b + 1, c + 1; z)$$

In our work the hypergeometric function is encountered primarily as a way to express definite integrals that describe measured quantities. The following example occurs in several situations of interest.

1. $\displaystyle\int_0^1 x^{b-1}(1 - x)^{c-b-1}(1 - xz)^{-a}\,dx = \frac{\Gamma(b)\Gamma(c - b)}{\Gamma(c)}\,_2F_1(a, b, c; z)$

$$\Re(c) > \Re(b) > 0$$

Sometimes we need integrals of the hypergeometric functions.

2. $\displaystyle\int_0^z \,_2F_1(a, b, c; x)\,dx = \frac{c - 1}{(a - 1)(b - 1)}[\,_2F_1(a - 1, b - 1, c - 1; z) - 1]$

3. $\displaystyle\int_0^1 x^{c-1}(1 - x)^{\mu-c-1}\,_2F_1(a, b, c; xz)\,dx = \frac{\Gamma(c)\Gamma(\mu - c)}{\Gamma(\mu)}\,_2F_1(a, b, \mu; z)$

$$|\arg(1 - z)| < \pi, \qquad z \neq 1, \qquad \Re(\mu) > \Re(c) > 0$$

The function $_2F_1(a, b, c; x)$ discussed above is the most important member of a *family* of hypergeometric functions. Some integrations are expressed in terms of a less familiar member of this family, namely

$$_1F_2(a, b, c; x)$$

The reversal of indices means that it has the following series expansion:

$$_1F_2(a, b, c; z) = 1 + \frac{a}{bc}\frac{z}{1!} + \frac{a(a + 1)}{b(b + 1)c(c + 1)}\frac{z^2}{2!} + \cdots$$

This series is valid for $|z| < 1$ and the function can be calculated outside the unit circle by continuation. Other properties of the hypergeometric function are found in the following standard references.

1. M. Abramowitz and I. A. Stegun, *Handbook of Mathematical Functions* (Dover, New York, 1964), 556–565.
2. W. Magnus and F. Oberhettinger, *Formulas and Theorems for the Special Functions of Mathematical Physics* (Chelsea, New York, 1949), 7–11.
3. E. T. Whittaker and G. N. Watson, *A Course of Modern Analysis* (Cambridge University Press and Macmillan, Cambridge, 1947), 280–301.

Appendix I

Aperture Averaging

It is often necessary to analyze the influence of aperture averaging on electromagnetic signals that have traveled through random media. The measured quantities can be expressed in terms of the correlation of the desired property averaged over the surface of the aperture of the receiver. In the vast majority of optical and microwave receivers circular apertures are used and this simplifies the analysis considerably. The cylindrical coordinates identified in Figure 3.8 provide the natural description for the required surface integrals and the separation between typical points on the receiving surface. The quantity to be averaged is related to the spatial correlation of the measured quantity by

$$\mathcal{J}(\kappa, a) = \frac{1}{\pi^2 a^4} \int_0^a r_1 \, dr_1 \int_0^a r_2 \, dr_2 \int_0^{2\pi} d\phi_1 \int_0^{2\pi} d\phi_2$$
$$\times C\left(\kappa\sqrt{r_1^2 + r_2^2 - 2r_1 r_2 \cos(\phi_1 - \phi_2)}\right)$$

where κ is an inverse scale length.

It is surprising that this four-fold integration can be reduced to a single integral without specifying the correlation function. There are several ways to do so. The simplest derivation[1] is to represent the spatial correlation as a Fourier Bessel transform:[2]

$$C(z) = \int_0^\infty u \, J_0(uz) f(u) \, du$$

On substituting this expression for the correlation function into the aperture

---

[1] V. I. Tatarskii, *Wave Propagation in a Turbulent Medium* (Dover, New York, 1967), 233.
[2] G. N. Watson, *A Treatise on the Theory of Bessel Functions* (Cambridge University Press, Cambridge, 1952), 576 *et seq.*

average we have

$$J(\kappa, a) = \int_0^\infty u f(u) \, du \, \frac{1}{\pi^2 a^4} \int_0^a r_1 \, dr_1 \int_0^a r_2 \, dr_2 \int_0^{2\pi} d\phi_1 \int_0^{2\pi} d\phi_2$$
$$\times J_0\!\left(u\kappa \sqrt{r_1^2 + r_2^2 - 2r_1 r_2 \cos(\phi_1 - \phi_2)} \right)$$

The four-fold integration was evaluated in Section 4.1.8 in order to describe phase averaging. The result depends only on the product κa and is given by (4.33):

$$J(\kappa a) = \int_0^\infty u f(u)\!\left(\frac{2J_1(u\kappa a)}{u\kappa a} \right)^2 du$$

The orthogonality of the Bessel function indicated in Appendix D allows the transform function $f(u)$ to be expressed in terms of $C(z)$ by a similar relation:

$$\int_0^\infty dz \, z J_0(z\zeta) C(z) = \int_0^\infty u f(u) \, du \int_0^\infty dz \, z J_0(z\zeta) J_0(uz)$$
$$= \int_0^\infty u f(u) \frac{\delta(u - \zeta)}{u} \, du$$
$$= f(\zeta)$$

We introduce this expression for $f(u)$ and reverse the order of integration:

$$J(\kappa a) = \frac{4}{\kappa^2 a^2} \int_0^\infty dz \, z C(z) \int_0^\infty \frac{1}{u} J_1^2(u\kappa a) J_0(uz) \, du$$

The integral involving three Bessel functions can be done analytically:

$$\int_0^\infty \frac{1}{x} J_1^2(x) J_0(xw) \, dx = \begin{cases} \dfrac{1}{\pi}\!\left[\arccos\!\left(\dfrac{w}{2}\right) - \dfrac{w}{4}\sqrt{4 - w^2} \right] & w < 2 \\[2ex] 0 & w > 2 \end{cases}$$

The aperture average is therefore presented as a single integral over the spatial correlation function:

$$J(\kappa a) = \frac{4}{\pi} \int_0^2 \left[\arccos\!\left(\frac{w}{2}\right) - \frac{w}{2}\sqrt{1 - \frac{w^2}{4}} \right] C(\kappa a w) w \, dw$$

This result is related to the problem of finding the common area of two overlapping circles.[3] An equivalent solution has two parametric integrations but is often easier to use:[4]

$$\mathcal{J}(\kappa a) = \frac{8}{\pi} \int_0^1 ds \sqrt{1 - s^2} \int_0^{2s} C(\kappa a w) w\, dw$$

[3] D. L. Fried, "Aperture Averaging of Scintillation," *Journal of the Optical Society of America*, **57**, No. 2, 169–175 (February 1967).

[4] E. Levin, R. B. Muchmore and A. D. Wheelon, "Aperture-to-Medium Coupling on Line-of-Sight Paths: Fresnel Scattering," *IRE Transactions on Antennas and Propagation*, **AP-7**, No. 2, 142–146 (April 1959).

Appendix J

Vector Relations

1. $\mathbf{a} \cdot \mathbf{b} \times \mathbf{c} = \mathbf{b} \cdot \mathbf{c} \times \mathbf{a}$
   $= \mathbf{c} \cdot \mathbf{a} \times \mathbf{b}$

2. $\mathbf{a} \times (\mathbf{b} \times \mathbf{c}) = (\mathbf{a} \cdot \mathbf{c})\mathbf{b} - (\mathbf{a} \cdot \mathbf{b})\mathbf{c}$

3. $(\mathbf{a} \times \mathbf{b}) \cdot (\mathbf{c} \times \mathbf{d}) = \mathbf{a} \cdot \mathbf{b} \times (\mathbf{c} \times \mathbf{d})$
   $= \mathbf{a} \cdot [(\mathbf{b} \cdot \mathbf{d})\mathbf{c} - (\mathbf{b} \cdot \mathbf{c})\mathbf{d})]$
   $= (\mathbf{a} \cdot \mathbf{c})(\mathbf{b} \cdot \mathbf{d}) - (\mathbf{a} \cdot \mathbf{d})(\mathbf{b} \cdot \mathbf{c})$

4. $(\mathbf{a} \times \mathbf{b}) \times (\mathbf{c} \times \mathbf{d}) = (\mathbf{a} \times \mathbf{b} \cdot \mathbf{d})\mathbf{c} - (\mathbf{a} \times \mathbf{b} \cdot \mathbf{c})\mathbf{d}$

5. $\nabla(\phi + \psi) = \nabla\phi + \nabla\phi$

6. $\nabla(\phi\psi) = \phi\,\nabla\psi + \psi\,\nabla\phi$

7. $\nabla \cdot (\mathbf{a} + \mathbf{b}) = \nabla \cdot \mathbf{a} + \nabla \cdot \mathbf{b}$

8. $\nabla \times (\mathbf{a} + \mathbf{b}) = \nabla \times \mathbf{a} + \nabla \times \mathbf{b}$

9. $\nabla \cdot (\phi\mathbf{a}) = \mathbf{a} \cdot \nabla\phi + \phi\,\nabla \cdot \mathbf{a}$

10. $\nabla \times (\phi\mathbf{a}) = \nabla\phi \times \mathbf{a} + \phi\,\nabla \times \mathbf{a}$

11. $\nabla(\mathbf{a} \cdot \mathbf{b}) = (\mathbf{a} \cdot \nabla)\mathbf{b} + (\mathbf{b} \cdot \nabla)\mathbf{a} + a \times (\nabla \times \mathbf{b}) + \mathbf{b} \times (\nabla \times \mathbf{a})$

12. $\nabla \cdot (\mathbf{a} \times \mathbf{b}) = \mathbf{b} \cdot \nabla \times \mathbf{a} - \mathbf{a} \cdot \nabla \times \mathbf{b}$

13. $\nabla \times (\mathbf{a} \times \mathbf{b}) = \mathbf{a}\,\nabla \cdot b - \mathbf{b}\,\nabla \cdot \mathbf{a} + (\mathbf{b} \cdot \nabla)\mathbf{a} - (\mathbf{a} \cdot \nabla)\mathbf{b}$

14. $\nabla \times \nabla \times \mathbf{a} = \nabla\nabla \cdot \mathbf{a} - \nabla^2\mathbf{a}$

15. $\nabla \times \nabla\phi = 0$

16. $\nabla \cdot \nabla \times \mathbf{a} = 0$

17. $(\nabla\psi \cdot \nabla)\nabla\psi = \frac{1}{2}\nabla(\nabla\psi)^2$

18. $\nabla^2(FG) = F\,\nabla^2 G + 2\nabla F \cdot \nabla G + G\,\nabla^2 F$

Appendix K

The Gamma Function

The gamma function occurs frequently in this work. It often crops up in evaluating integrals or Laplace transforms. Mellin transforms provide a powerful way to analyze propagation in random media[1] and they are usually expressed in terms of gamma functions, which are defined by the integral

$$\Gamma(z) = \int_0^\infty x^{z-1} \exp(-x)\, dx \qquad \Re(z) > 0$$

The gamma function is defined for complex values of the argument[2] but we usually encounter it for real values of z. It is finite with well-defined derivatives unless the argument is zero or a negative integer, in which case it has simple poles. This is evident from Figure K.1. When z is a positive integer it is related to the *factorial*:

$$\Gamma(n) = (n-1)! = 1 \times 2 \times 3 \times \cdots (n-1)$$

A brief table of $\Gamma(z)$ for noninteger values of z is given at the end of this appendix (Table K.1), together with some useful mathematical constants. Complete tabulations of this function are available in standard references.[3] The following special values are helpful:

$$\Gamma(\tfrac{1}{2}) = \sqrt{\pi}$$

$$\Gamma(1) = 1$$

$$\Gamma(\tfrac{3}{2}) = \tfrac{1}{2}\sqrt{\pi}$$

$$\Gamma(2) = 1$$

[1] R. J. Sasiela, *Electromagnetic Wave Propagation in Turbulence* (Springer-Verlag, Berlin, 1994).
[2] P. M. Morse and H. Feshbach, *Methods of Theoretical Physics: Part I* (McGraw-Hill, New York, 1953), 419–425.
[3] M. Abramowitz and I. A. Stegun, *Handbook of Mathematical Functions* (Dover, New York, 1964), 255–293.

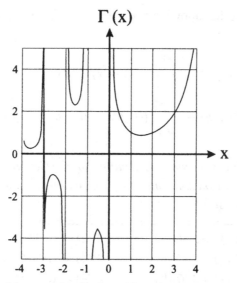

Figure K.1: A plot of the gamma function for positive and negative real values of the argument.

The *recurrence* formula can be used to extend Table K.1 to other positive and negative values:

$$\Gamma(z+1) = z\Gamma(z)$$

The *duplication* formula can also be used to extend Table K.1 to larger values:

$$\Gamma(2z) = \frac{1}{\sqrt{\pi}} 2^{2z-1} \Gamma(z) \Gamma\left(z + \frac{1}{2}\right)$$

The following *reflection* formula is often used to simplify expressions that involve several gamma functions:

$$\Gamma(z)\Gamma(1-z) = \frac{\pi}{\sin(\pi z)}$$

For very large values one can use Sterling's approximation:

$$\lim_{z \to \infty} [\Gamma(z)] = \sqrt{2\pi} z^{z-\frac{1}{2}} e^{-z}\left(1 + \frac{1}{12z} + \frac{1}{288z^2} + \cdots\right) \qquad |\arg z| < \pi$$

The beta function is closely related to the gamma function:

$$B(a, b) = \frac{\Gamma(a)\Gamma(b)}{\Gamma(a+b)} \qquad \Re(a) > 0 \text{ and } \Re(b) > 0$$

This connection allows the following two common definite integrals to be expressed in terms of $\Gamma(a)$ and $\Gamma(b)$:

$$B(a, b) = \int_0^1 x^{a-1}(1 - x)^{b-1}\, dx$$

$$B(a, b) = 2\int_0^{\frac{\pi}{2}} (\sin\theta)^{2a-1}(\cos\theta)^{2b-1}\, d\theta$$

Table K.1: *Numerical values for the Gamma function and commonly used mathematical constants*

$\Gamma(\frac{1}{6}) = 5.56632$	$\Gamma(2) = 1.00000$
$\Gamma(\frac{1}{4}) = 3.62561$	$\Gamma(\frac{7}{3}) = 1.19064$
$\Gamma(\frac{1}{3}) = 2.67894$	$\Gamma(\frac{17}{6}) = 1.72454$
$\Gamma(\frac{2}{3}) = 1.35412$	$\Gamma(\frac{11}{3}) = 4.01220$
$\Gamma(\frac{1}{2}) = 1.77245$	$\Gamma(\frac{25}{6}) = 7.42606$
$\Gamma(\frac{3}{4}) = 1.22542$	$\Gamma(-\frac{1}{2}) = -3.54490$
$\Gamma(\frac{5}{6}) = 1.12879$	$(2)^{\frac{1}{3}} = 1.25992$
$\Gamma(1) = 1.00000$	$(\pi)^{\frac{1}{3}} = 1.46459$
$\Gamma(\frac{7}{6}) = 0.92772$	$(2\pi)^{\frac{1}{3}} = 1.84527$
$\Gamma(\frac{4}{3}) = 0.89298$	$(10)^{\frac{1}{3}} = 2.15443$
$\Gamma(\frac{5}{3}) = 0.90275$	$(2)^{\frac{1}{6}} = 1.12246$
$\Gamma(\frac{11}{6}) = 0.94066$	$(2\pi)^2 = 39.47842$

Appendix L

Green's Function

Green's function satisfies a wave equation driven by a *point source* with unit strength:

$$\left(\nabla_r^2 + k^2\right)G(\mathbf{R}, \mathbf{r}) = -\delta(\mathbf{R} - \mathbf{r})$$

We can solve this equation by introducing its three-dimensional Fourier transformation:

$$G(\mathbf{R}, \mathbf{r}) = \frac{1}{8\pi^3} \int d^3\eta \, \mathcal{G}(\eta) \exp[i\eta \cdot (\mathbf{R} - \mathbf{r})]$$

The vector delta function describes a point source. It can be represented as the transform of one, as discussed in Appendix F:

$$\delta(\mathbf{R} - \mathbf{r}) = \frac{1}{8\pi^3} \int d^3\eta \exp[i\eta \cdot (\mathbf{R} - \mathbf{r})]$$

An algebraic equation for the transform of Green's function emerges when we substitute these expressions into the differential equation for $G(\mathbf{R}, \mathbf{r})$ given above:

$$\mathcal{G}(\eta) = \frac{1}{\eta^2 - k^2}$$

This solution provides an integral representation for Green's function:

$$G(\mathbf{R}, \mathbf{r}) = \frac{-1}{8\pi^3} \lim_{\epsilon \to 0} \int d^3\eta \, \frac{\exp[i\eta \cdot (\mathbf{R} - \mathbf{r})]}{k^2 - \eta^2 + i\epsilon}$$

The infinitesimal imaginary term in the denominator ensures that $G(\mathbf{R}, \mathbf{r})$ represents an outgoing wave connecting $\mathbf{r}$ to $\mathbf{R}$. To obtain an explicit description we cast the

425

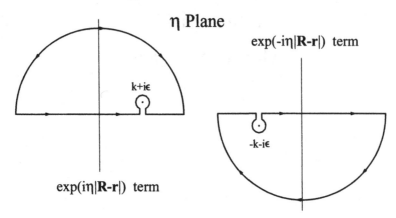

Figure L.1: Integration contours for evaluating terms in the integral representation of the spherical Green's function for an unbounded region.

η integration in spherical coordinates centered on the vector $\mathbf{R} - \mathbf{r}$:

$$G(\mathbf{R}, \mathbf{r}) = \frac{-1}{8\pi^3} \int_0^\infty d\eta\, \eta^2 \int_0^\pi d\theta \sin\theta \int_0^{2\pi} d\phi\, \frac{\exp[i\eta \cos(\theta\, |\mathbf{R} - \mathbf{r}|)]}{k^2 - \eta^2 + i\epsilon}$$

$$= \frac{i}{4\pi^2 |\mathbf{R} - \mathbf{r}|} \int_0^\infty \frac{d\eta\, \eta}{k^2 - \eta^2 + i\epsilon} [\exp(i\eta|\mathbf{R} - \mathbf{r}|) - \exp(-i\eta|\mathbf{R} - \mathbf{r}|)]$$

The remaining integration can be evaluated as contour integrals in the complex-η plane. Since the integrand is symmetrical about the origin, the integration can be extended from $-\infty$ to $+\infty$ and the appropriate closing contours are shown in Figure L.1. The small imaginary component correctly guides one around the poles in the denominator and the result is the same for both terms:

$$G(\mathbf{R}, \mathbf{r}) = \frac{\exp(ik|\mathbf{R} - \mathbf{r}|)}{4\pi |\mathbf{R} - \mathbf{r}|}$$

We shall use this form and the integral representation interchangeably.

Appendix M

The Method of Cumulant Analysis

The ensemble average of a complex exponential involving the random variable z can be expressed in terms of its cumulants as follows:

$$\langle \exp(i\lambda z) \rangle = \exp\left(\sum_{n=1}^{\infty} K_n \frac{(i\lambda)^n}{n!} \right)$$

where

$$K_1 = \langle z \rangle$$

$$K_2 = \langle z^2 \rangle - \langle z \rangle^2$$

$$K_3 = \langle z^3 \rangle - 3\langle z^2 \rangle\langle z \rangle + 2\langle z \rangle^3$$

$$K_4 = \langle z^4 \rangle - 4\langle z^3 \rangle\langle z \rangle - 3\langle z^2 \rangle^2 + 12\langle z \rangle^2\langle z^2 \rangle - 6\langle z \rangle^4$$

$$K_5 = \langle z^5 \rangle + 24\langle z \rangle^5 - 60\langle z \rangle^3\langle z^2 \rangle + 20\langle z^3 \rangle\langle z \rangle^2 + 30\langle z^2 \rangle^2\langle z \rangle$$
$$- 5\langle z^4 \rangle\langle z \rangle - 10\langle z^3 \rangle\langle z^2 \rangle$$

These expressions are valid for any random variable z and do not assume that z is distributed as a Gaussian random variable. They can be inverted to express the moments of z in terms of its cumulants as follows:

$$\langle z \rangle = K_1$$

$$\langle z^2 \rangle = K_1^2 + K_2$$

$$\langle z^3 \rangle = K_1^3 + 3K_1 K_2 + K_3$$

$$\langle z^4 \rangle = K_1^4 + 6K_1^2 K^2 + 3K_2^2 + 4K_1 K_3 + K_4$$

$$\langle z^5 \rangle = K_1^5 + 10K_1^3 K_2 + 10K_1^2 K_3 + 15K_1 K_2^2 + 5K_1 K_4 + 10K_2 K_3 + K_5$$

Further invaluable results are presented in the following excellent book which is available only in Russian:

A. A. Malakhov, *Cumulant Analysis of Random Non-Gaussian Processes and Their Transformations* (Soviet Radio, Moscow, 1978).

The following standard reference also contains many results of cumulant analysis:

A. Stuart and J. K. Ord, *Kendall's Advanced Theory of Statistics,* 5th edition, vol. 1 (Oxford University Press, New York, 1987).

Appendix N

Diffraction Integrals

1. $\dfrac{1}{(2\pi)^3} \displaystyle\int d^3 p \, \dfrac{\exp(i\mathbf{p}\cdot\mathbf{R})}{k^2 - p^2 + i\epsilon} = \dfrac{1}{4\pi R}\exp(ikR)$

2. $\dfrac{1}{(2\pi)^3} \displaystyle\int d^3 p \, \dfrac{\exp(i\mathbf{p}\cdot\mathbf{R})}{\left(k^2 - p^2 + i\epsilon\right)^2} = \dfrac{-i}{8\pi}\dfrac{\exp(ikR)}{k}$

3. $\dfrac{1}{(2\pi)^3} \displaystyle\int d^3 p \, \dfrac{\exp(i\mathbf{p}\cdot\mathbf{R})}{\left(k^2 - p^2 + i\epsilon\right)^3} = \dfrac{-1}{32\pi}\left(\dfrac{R}{k^2} + \dfrac{i}{k^3}\right)\exp(ikR)$

4. $\dfrac{1}{(2\pi)^3} \displaystyle\int d^3 p \, \dfrac{\exp(i\mathbf{p}\cdot\mathbf{R})}{\left(k^2 - p^2 + i\epsilon\right)^4} = \dfrac{1}{192\pi}\left(-\dfrac{3R}{k^4} + \dfrac{iR^2}{k^3} - \dfrac{3i}{k^5}\right)\exp(ikR)$

5. $\dfrac{1}{(2\pi)^3} \displaystyle\int d^3 p \, \dfrac{\exp(i\mathbf{p}\cdot\mathbf{R})}{\left(k^2 - p^2 + i\epsilon\right)^5} = \dfrac{1}{8 \times 192\pi}$

$$\times \left(\dfrac{R^3}{k^4} + 6i\dfrac{R^2}{k^5} - 15\dfrac{R}{k^6} - \dfrac{15i}{k^7}\right)\exp(ikR)$$

Appendix O

Feynman Formulas

1. $\dfrac{1}{ab} = \displaystyle\int_0^1 du\, [(1-u)a + bu]^{-2}$

2. $\dfrac{1}{a^2 b} = 2\displaystyle\int_0^1 du\,(1-u)[a(1-u) + bu]^{-3}$

3. $\dfrac{1}{a^3 b} = 3\displaystyle\int_0^1 du\,(1-u)^2[a(1-u) + bu]^{-4}$

4. $\dfrac{1}{a^2 b^2 c} = 24\displaystyle\int_0^1 du\,(1-u)\int_0^u dv\,(u-v)[a(1-u) + b(u-v) + cv]^{-5}$

5. $\dfrac{1}{abc} = 2\displaystyle\int_0^1 du\int_0^u dv\,[a(1-u) + b(u-v) + cv]^{-3}$

6. $\dfrac{1}{a^2 bc} = 6\displaystyle\int_0^1 du\,(1-u)\int_0^u dv\,[a(1-u) + b(u-v) + cv]^{-4}$

7. $\dfrac{1}{a^3 bc} = 12\displaystyle\int_0^1 du\,(1-u)^2\int_0^u dv\,[a(1-u) + b(u-v) + cv]^{-5}$

8. $\dfrac{1}{abcf} = 6\displaystyle\int_0^1 du\int_0^u dv\int_0^v dw\,[a(1-u) + b(u-v) + c(v-w) + fw]^{-4}$

9. $\dfrac{1}{a^2 bcf} = 24\displaystyle\int_0^1 du\,(1-u)\int_0^u dv\int_0^v dw$
$$\times\,[a(1-u) + b(u-v) + c(v-w) + fw]^{-5}$$

Author Index

431

434 *Author Index*

Thompson, M. C., 117, 168, 186, 199, 212, 246, 277, 313, 354, 357, 366
Time, N. S., 186, 315, 355
Titus, J. M., 119
Tsao, C. K. H., 212
Tsvang, L. R., 132, 164, 315
Tsvyk, R. Sh., 118
Tur, M., 212
Twiss, R. Q., 119

Uscinski, B. J., 282, 313
Umeki, R., 121, 213

Van Vleck, J. H., 256, 276
Vernin, J., 147, 148, 153, 156, 165, 166, 213
Vetter, M. J., 313, 366
Vilar, E., 187, 356, 357, 358
Vogel, W. J., 357
Vogler, L. E., 357
von Karman, Th., *see* Subject Index: von Karman model

Walker, J. G., 357
Wang, C. N., 358
Wang, G. Y., 356
Wang, Ting-i., 155, 166

Waterman, A. T., 133, 165, 186, 198, 212
Watson, G. N., 13, 31, 166, 395, 410, 417, 418
Watterson, J., xx
Weitz, A., 212
Wells, P. I., 313, 366
Wheelon, A. D., 211, 314, 356, 357, 392, 395, 420
Wheelon, M. A., xix
Wheelon, P. G., xix
Whitney, H. E., 120, 357
Whittaker, E. T., 410, 417
Wilson, J. J., 187, 213
Wolf, E., 116
Wood, L. E., 117, 186, 212, 246, 354
Worthington, D. T., 212
Wright, N. J., 315

Yamada, M., 358
Yeh, K. C., 120, 121, 164, 187, 211, 213
Yura, H. T., xix, 24, 26, 27, 32, 214, 247, 275, 276, 279, 288, 313, 314, 315
Young, A. T., 96, 119, 180, 187, 341, 357

Zavorotnyi, V. U., xix, 356
Zhuk, I. N., 120
Zhukova, L. N., 119, 205, 213
Zieske, P., 155, 156

Subject Index